Advances in 21st Century Human Settlements

Indexed by SCOPUS

This Series focuses on the entire spectrum of human settlements – from rural to urban, in different regions of the world, with questions such as: What factors cause and guide the process of change in human settlements from rural to urban in character, from hamlets and villages to towns, cities and megacities? Is this process different across time and space, how and why? Is there a future for rural life? Is it possible or not to have industrial development in rural settlements, and how? Why does 'urban shrinkage' occur? Are the rural areas urbanizing or is that urban areas are undergoing 'ruralisation' (in form of underserviced slums)? What are the challenges faced by 'mega urban regions', and how they can be/are being addressed? What drives economic dynamism in human settlements? Is the urban-based economic growth paradigm the only answer to the quest for sustainable development, or is there an urgent need to balance between economic growth on one hand and ecosystem restoration and conservation on the other – for the future sustainability of human habitats? How and what new technology is helping to achieve sustainable development in human settlements? What sort of changes in the current planning, management and governance of human settlements are needed to face the changing environment including the climate and increasing disaster risks? What is the uniqueness of the new 'socio-cultural spaces' that emerge in human settlements, and how they change over time? As rural settlements become urban, are the new 'urban spaces' resulting in the loss of rural life and 'socio-cultural spaces'? What is leading the preservation of rural 'socio-cultural spaces' within the urbanizing world, and how? What is the emerging nature of the rural-urban interface, and what factors influence it? What are the emerging perspectives that help understand the human-environment-culture complex through the study of human settlements and the related ecosystems, and how do they transform our understanding of cultural landscapes and 'waterscapes' in the twenty-first Century? What else is and/or likely to be new vis-à-vis human settlements – now and in the future? The Series, therefore, welcomes contributions with fresh cognitive perspectives to understand the new and emerging realities of the 21st Century human settlements. Such perspectives will include a multidisciplinary analysis, constituting of the demographic, spatio-economic, environmental, technological, and planning, management and governance lenses.

If you are interested in submitting a proposal for this series, please contact the Series Editor, or the Publishing Editor:

Bharat Dahiya (bharatdahiya@gmail.com) or

Loyola D'Silva (loyola.dsilva@springer.com)

Limao Zhang • Renfei He
Qiong Liu • Paolo Gardoni

Resilience in Urban Transit Systems

Springer

Limao Zhang
School of Civil and Hydraulic Engineering
Huazhong University of Science and Technology
Wuhan, Hubei, China

Qiong Liu
Department of Architecture and Civil Engineering
City University of Hong Kong
Hong Kong, China

Renfei He
School of Civil and Environmental Engineering
Nanyang Technological University
Singapore, Singapore

Paolo Gardoni
Department of Civil and Environmental Engineering
University of Illinois Urbana-Champaign
Urbana, IL, USA

ISSN 2198-2546 ISSN 2198-2554 (electronic)
Advances in 21st Century Human Settlements
ISBN 978-981-95-4321-2 ISBN 978-981-95-4322-9 (eBook)
https://doi.org/10.1007/978-981-95-4322-9

This Springer imprint is published by the registered company Springer Nature Singapore Pte Ltd.
The registered company address is: 152 Beach Road, #21-01/04 Gateway East, Singapore 189721, Singapore

Preface

Urban Transit Systems (UTS) have become indispensable infrastructures that sustain the daily mobility needs of rapidly expanding cities throughout the world. Their continuous development has greatly improved urban accessibility and economic vitality. However, with the growing complexity of urban environments and the increasing frequency of natural hazards and other unexpected disruptions, UTS face significant challenges in maintaining safe and reliable operations. Incidents such as floods and fires can severely disrupt services, cause substantial economic losses, and even threaten passenger safety. In this context, the concept of resilience is understood as the capacity of a system to prepare for, absorb, recover from, and adapt to disturbances. It offers a valuable perspective for enhancing disaster management in UTS. Strengthening the resilience of UTS helps cities safeguard public mobility, protect critical infrastructure, and ensure long-term sustainable urban development.

This book provides an integrated and up-to-date overview of resilience thinking applied to UTS, with a clear structure reflecting the logical progression from understanding vulnerabilities to implementing enhancement strategies. It starts with "Introduction to Resilience of Urban Transit System" in Chap. 1. Then the main contents of the book are organized into three interrelated parts. Part I, System Vulnerability Modeling, introduces advanced modeling approaches, including "Disaster Chains Modeling" (Chap. 2), "Cascading Failure Modeling" (Chap. 3), and "Evolutionary Dynamics Modeling" (Chap. 4), to capture the complex behaviors and interdependencies of transit systems under stress. Part II, System Vulnerability Evaluation, focuses on practical hazard assessments and risk analyses, covering "Flood Hazard Analysis" (Chap. 5), "Fire Hazard Analysis" (Chap. 6), and "Passenger Evacuation Analysis" (Chap. 7). Part III, System Resilience Enhancement, presents planning and optimization solutions for enhancing system resilience, including "Facility Operation Planning" for flood response (Chap. 8), "Emergency Rescue Planning" for fire incidents (Chap. 9), "Resource Preparedness Planning" (Chap. 10), and "System Repair Planning" to enable swift recovery (Chap. 11). Finally, the main results are summarized in "Conclusions and Future Directions" (Chap. 12). Together, these works demonstrate the significant potential of resilience modeling, evaluation, and enhancement to build safer, more adaptive, and more sustainable UTS, ensuring cities remain functional, reliable, and better prepared for future uncertainties.

This book expects to: (1) reveal the hidden complexities and interdependencies within UTS that affect their resilience performance; (2) present analytical models, evaluation methods, and practical planning strategies for assessing and strengthening transit resilience; (3) showcase real-world applications and evidence-based practices to demonstrate how resilience thinking can be embedded in transit operations and management; (4) promote a deeper understanding of current advancements and inspire further research and implementation of resilience-oriented solutions in UTS.

This book is a valuable reference for scientists, engineers, and graduate students in various disciplines, including but not limited to transportation engineering, intelligent construction, urban planning, disaster management, systems engineering, etc.

Wuhan, China, Singapore — Limao Zhang
Renfei He
Hong Kong, China — Qiong Liu
Champaign, USA — Paolo Gardoni
September 2025

Acknowledgment

It is our sincere desire to thank the anonymous peer reviewers for this work. We would like to extend our sincere gratitude to our esteemed colleagues for their invaluable contributions to this scholarly endeavor. Our sincere appreciation goes to Prof. Lieyun Ding and Prof. Xianguo Wu from Huazhong University of Science and Technology for their unwavering support throughout this research work. Special recognitions are extended to Dr. Hongyu Chen, Dr. Kai Guo, Dr. Yang Zhang, Dr. Ying Wang, Dr. Ankang Ji, and Dr. Lulu Wang for their rigorous feedback and stimulating discussions. Their combined knowledge and steadfast commitment proved instrumental in advancing the academic depth and presentation standards of this book. This work is supported in part by the National Natural Science Foundation of China (Nos. 72571110, 72271101).

Competing Interests The authors have no competing interests to declare that are relevant to the content of this manuscript.

Acknowledgment

[illegible]

Contents

Chapter 1
Introduction to Resilience of Urban Transit Systems

1.1 What Is Resilience

Resilience is a multidisciplinary concept with origins deeply rooted in several scientific fields. The term itself derives from the Latin word *resilire*, meaning "to rebound" or "to spring back" [1]. In early scientific usage, resilience was primarily used in material science and engineering to describe a material's ability to absorb energy and return to its original shape after deformation. The concept evolved significantly when Holling [2] introduced resilience to ecology, defining it as the capacity of an ecosystem to absorb disturbances and reorganize while undergoing change so as to retain essentially the same function, structure, and feedback. Since then, the idea of resilience has been employed across various domains, including psychology, socio-ecological systems, disaster risk management, and infrastructure planning [3].

Different scholars and organizations have proposed and enriched the definition of resilience to fit distinct contexts. In engineering, resilience is often described as the ability of a system to withstand a disruptive event and return to a steady operational state quickly, aligning with what is termed "engineering resilience" [4]. Ecological and socio-ecological resilience broaden this perspective by considering systems as dynamic and adaptive, capable of transforming in response to stressors, which is more aligned with "evolutionary resilience" [5]. The U.S. National Academy of Sciences (NAS) defines resilience in the context of critical infrastructure as "the ability to prepare and plan for, absorb, recover from, and more successfully adapt to adverse events" [6]. In addition, resilience was defined as the ability of a system to sustain disruptions without interrupting system performance and recover fully and quickly if the system is interrupted, which is according to the American Society of Mechanical Engineers [7]. Heinimann [8] introduced a framework for measuring the resilience of socio-technical infrastructure systems, consisting of eight functions. This framework, referred to as AR6A, considered the entirety

L. Zhang et al., *Resilience in Urban Transit Systems*, Advances in 21st Century Human Settlements, https://doi.org/10.1007/978-981-95-4322-9_1

of the system's performance, including preparedness and attentiveness activities prior to an event, as well as post-event behaviors such as remembering, strengthening, and adaptive functions for future events. The City Resilience Framework, developed by the Rockefeller Foundation and Arup [9], defined urban resilience as the capability of individuals, communities, or systems to survive, adapt, and grow under various chronic stresses and acute shocks. Additionally, the framework proposes seven features of resilience, including robustness, redundancy, flexibility, resources, reflection, inclusion, and integration. These definitions reflect a shared recognition that resilience is not a static trait but an evolving capacity to deal with uncertainty, nonlinearity, and complexity.

Drawing from extensive literature and focusing on the resilience of an engineering system, this section adopts the definition of resilience made by NAS, which consists of four core capabilities:

1. **Preparedness.** Preparedness refers to the ability of the system to be prepared for possible disruptions. To prepare for disruptions like accidents or disasters, it is crucial to identify the hazard's type and occurrence probability, estimate possible consequences, forecast, and monitor the disruption events. Preparedness is also embodied in the preparation for emergencies, which involves stock buildup, emergency plan establishment, staff training, and so forth. Effective preparedness reduces the likelihood of severe system disruption and enables a quicker mobilization of resources when events occur.
2. **Absorbability.** Absorbability is the capacity of a system to withstand shocks without losing its essential functions. This means that it should be robust to different types of disruptions or have built-in redundancies that allow them to continue functioning even in the face of such disruptions. For example, in engineering, this might involve strengthening physical structures, diversifying critical supply chains, or incorporating backup systems that ensure continuity under stress.
3. **Recoverability.** Even with the best planning and design, systems can still be disrupted. Therefore, it is important for these systems to be able to recover from disruptions. This involves having response mechanisms in place and thus implementing optimal restoration plans. Note that recoverability is not solely about speed; it also entails the cost-effectiveness and completeness of the process. A resilient system not only returns to functionality quickly but does so in a manner that minimizes resource consumption and long-term vulnerability.
4. **Adaptability.** Adaptability is the capacity of the system to adjust its structures, operations, or strategies in response to new conditions and lessons learned from past disturbances. It goes beyond merely recovering to the previous state; it implies transformation to improve future resilience. This requires the system operators to identify new risks and emerging threats and quickly adjust their plans and operations to mitigate them. Adaptability is crucial for complex urban systems that must respond to evolving hazards such as climate change, rapid urbanization, and technological shifts.

Together, these four capacities—preparedness, absorbability, recoverability, and adaptability—provide a comprehensive lens to evaluate and enhance resilience in practice. They highlight that resilience is not an isolated attribute but a dynamic process requiring continuous investment, monitoring, and improvement. The four capacities can also be visualized via the "resilience curve" (Fig. 1.1), which illustrates how a system's performance changes over time in response to a disruption. Specifically, the curve can be divided into four phases that correspond to the four capacities: preparedness ensures that a system is well-equipped before a disruption occurs; absorbability describes the system's ability to withstand the immediate impact; recoverability indicates how effectively the system can regain its performance after the impact; and adaptability reflects the system's capacity to adjust and improve in response to new conditions. This phased representation enables stakeholders to identify gaps in each stage and to implement targeted measures to strengthen overall system resilience. Based on the resilience curve, various quantitative indicators have been proposed in the literature to measure system resilience, such as calculating the integral of the performance curve between the time of disruption (t_{event}) and the time when the system reaches a new stable state (t_{post}) [10].

Today, resilience has become an important framework in urban research, climate change adaptation, and sustainable development. Unlike traditional risk management or reliability engineering, which mainly focus on predicting and mitigating known hazards, resilience addresses the inherent uncertainty and adaptability of complex systems. A clear understanding of its evolution and multidisciplinary perspectives is crucial for effectively applying resilience to urban infrastructure.

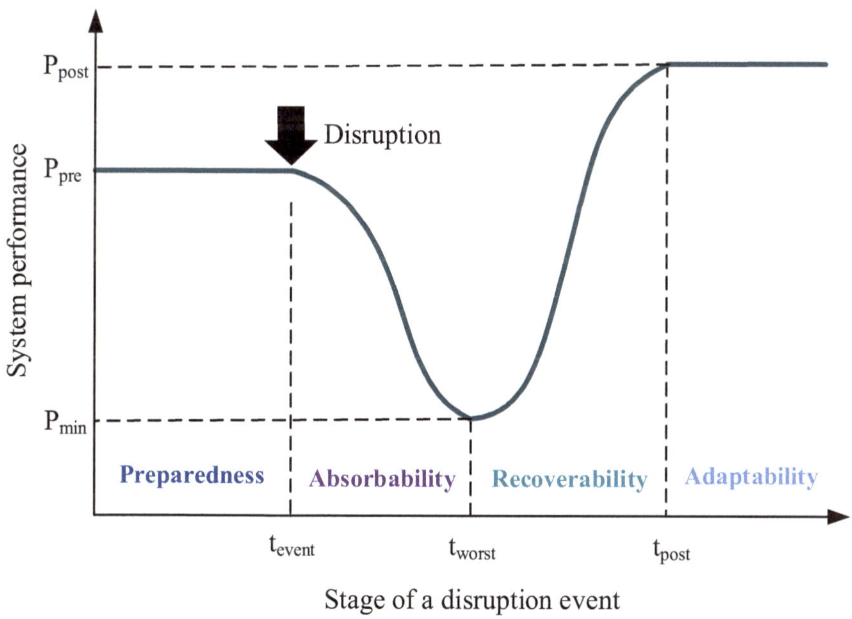

Fig. 1.1 Resilience curve

Modern urban infrastructure is highly interconnected and embedded within broader socio-technical and environmental contexts, making them susceptible to various disruptions such as natural disasters, technical failures, human errors, and targeted attacks. Resilience thinking helps cities better withstand unexpected events, maintain essential services, and adapt to changing conditions. By incorporating preparedness, absorbability, recoverability, and adaptability into the design, operation, and governance of critical infrastructure, cities can minimize service interruptions, shorten recovery times, and enhance overall sustainability and quality of people's lives.

1.2 What Is Urban Transit Systems?

Urban transit systems encompass a variety of transport modes designed to facilitate the efficient movement of people within cities. These systems typically include metro or subway networks, light rail transit, buses, and trams, all of which work together to provide comprehensive and accessible mobility solutions for urban residents. Among these modes, the metro system has emerged as the most prominent, widely used, and rapidly expanding form of urban transit in many metropolitan areas worldwide. Due to its high capacity, speed, and reliability, the metro system plays a central role in supporting daily commuting and mitigating urban congestion. Therefore, in this book, particular emphasis is placed on the metro system as a representative example of urban transit systems. For clarity and consistency, the terms "urban transit system" and "urban metro system" will be used interchangeably in the following sections.

1.2.1 Development of Urban Metro System

Urban metro systems, now a defining feature of modern cities, have undergone a remarkable transformation since their inception in the nineteenth century. Their evolution can be broadly divided into three distinct phases, each shaped by technological advances, urban growth, and shifting socio-economic contexts:

- **Early development phase (1863–1924).** The origin of the metro can be traced back to 1863, when London opened the world's first underground railway, the Metropolitan Railway, with steam-powered trains running along a 6.4 km line to address severe street congestion at the time [11]. Shortly after, innovations spread: Germany showcased the first electric tramway at the Berlin Industrial Exhibition in 1881, and by 1890, London introduced its first electric underground line, marking the start of electrification that soon replaced steam traction [12]. During this period, Europe led the pace—13 major cities constructed initial metro lines to serve rapidly industrializing urban centers. On the European mainland, Budapest's underground line opened in 1896, and Paris launched its Métro in 1900, from which the term "metro" originated. Asia followed with Japan's Ginza Line in Tokyo, which began service in 1927, laying a foundation for metro transit across the continent [13].

- **Stagnation and redevelopment (1924–1969).** Despite the early momentum of metro construction, the interwar period and World War II severely constrained further development. High construction costs, shifting transportation preferences towards private automobiles, and widespread wartime destruction caused many planned metro projects to be postponed or cancelled altogether. A few cities—most notably Tokyo, Osaka, and Moscow—continued modest expansions [14, 15]. In the post-war decades, rapid urban population growth and rising motorization led to worsening traffic congestion, air pollution, and energy consumption, prompting city governments to reconsider underground rail as a sustainable alternative to road-based transport. To support post-war reconstruction and economic revitalization, many cities in Western Europe and Asia renewed investments in their metro networks, expanding existing lines and planning new ones to meet the demands of modern urban life [16].
- **Rapid growth and modern expansion (1970–present).** Since the 1970s, metro systems have entered an era of accelerated global expansion, becoming central to sustainable urban mobility strategies. Factors such as intensified urbanization, the need for efficient mass transit, and heightened environmental awareness have driven governments to prioritize metro development. Technological innovations—such as fully electric rolling stock, advanced signaling systems, and automated train operations—have further enhanced capacity, safety, and operational efficiency.

In the present phase, the Asia-Pacific region has emerged as the global leader, with unprecedented construction scale and network density. China's metro boom is particularly striking: after decades of gradual development since the opening of the Beijing Subway in 1969, the country's metro length skyrocketed in the past two decades. By 2024, China operated over 12,000 km of metro lines across over 50 cities, accounting for more than half of the world's total urban rail length. Major cities like Beijing and Shanghai have metro systems exceeding 800 km each, with annual ridership figures in the billions. Policy thresholds, such as minimum population and GDP benchmarks, have guided which cities qualify for metro investment, ensuring financial feasibility and network sustainability. National development plans, including successive Five-Year Plans, have kept metro construction a high priority, backed by massive infrastructure funding [17, 18].

Globally, other regions have also witnessed robust growth. Europe continues to modernize and expand legacy networks in cities like London, Paris, and Moscow. Emerging economies, including India, Thailand, and parts of Latin America, are rapidly building metro lines to address urban mobility and air quality challenges. According to recent industry reviews, the total global metro network length surpassed 22,000 km in 2024, with Asia alone contributing over 70% [19].

The last two decades have also seen a profound shift toward digitalization and automation. Modern metro systems increasingly adopt Communications-Based Train Control (CBTC), driverless operation, real-time passenger information systems, and smart ticketing. In China, smart metro development integrates mobile payment, passenger flow monitoring, and AI-based predictive maintenance, setting global benchmarks for operational efficiency. In the near future, cities worldwide will carry out further network extensions, increased automation, and deeper integration with other transit modes to create seamless, multimodal urban mobility ecosystems.

In summary, from a pioneering solution to urban congestion in nineteenth-century London to today's massive, smart, and sustainable networks throughout the world, metro systems have evolved through phases of innovation, stagnation, and unprecedented growth. This historical perspective sets the foundation for understanding the broader significance of urban metro systems, which will be explored further in the next section.

1.2.2 Significance of Urban Metro System

Urban metro systems have become an indispensable backbone of urban mobility in modern cities, providing a lot of benefits that extend well beyond simple transportation. Their significance can be examined from multiple perspectives, including transportation efficiency, economic vitality, environmental sustainability, social equity, and urban spatial optimization [20].

1. **Transportation efficiency.** At the most fundamental level, metro systems address one of the most pressing challenges faced by rapidly urbanizing areas: high-volume passenger movement within dense urban cores. Unlike conventional road-based transport, metros operate on exclusive right-of-way tracks, free from road congestion. This enables them to provide frequent, punctual, and high-capacity services regardless of surface traffic conditions. A single metro line can transport tens of thousands of passengers per hour per direction, far surpassing the capacity of buses or private cars. Hence, in megacities, metro lines serve as the main arteries of daily commuting, alleviating severe rush-hour traffic and ensuring urban centers remain accessible and functional.
2. **Economic growth.** Metro systems play a key role in driving urban economic growth, especially through their impact on land use and property values. New metro lines often raise land and rental prices around stations, making these areas more attractive for businesses and residential developments. According to Wang, Feng [21], in Shanghai, rental prices increased by about 0.4% for every 100 m reduction in distance from the nearest metro transport. In addition, metro systems greatly improve connections between where people live, work, and do business, which shortens travel time and makes it easier for companies to reach more workers and customers. This will bring about broader economic benefits [20].

3. **Environmental sustainability.** Metro systems significantly contribute to urban environmental sustainability by promoting cleaner, more efficient travel and reducing the negative impacts of car use. Studies in both developed and developing countries show that metro networks help cut greenhouse gas emissions and improve air quality by encouraging a shift from private cars to electric-powered public transport. For example, research in China and Brazil found that new metro operations reduced local concentrations of pollutants such as CO and PM2.5 [22, 23]. Besides, by shifting a large share of commuters from private vehicles and buses to underground trains, metro systems help reduce surface road traffic noise, which is a major source of urban noise pollution [24].
4. **Social equity.** Metro systems also promote social equity by providing affordable, reliable, and accessible mobility options to diverse population groups, including low-income residents, students, the elderly, and people with disabilities. Unlike cars, which require significant upfront and ongoing costs, metro fares are generally subsidized and remain stable, making them an essential mode of transport for those without private vehicles. Moreover, metros enhance spatial equity by connecting remote residential areas with employment centers and essential services. This connectivity reduces social segregation and supports balanced urban development. In many cities, metro expansions have been aligned with policies aiming to bridge the urban-rural gap and stimulate development in less developed districts [20].
5. **Spatial efficiency.** A unique advantage of metro systems lies in their ability to optimize scarce urban surface land by utilizing underground space. This underground space not only frees up valuable surface space for pedestrian areas, green spaces, and mixed-use developments but also facilitates the creation of multi-layered underground complexes integrating transit stations, shopping malls, and public facilities. The trend of coupling metro stations with underground commercial spaces, parking facilities, and pedestrian tunnels has proven highly effective in maximizing the economic returns of metro infrastructure while simultaneously improving the quality of urban public spaces above ground [25].

In summary, the metro system is far more than a mode of transportation. It is a powerful driver of economic growth, environmental sustainability, social equity, and spatial efficiency. All these benefits make it an essential part of today's cities and lay the groundwork for applying resilience to the management of urban transit systems.

1.3 How Resilience Benefits Urban Transit Systems

Despite their vital role in ensuring smooth urban mobility, metro systems face a wide range of risks that can threaten their safe and uninterrupted operation. These risks arise from both natural and human-made hazards, which can disrupt services, endanger passengers, and cause severe economic and social impacts. Understanding the main types of risks that metro systems commonly encounter is an important step toward strengthening their resilience. Here, five key categories of risks, including rainstorm flooding, fire hazards, earthquake disasters, terrorism attacks, and technical failures and cyber threats, are identified and analyzed below:

1. **Rainstorm flooding.** Urban metro systems are particularly vulnerable to rainstorm flooding due to their extensive underground tunnels, low-lying stations, and often limited drainage capacity. Heavy rainfall can quickly overwhelm pumping systems and flood entrances, tracks, and critical equipment rooms. Besides, it is difficult for people to evacuate from metro systems during flood events since the exits for escaping are limited and the evacuation routes are also the intrusion paths of floodwater (people need to evacuate against the flow). A tragic example, as illustrated in Fig. 1.2a, occurred during the July 2021 Zhengzhou flood in China, where extreme rainfall led to severe inundation of Metro Line 5 and caused a total of 14 casualties [27].
2. **Fire hazards.** As shown in Fig. 1.2b, fire is one of the most feared risks in metro systems because of the enclosed stations and tunnels as well as high passenger density, which makes smoke spread rapidly and evacuation can be challenging. Electrical faults, flammable materials, or intentional arson can lead to fire hazards in metro systems. The 2003 Daegu subway fire in South Korea remains one of the deadliest metro disasters, claiming 192 lives when a deliberately set fire

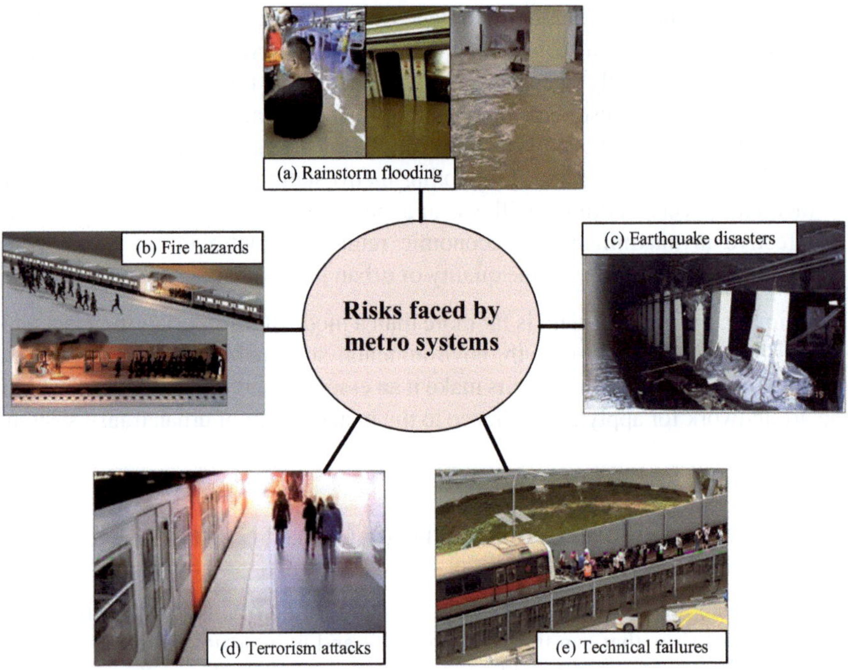

Fig. 1.2 Risks of metro systems: (**a**) rainstorm flooding (source: China Press); (**b**) fire hazard (source: Chosun Daily); (**c**) earthquake disaster [26]; (**d**) terrorism attack (source: Wikipedia); (**e**) technical failures (source: Lianhe Zaobao)

filled trains and stations with toxic smoke [28]. Such incidents highlight the critical need for fire detection and robust emergency procedures to prevent small ignition points from escalating into large-scale fire hazards.

3. **Earthquake disasters.** In regions prone to seismic activity, metro systems face the threat of structural damage, tunnel collapse, and service paralysis during major earthquakes. Underground stations and tunnels can suffer from ground shaking, soil liquefaction, and aftershocks. A striking example is the 1995 Great Hanshin Earthquake in Kobe, Japan. This magnitude 7.3 earthquake damaged a total of ten subway stations and most of the 21.4 km-long subway tunnels, with Daikai Station suffering the worst structural failure (see Fig. 1.2c) [29]. Earthquakes also pose long-term risks by weakening infrastructure and requiring costly inspections and retrofitting.
4. **Terrorism attacks.** Metro systems, due to their high passenger volumes and open-access design, are attractive targets for terrorism and malicious acts. Past attacks have caused significant loss of life and shaken public confidence in urban transit. Notably, the 2005 London Underground bombings killed 52 people and injured hundreds, while the 2016 Brussels bombings targeted both the airport and metro (see Fig. 1.2d), paralyzing city mobility and security systems. These events underscore the persistent need for vigilant surveillance, rapid emergency response, and robust security work in metro operations.
5. **Technical failures and cyber threats.** Modern metros rely heavily on complex digital control systems, including automated signaling, real-time monitoring, and driverless operations. While these technologies increase efficiency, they also introduce new failure points and security concerns. Technical failures, such as signal breakdowns or power outages, can cause the entire line to shut down, leaving thousands of commuters stranded. For instance, in September 2024, a metro train in Singapore malfunctioned (as shown in Fig. 1.2e) and damaged cables and triggered a power trip, causing a suspension of train services for about 1 week. Additionally, as cyber threats grow, metro operators must guard against hacking attempts that could compromise train control, passenger information, or even critical safety mechanisms, posing serious operational risks.

Faced with the various risk events described above, resilience has become essential for modern urban transit systems. Unlike traditional safety and reliability management, which mainly focuses on avoiding specific failures, resilience thinking provides a broader and more realistic approach to dealing with unexpected incidents and cascading disruptions. This perspective recognizes that not every incident can be predicted or prevented in advance. Extreme weather, sudden equipment breakdowns, or deliberate attacks can still happen despite strict safeguards. Therefore, instead of relying solely on static risk controls, resilience emphasizes how a system can prepare in advance, absorb the impact when disruptions occur, recover operations quickly, and adapt to new challenges over time. By adopting this perspective, metro operators and city planners can strengthen the system's capacity to withstand uncertainty and ensure that urban mobility remains reliable even in the face of growing complexity and unpredictable threats. To be more specific, enhancing the resilience of urban transit systems can bring the following key benefits:

- **Service robustness and continuity.** Enhancing resilience helps metro systems maintain steady service even under stress. By reinforcing infrastructure, improving real-time monitoring, and optimizing emergency decision-making, operators can keep key metro lines running during floods, technical failures, or other emergencies. This lowers the risk of total shutdowns and ensures that passengers still have access to reliable transport, which is crucial during peak travel times or citywide events.
- **Rapid Recovery and Minimal Downtime.** No transit system can avoid all interruptions, but a more resilient metro can bounce back faster after incidents. Clear emergency plans, well-trained staff, and pre-positioned resources allow operators to solve problems quickly and restore normal service sooner. Faster recovery minimizes passenger inconvenience and enables businesses and communities to rely on stable transport even after serious disruptions such as earthquakes or major system faults.
- **Passenger Safety and Trust.** Resilience-based measures directly improve how safely and confidently people use metro services. Effective hazard detection, reliable evacuation routes, and clear communication during incidents reduce the risk of casualties and panic. When passengers trust that operators can handle unexpected events properly, public confidence in the system grows, encouraging more people to choose metro transport over private vehicles.
- **Economic Stability and Social Well-being.** Disruptions of urban transit systems can cause significant economic losses, ranging from missed work to loss of revenue for nearby shops. By strengthening resilience, cities can limit these impacts by keeping people moving as smoothly as possible during unexpected incidents. Faster service recovery reduces downtime for businesses and helps maintain the daily flow of commuters and customers. In the long run, a resilient transit network supports local economies, reduces the social stress caused by large-scale shutdowns, and ensures that urban areas remain attractive, competitive, and livable for residents and businesses.
- **Support for Sustainable Urban Growth.** Investing in resilience aligns with long-term urban sustainability goals. As cities grow denser and face new risks from climate change and population pressures, a resilient urban transit system provides a backbone for low-carbon mobility and flexible adaptation. It helps cities cope with future uncertainties while delivering safe, reliable, and efficient public transport that meets evolving community needs.

In summary, these benefits show how resilience enhancement helps urban transit systems deal with diverse risks while keeping cities moving safely and sustainably. Therefore, it is necessary to systematically study how to model, evaluate, and enhance resilience in metro networks, which form the basis for the content of this book.

1.4 Organization of the Book

This chapter (Chap. 1) briefly introduces the concept of resilience and reviews the development and significance of urban transit systems. The substantial benefits of resilience in urban transit systems are summarized.

Chapter 2 "Disaster Chains Modeling" describes a hybrid modeling approach that integrates disaster chain theory and complex network analysis to assess vulnerability in urban transit systems. It first reviews traditional methods such as GIS mapping, ISM, and ANP, noting their limitations in capturing transmission dependencies. To address this, the chapter constructs a weighted complex network where disaster events are nodes and their causal links are directed edges, with weights reflecting occurrence frequencies. Analyzing node degrees, eigenvector centrality, and betweenness identifies critical disaster nodes and links. The model is further validated using real-world disaster data, expert interviews, and empirical cases to show how cascading failures propagate in metro operations. Vulnerability metrics at micro and macro levels help pinpoint weak points and optimize risk mitigation strategies. This framework provides a systematic tool for safety engineers and planners to understand disaster evolution, allocate emergency resources effectively, and ultimately strengthen the resilience of urban transit networks.

Chapter 3 "Cascade Failure Modeling" describes a hybrid complex network approach to evaluate how cascading failures impact the vulnerability of urban transit systems. It first reviews static and dynamic network-based simulation methods and highlights the limitations of ignoring cascading effects. To address this, the chapter develops a simulation framework that models passenger load redistribution and tolerance capacity during node failures. Five attack strategies, including degree-based, betweenness-based, and random attacks, are tested to reveal how failures propagate. A detailed case study on Wuhan Metro demonstrates that intentional attacks targeting critical interchange stations cause severe capacity loss, while appropriate tolerance parameters and passenger transfer rates can mitigate cascading effects. This method provides planners with a practical tool to identify key nodes, optimize network robustness, and support informed decision-making for future metro expansions.

Chapter 4 "Evolutionary Dynamics Modeling" describes a probabilistic system dynamics (SD) approach integrated with Monte Carlo (MC) simulation to assess long-term vulnerability in urban metro systems. The chapter reviews existing static and dynamic methods, noting their limitations in handling uncertainty and temporal evolution. It then develops a hybrid framework that models causal feedback among workforce, organizational, equipment, and environmental subsystems, while MC simulation captures global uncertainty. A detailed case study on the Wuhan Metro demonstrates how the model predicts vulnerability trends over 25 years, identifies sensitive factors like ticket pricing and safety training, and tests alternative management strategies. Scenario analyses show that increasing safety investment and maintenance frequency can effectively delay vulnerability growth and enhance system resilience. This approach offers decision-makers a robust tool for long-term planning and proactive risk mitigation in metro operations.

Chapter 5 "Flood Hazard Analysis" develops an evidential reasoning (ER)-based optimality approach to assess and mitigate the flood risks faced by metro stations, particularly under the growing threats of climate change and rapid urbanization. The chapter first analyzes the shortcomings of traditional numerical simulations and multi-criteria decision analysis (MCDA) methods, especially the uncertainty of subjective parameter settings. By treating subjective parameters as intervals and applying an evolutionary optimization algorithm, the proposed method determines the worst-case risk expectation to provide conservative risk estimates. Furthermore, a new combined gradient sensitivity indicator is introduced to identify critical risk factors and guide both short-term and long-term mitigation strategies. A case study on Shanghai metro stations verifies the approach's practicality, showing consistent results with past flood events and current government countermeasures.

Chapter 6 "Fire Hazard Analysis" introduces a systematic framework combining the ASET/RSET method, numerical simulations, and multi-attribute decision analysis (MADA) to evaluate and control fire safety risks in metro stations. Recognizing the severe consequences of fires in underground spaces with dense crowds, the chapter elaborates on calculating the Safety Index (SI) based on the balance between Available Safe Egress Time (ASET) and Required Safe Egress Time (RSET). To overcome the limitations of traditional two-zone models, a more realistic 3D simulation using Fire Dynamics Simulator (FDS) is applied to predict smoke spread, temperature, CO concentration, and visibility. The methodology is validated through a case study of a Wuhan metro station under multiple fire scenarios, revealing critical factors that influence evacuation safety and demonstrating how worst-case scenarios can be effectively identified and mitigated.

Chapter 7 "Passenger Evacuation Analysis" proposes a hybrid framework integrating simulation-based modeling and random forest meta-modeling, systematically analyzing the vulnerability of passenger evacuation in metro stations, and supports the evacuation strategy planning under multiple scenarios. Utilizing AnyLogic simulation software and Random Forest algorithms, the research accurately predicts evacuation performance indicators such as evacuation time, passenger density, and renovation cost, validated through a detailed case study of Singapore's Jurong East MRT station. The chapter identifies key influencing factors, including evacuee behaviors, station structures, and emergency guidance, which significantly affect evacuation outcomes. The framework effectively compares diverse evacuation strategies under different scenarios, including constrained emergency conditions, highlighting trade-offs between different evacuation metrics and supporting more informed and adaptive evacuation planning for metro stations.

Chapter 8 "Facility Operation Planning" proposes a network-based optimization framework for pre-flood station closure and protection planning in metro systems. While existing research largely focuses on post-flood responses like passenger evacuation and system restoration, this chapter emphasizes the underexplored pre-flood preparations, which are crucial for minimizing flood impacts. The framework starts by constructing a time-weighted metro network that accounts for transfer walking times and then assesses network performance using an Integrated Network Performance (INP) metric that combines efficiency and passenger utility. A cost-performance-based greedy algorithm (CPGA) is developed to determine which high-risk stations should receive protective resources and which should be closed, given limited resources. The framework is tested on Shanghai's metro network, demonstrating its effectiveness in enhancing network resilience and informing practical flood management decisions.

Chapter 9 "Emergency Rescue Planning" addresses the optimization of metro fire emergency responses by developing an integrated simulation and multi-objective optimization framework. Recognizing that current responses often rely on experience and may suffer from inefficiencies, the chapter employs Hierarchical Timed Colored Petri Nets (HTCPN) to accurately model the complex interactions among emergency responders, passengers, and evolving fire conditions. The Skyline operator and a fuzzy-based mechanism are then used to identify Pareto-optimal solutions that balance two conflicting objectives: minimizing response time and minimizing firefighter deployment. A case study of a fire at Wuhan's Simenkou metro station shows that the proposed approach significantly reduces both response time and resource use, providing robust decision support for real-time emergency management and contributing to greater resilience and safety in metro operations.

Chapter 10 "Resource Preparedness Planning" presents a multi-condition bilevel optimization (MBO) model for the integrated planning of flood control resource warehouse location and inventory in metro systems. While existing emergency facility studies often overlook the unique operational needs of interconnected metro networks during floods, this chapter addresses the critical challenge of optimizing the location and inventory for resource warehouses tailored to urban transit systems. The methodology begins with a time-weighted metro network model and evaluates system performance using Average Passenger Utility (APU). A GIS-based spatio-temporal storm flood simulation is conducted to estimate inundation depths and resource demands at metro stations. On this basis, the upper-level of the MBO model optimizes warehouse locations and inventory to maximize the weighted average APU across multiple flood conditions, while the lower-level allocates resources to maximize protected passenger flow under each flood condition. A genetic-algorithm-based nested method is developed to solve the model. A case study on the Shanghai metro system demonstrates the model's ability to produce a robust warehouse location-inventory plan outperforming single-condition optimizations, which offers practical insights for metro flood preparedness and resilience.

Chapter 11 "System Repair Planning" introduces a data-driven, simulation-based framework for optimizing urban transit system's recovery strategies under natural disaster-induced disruptions. This chapter proposes a novel two-layer network model that explicitly separates the infrastructure and service layers, with link flow (LF) derived from smart card data representing localized functionality. A fuzzy-set-based mapping method is developed to estimate time-varying LF with uncertainty, enabling high-resolution tracking of service degradation. Multiple node attack protocols—random, simultaneous, and sequential—are simulated to assess network absorptive resilience, while two recovery strategies are compared: one guided by dynamic LF, and the other by traditional recovery cost (RC) metrics. A case study on the Hangzhou metro system reveals that the LF-based strategy outperforms RC in both speed and robustness of service restoration across 18 attack-repair scenarios.

Chapter 12 "Conclusions and Future Directions" summarizes the conclusions and points out future directions. It highlights the book's comprehensive framework that integrates systematic vulnerability modeling, multi-hazard risk evaluation, and resilience-oriented planning to enhance the robustness and adaptability of urban transit systems. Several future directions are put forward, including applying advanced information technologies, improving model portability, robustness, security and safety, and strengthening policy and stakeholder collaboration. These pave a more affordable and effective way to build intelligent, adaptive, and resilient urban transit systems in the face of growing risks and uncertainties.

References

1. Amirzadeh M, Sobhaninia S, Sharifi A (2022) Urban resilience: a vague or an evolutionary concept? Sustain Cities Soc 81:103853
2. Holling CS. Resilience and stability of ecological systems. International Institute for Applied Systems Analysis Laxenburg; 1973
3. Büyüközkan G, Ilıcak Ö, Feyzioğlu O (2022) A review of urban resilience literature. Sustain Cities Soc 77:103579
4. Yodo N, Wang P (2016) Engineering resilience quantification and system design implications: a literature survey. J Mech Des 138:111408
5. Davoudi S, Brooks E, Mehmood A (2013) Evolutionary resilience and strategies for climate adaptation. Plan Pract Res 28:307–322
6. Rus K, Kilar V, Koren D (2018) Resilience assessment of complex urban systems to natural disasters: a new literature review. Int J Disaster Risk Reduct 31:311–330
7. Institute AIT (2009) All hazards risk and resilience prioritizing critical infrastructure using the RAMCAP plus approach. ASME Press
8. Heinimann HR (2016) A generic framework for resilience assessment. An edited collection of authored pieces comparing, contrasting, and integrating risk and resilience with an emphasis on ways to measure resilience, p 90
9. Index CR (2014) City resilience framework. The Rockefeller Foundation and ARUP, p 928
10. Poulin C, Kane MB (2021) Infrastructure resilience curves: performance measures and summary metrics. Reliab Eng Syst Saf 216:107926
11. Galviz CL (2013) Metropolitan railways: urban form and the public benefit in London and Paris c. 1850–1880. Lond J 38:184–202
12. Ovenden M (2019) London underground by design. Penguin UK
13. Acharya SR, Parumog-Pernia M, Morichi S (2012) Evolving concepts in urban transport. In: Transport development in Asian megacities: a new perspective. Springer, pp 15–31
14. Dushkina N (2020) The post-war metro in Moscow and in the former USSR. Values and significance. ICOMOS–Hefte des Deutschen Nationalkomitees 74:66–72
15. Levinson HS, G. AJ, Hoey WF (2012) Light rail since world war II: abandonments, survivals, and revivals. J Urban Technol 19:65–79
16. Zerdoun C (2016) Under ground: subways and metros of the world. Firefly Books
17. Lin D, Nelson JD, Beecroft M, Cui J (2021) An overview of recent developments in China's metro systems. Tunn Undergr Space Technol 111:103783

18. Hou X, Feng C, Ye H, Zuo C (2024) Overview of urban rail transit operations in China's mainland in 2024. Urban Rapid Rail Transit 38:13–19. (in Chinese)
19. Han B, Li Z, Jin Y, Lu F, Hu J, Huang S et al (2025) Statistical analysis of urban rail transit operations worldwide in 2024: a review. Urban Rapid Rail Transit 38:1–12. (in Chinese)
20. Lin D, Broere W, Cui J (2022) Metro systems and urban development: impacts and implications. Tunn Undergr Space Technol 125:104509
21. Wang Y, Feng S, Deng Z, Cheng S (2016) Transit premium and rent segmentation: a spatial quantile hedonic analysis of Shanghai Metro. Transp Policy 51:61–69
22. Wei H (2019) Impacts of China's national vehicle fuel standards and subway development on air pollution. J Clean Prod 241:118399
23. Xiao D, Li B, Cheng S (2020) The effect of subway development on air pollution: evidence from China. J Clean Prod 275:124149
24. Broere W (2016) Urban underground space: solving the problems of today's cities. Tunn Undergr Space Technol 55:245–248
25. Cui J, Broere W, Lin D (2021) Underground space utilisation for urban renewal. Tunn Undergr Space Technol 108:103726
26. Lu C-C, Hwang J-H (2019) Nonlinear collapse simulation of Daikai Subway in the 1995 Kobe earthquake: necessity of dynamic analysis for a shallow tunnel. Tunn Undergr Space Technol 87:78–90
27. He R, Tiong RLK, Yuan Y, Zhang L (2024) Enhancing resilience of urban underground space under floods: current status and future directions. Tunn Undergr Space Technol 147:105674
28. Cong W, Shi L, Shi Z, Peng M, Yang H, Cheng X (2022) Numerical study on the ceiling gas temperature in a subway train with different fire locations. Build Simul 15:549–560
29. Hu J, Wen W, Zhai C, Pei S (2024) A comprehensive review of resilience of urban metro systems: a perspective from earthquake engineering. Tunn Undergr Space Technol 152:105920

18. Hou X, Peng C, Ye H, Zuo C (2024) Overview of urban rail transit operations in China's main-land in 2024. Urban Rapid Rail Transit 38:[illegible] (in Chinese)
19. [illegible]

Part I
System Vulnerability Modeling

Chapter 2
Disaster Chains Modeling

2.1 Introduction

With rapid population growth, increasing congestion, and pollution resulting from rising transportation demand, governments in both developed and developing countries have drawn increasing attention. As a form of large-scale public transportation, urban transit systems (UTS) offer advantages such as high capacity, operational efficiency, and reduced emissions. These characteristics make them an effective solution for alleviating pressure on conventional transportation networks [1]. However, the UTS is a complex network composed of numerous interconnected elements, activities, and events. The high level of interdependency among system components makes metro operations vulnerable to various disaster risks and cascading failures [2]. Therefore, developing effective methods for modeling and assessing system vulnerability is essential for disaster risk reduction.

Several methodologies have been developed to assess vulnerability in UTS, including the use of geographical information systems (GIS) for spatial identification of critical infrastructure, interpretative structural modeling (ISM) for exploring hierarchical relationships among vulnerability factors, and the analytical network process (ANP) for evaluating interdependent components from a multi-criteria perspective [3]. These approaches have advanced the understanding of transit vulnerability and informed disaster risk reduction strategies during system operations. However, most studies primarily emphasize the identification and weighting of external vulnerability factors, with limited attention given to transmission dependencies and internal network structures in disaster scenarios. To address this gap, complex network theory offers a promising alternative for modeling and assessing vulnerability in UTS.

Real-world systems are neither entirely regular nor entirely random and thus cannot be fully described by traditional network theory. To address this limitation, complex network theory incorporates topological features of both scale-free and

L. Zhang et al., *Resilience in Urban Transit Systems*, Advances in 21st Century Human Settlements, https://doi.org/10.1007/978-981-95-4322-9_2

small-world networks [4]. In a complex network, nodes represent distinct entities or concepts, while directed or undirected edges denote the relationships between them [5]. This structure allows the representation of complex systems in a mathematical format, enabling topological analysis. Various network-based metrics, such as node degree, edge betweenness, and average path length, have been developed to quantify structural properties and assess network vulnerability [6].

Given its ability to capture global network properties, complex network theory is well-suited for systematically modeling and assessing the vulnerability of UTS. Several studies have applied this approach to investigate vulnerability in public transportation networks [7–9]. However, these studies primarily focus on the physical structure of metro systems, specifically the stations and their interconnections, while paying limited attention to external disaster sources such as hazardous human behaviors, natural hazards, equipment malfunctions, and management errors. In practice, failures in transit systems rarely occur in isolation. Instead, they often trigger disaster chains that can lead to severe system-wide disruptions [10].

To address these challenges, this chapter develops a hybrid approach that integrates disaster chains and complex networks to model and assess vulnerability in UTS. A weighted complex network is established by combining various disaster chains and their frequencies in the operation of the urban transit system. Several measures are proposed to quantify the vulnerability magnitude of the disaster nodes, edges, and network from both micro-level and macro-level perspectives. The main objective of this research is to propose a novel network analysis approach to effectively identify the vulnerability of disaster events and links in the urban transit system and then to discover the best strategy for improving the resilience of the transit system against disasters.

2.2 Disaster Chain in UTS

Improving emergency management capabilities relies heavily on understanding the formation and progression of disaster chains. Disaster chain modeling allows the quantitative examination of propagation processes, resource allocation needs, and systemic vulnerabilities, offering critical support for preparedness and mitigation planning. Considering the complexity of UTS and the rapid escalation typical of disaster chains, researchers have introduced a variety of analytical frameworks. Notably, probability-based modeling, remote sensing, and complex network modeling methods have been recognized for their effectiveness in capturing the dynamic and multifaceted nature of disasters.

1. **Probability modeling methods.** Probability-based modeling methods focus on the quantitative representation of the relationships among disaster events, a capability that is critical for modeling disaster propagation scenarios. Bayesian networks advance this methodology further by incorporating probabilistic inference mechanisms to effectively manage uncertainties and assess intricate depen-

dencies among disaster events [11]. Additionally, neural networks employ historical data to predict the likelihood and impact of secondary disasters [12], while expert scoring systems integrate expert opinions to evaluate the probabilities of disaster occurrences [13]. However, when applied to metro transit scenarios, these methods often oversimplify real-world interactions, inadequately addressing complex interdependencies and operational constraints.

2. **Remote sensing methods.** Remote sensing methods leverage satellite technology to detect, monitor, and evaluate disaster events rapidly. The advent of advanced geospatial observation techniques has enabled comprehensive assessments of disasters over extensive spatial and temporal scales [14]. Remote sensing data contributes significantly to disaster management by providing timely information regarding disaster impacts and their geographical extents. However, current applications predominantly target geological disasters, such as landslides and earthquakes, with limited exploration in more dynamic and confined environments like metro systems. Therefore, while promising, these methods currently offer restricted applicability for modeling disaster chain propagation within urban transit infrastructures.
3. **Complex network modeling methods.** Complex network theory, conversely, provides considerable promise for capturing the intricacies of disaster chain dynamics within metro systems. Representing disaster chains as complex networks involves modeling each disaster event as a node, with directed edges illustrating causal or consequential relationships between disasters [15]. This kind of representation facilitates the analysis of both disaster propagation paths and system-wide vulnerability assessments, accommodating the complex topology and dynamic interactions characteristic of disaster chains [16]. Complex network approaches excel at integrating diverse disaster events, frequencies, and relationships into unified analytical frameworks, providing robust quantitative measures for disaster resilience planning and emergency response optimization.

In summary, disasters in UTS often escalate rapidly and involve intricate interactions, requiring comprehensive vulnerability assessments. However, current methodologies primarily focus on isolated disasters, neglecting integrated assessments of disaster propagation dynamics and operational constraints. To address this gap, we develop a hybrid modeling approach that combines disaster chain theory and weighted complex network analysis, systematically quantifying disaster vulnerabilities at micro- (individual nodes and edges) and macro-levels (network topology), thus improving vulnerability evaluation and decision-making under complex and dynamic UTS.

2.3 Disaster Chain Modeling for Vulnerability Assessment

To account for the transmission dependencies and internal structures of disaster events within UTS, this chapter integrates disaster chain analysis with complex network theory. The workflow of the proposed approach, as shown in Fig. 2.1, consists of three main stages: disaster chain analysis, complex network construction, and vulnerability analysis and assessment. The resulting complex disaster network offers an effective means of identifying high-risk disaster events and critical links. By analyzing the vulnerability of this network, decision-makers can be supported in reducing disaster risk and enhancing the allocation of emergency resources.

2.3.1 *Disaster Chain Analysis*

The construction of the complex network is grounded in the identification of risk factors and disaster chains. To ensure the completeness and reliability of the network structure, disaster chains within UTS must be systematically analyzed. Currently, there is no widely accepted framework for illustrating the internal relationships among disaster events in these systems. This stage of the methodology focuses on identifying disaster events and their transmission pathways, with disaster chains serving as the foundation for relationship discovery.

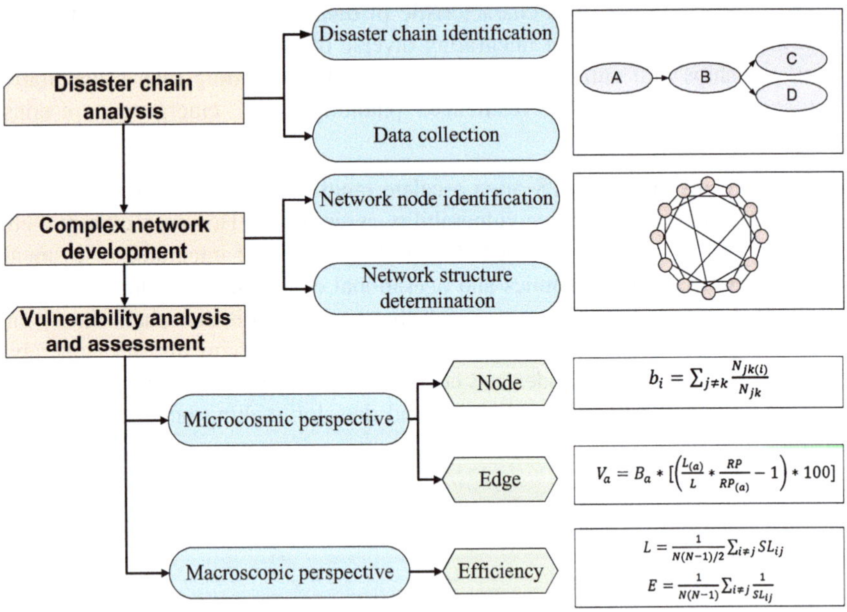

Fig. 2.1 Workflow of the developed complex network approach

Based on the origin of the initiating event, disaster chains in UTS can be categorized into four primary types: those triggered by human behavior, equipment failures, environmental hazards, and management deficiencies. Figure 2.2 presents a simplified example of a disaster chain involving a fire incident. In this scenario, unsafe passenger behavior plays a key role in initiating a man-made fire. The direct cause may be the passengers bringing combustibles into the station or igniting existing flammable materials. The root cause, however, lies in inadequate safety management. Historical records indicate that such incidents can lead to human injuries, stampedes, property damage, and significant operational delays.

The data collection in this study aims to quantify the correlations among disaster events and understand their frequency of occurrence within UTS. To enhance the comprehensiveness of the disaster chain analysis, three types of data sources are utilized: historical records of transit-related disasters, engineering reports, academic literature, and expert interviews with professionals possessing long-term experience in metro system operations.

2.3.2 Complex Network Development

The complex network represents an advancement of random network theory, integrating principles of topology and randomness to characterize real-world networks that lie between complete order and total randomness. This framework is widely applied in scientific research and is particularly effective in describing interpersonal relationships and interactions within complex systems. Two fundamental properties of complex networks are the small-world and scale-free characteristics. The small-world property highlights the network's clustering tendency and transitivity, reflecting tightly connected groups and short path lengths between nodes. Complex networks generally exhibit a higher clustering coefficient than random networks. Figure 2.3 compares regular, random, and small-world networks, illustrating how small-world networks combine structural features of both.

The disaster transmission mechanism in UTS demonstrates a structure similar to the small-world model. It displays both regularity and randomness in the

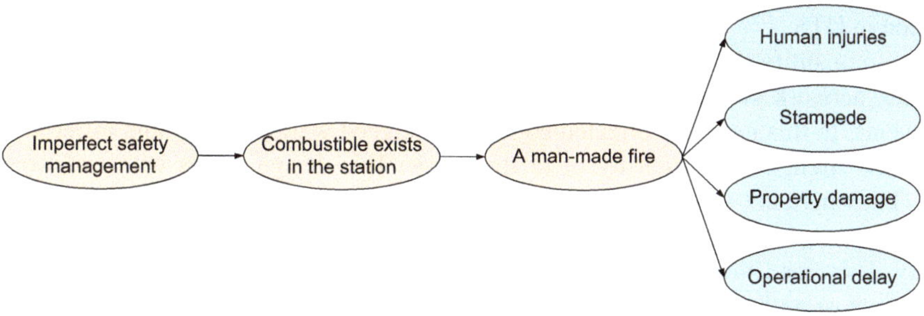

Fig. 2.2 An example of a disaster chain related to a fire accident

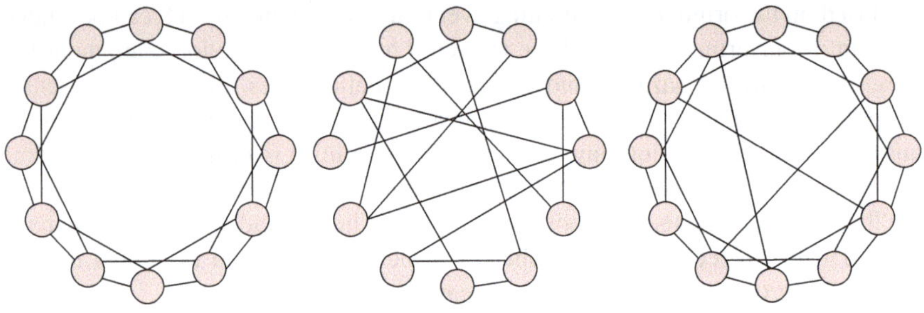

Fig. 2.3 Examples of (**a**) a regular network, (**b**) a random network, and (**c**) a small-world network

propagation of disaster events. Multiple transmission chains may connect two disaster nodes, among which at least one path typically enabling the most efficient information flow within the system. The scale-free property reflects a power-law degree distribution, where a small number of nodes (hubs) have disproportionately high connectivity. This characteristic allows the network to capture varying levels of node importance. In the case of the transit disaster transmission network, certain nodes exhibit significantly higher degrees than others, indicating their critical role in the structure and dynamics of the system.

The topological characteristics of complex networks contribute to their robustness, allowing them to maintain functionality under targeted or random attacks. In this study, a directed and unweighted complex network is constructed to model the transmission of disaster chains in UTS. This network captures both transitivity and robustness, enabling the simulation and interpretation of real-world disaster propagation within transit systems. In the constructed network, disaster events are represented as nodes, while the transmission relationships between events are represented as links. The complex network is developed based on previously identified disaster chains in the urban transit system. Prior to network construction, these disaster chains are treated as independent sequences of events. However, some disaster events appear in multiple chains. These recurring nodes serve as bridges, linking individual chains into an interconnected network. Chains containing identical disaster events can be integrated by merging the corresponding nodes, thereby forming a cohesive structure that reflects the interconnected nature of disaster propagation in UTS.

To quantify the importance and scale of disaster propagation, a weighted complex network is developed to represent the significance of connections between events. In this network, directed edges are assigned weights, reflecting differences in the strength and intensity of relationships. This design aligns with real-world UTS, where not all connections have the same capacity or level of impact. In this chapter, it is assumed that all disaster events result in consequences of equal severity. Therefore, the frequency of occurrence is used to represent the probability or weight of a directed edge in the complex network. Specifically, if a directed edge from node i to node j appears in the pth disaster chain ($p = 1, 2, \ldots, P$), the edge is assigned a weight of 1 for that instance. As more disaster chains include the same edge, the weights are accumulated accordingly. The number of times a directed edge from node i to node j appears across all disaster chains is defined by Eq. (2.1), and the corresponding edge weight, denoted as W_{ij}, is given in Eq. (2.2).

$$T_{ij(p)} = \begin{cases} 1 & \text{If the edge from node } i \text{ to } j \text{ lies in the } p\text{th disaster chain} \\ 0 & \text{Otherwise} \end{cases} \tag{2.1}$$

$$W_{ij} = \sum_{p=1}^{P} T_{ij(p)} \tag{2.2}$$

where $T_{ij(p)}$ represents whether the directed edge from node i to node j lies in the pth disaster chain ($p = 1,2\ldots.,P$), P is the total number of disaster chains discovered, and W_{ij} represents the weight of the directed edge from node i to node j.

Figure 2.4 presents a simplified example illustrating the transformation of disaster chains into a weighted complex network. As shown on the left side of the figure, three disaster chains are identified, with several common nodes (e.g., nodes 2 and 3) appearing across them. These shared nodes and edges are merged to construct an integrated complex disaster network, as depicted on the right. According to Eqs. (2.1) and (2.2), edge weights are calculated based on the frequency with which each directed edge appears in the disaster chains. These weights are displayed above the corresponding edges and are visualized by varying edge thickness to reflect the strength of connections.

2.3.3 *Vulnerability Analysis and Assessment*

Once the complex network is constructed, various types of vulnerability analysis can be conducted to enhance the understanding of disaster risks in UTS. Several commonly used metrics are employed to evaluate network vulnerability. At the node level, weighted degree and betweenness centrality indicate node vulnerability; at the edge level, edge betweenness reflects the vulnerability of connections. At the network level, average path length, network diameter, and weighted degree distribution are used to assess overall system vulnerability. Using these metrics, the

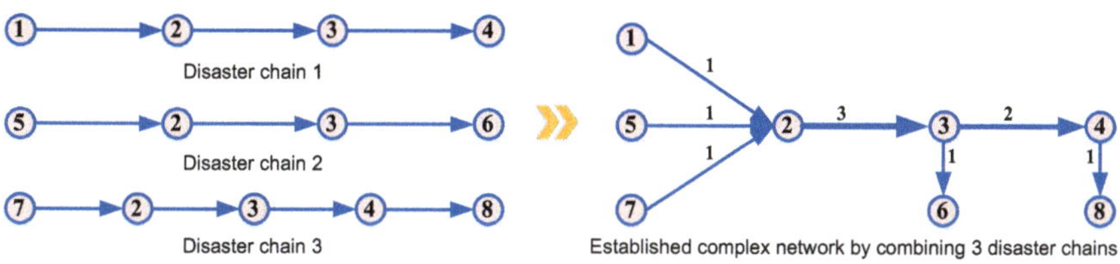

Fig. 2.4 A simple example demonstrates the transformation from disaster chains to the weighted complex network. (Note: The values above the edges indicate the edge weights in the established complex network)

weighted complex network of the urban transit system is analyzed from both local (micro-level) and global (macro-level) perspectives.

2.3.3.1 Microcosmic Analysis

In urban transit management, evaluating the importance and transmission potential of disaster events is critical for effective risk mitigation. This analysis focuses on identifying the significance of individual nodes and edges within the constructed complex network. Nodes with higher influence are more likely to trigger serious consequences during disruptions and, therefore, require greater attention during disaster preparedness.

Two key indicators are used to assess node vulnerability: node degree and betweenness centrality. The network is constructed and analyzed using the Pajek software package. In a directed complex network, node degree is divided into input and output degrees. The input degree refers to the number of nodes influencing a particular node, while the output degree denotes the number of nodes influenced by it. In this study, the total node degree is calculated as the sum of its weighted incoming and outgoing edges, providing a measure of the node's overall influence. This is defined as D_i and calculated using Eq. (2.3).

$$D_i = \sum_{j=1, j \neq i}^{N} \left(W_{ji} + W_{ij} \right) \tag{2.3}$$

where D_i is the value of the node degree of the node i, $i = 1, 2, \ldots, N$, W_{ij} is the weight of the directed edge from node i to node j (see Eq. (2.1)), and N is the total number of nodes in the complex network. Indeed, $\sum_{j=1, j \neq i}^{N} W_{ji}$ represents the node input degree while $\sum_{j=1, j \neq i}^{N} W_{ij}$ represents the node output degree for node i.

Eigenvector centrality (EI) is also employed to reflect the importance of a node based not only on its direct connections but also on the centrality of the nodes to which it is connected. In this context, edges connected to highly influential nodes contribute more to a node's eigenvector centrality. Based on the weighted adjacency matrix $W = [W_{ij}]_{N \times N}$, EI for the node i, represented by EI_i, is given by Eq. (2.4).

$$\mathrm{EI}_i = \frac{1}{\lambda} \sum_{j=1}^{N} W_{ij} \mathrm{EI}_j \tag{2.4}$$

where EI_i is the value of the eigenvector centrality of the node i, λ is a constant value, and N is the total number of nodes in the complex network.

Node betweenness is another measure to assess the betweenness centrality of the nodes in the network. It measures the fraction of the shortest path that passes through a node, relative to the total shortest path in the network. The node betweenness of the node i, represented by NB_i, is given by Eq. (2.5).

$$\mathrm{NB}_i = \sum_{j \neq k} \frac{N_{jk(i)}}{N_{jk}} \tag{2.5}$$

where NB_i is the value of the betweenness of the node i, $N_{jk(i)}$ is the number of the shortest path across the node i between the node j and node k, and N_{jk} is the number of the shortest path between the node j and k.

In addition to node-level indicators, edge vulnerability is also assessed to determine how critical certain links are to the network's stability. Edge betweenness, computed similarly to node betweenness using Eq. (2.6), measures the role of an edge as a bridge in information flow. High edge betweenness indicates greater importance in maintaining network connectivity. To evaluate the overall structure of the network, the average path length is used to reflect the average shortest distance between node pairs, indicating network dispersion. This is calculated as Eq. (2.7).

$$\mathrm{EB}_a = \sum_{i \neq j} \frac{N_{ij(a)}}{N_{ij}} \tag{2.6}$$

$$L = \frac{1}{N(N-1)} \sum_{i \neq j} \mathrm{SL}_{ij} \tag{2.7}$$

where EB_a is the value of the betweenness of the edge a, $N_{ij(a)}$ is the number of the shortest path that goes across the edge a between the node i and j, N_{ij} is the number of the shortest path between the node i and j, L is the average path length of the network, and SL_{ij} is the length of the shortest path between the node i and j.

To integrate both local and global effects, a composite edge vulnerability measure EV_a is introduced. This index considers both the edge's importance in information transmission and its contribution to network dispersion. Equation (2.8) provides the formula for calculating this combined measure.

$$\mathrm{EV}_a = \mathrm{EB}_a * \left[\left(\frac{L_{(a)}}{L} * \frac{\mathrm{RP}}{\mathrm{RP}_{(a)}} - 1 \right) * 100 \right] \tag{2.8}$$

where EV_a is the vulnerability of the edge a, $L_{(a)}$ is the average path length of the complex network with removing the edge a, RP is the number of reachable pairs of the complex network, and $\mathrm{RP}_{(a)}$ is the number of reachable pairs of a complex network.

2.3.3.2 Macroscopic Analysis

The macro-level analysis focuses on evaluating the overall structure and performance of the complex network, particularly in terms of its dispersion and efficiency. Two widely accepted metrics for this purpose are network diameter and global efficiency. Network diameter measures the extent of dispersion in the network and is defined as the longest of all the shortest paths between any two nodes. A larger diameter implies a more dispersed structure. The calculation method for network diameter D is provided in Eq. (2.9).

$$D = \max\left(\mathrm{SL}_{ij}\right) \tag{2.9}$$

Global efficiency assesses the effectiveness of information exchange across the entire network. Efficiency is inversely proportional to the shortest path length between nodes: the shorter the path, the higher the efficiency of communication. Therefore, global efficiency is computed as the average of the inverses of the shortest path lengths between all node pairs. The corresponding formula for global efficiency E is shown in Eq. (2.10).

$$E = \frac{1}{N\left(N-1\right)}\sum\nolimits_{i \neq j} \frac{1}{\mathrm{SL}_{ij}} \tag{2.10}$$

where SL_{ij} is the length of the shortest path between the node i and j.

2.4 Experimental Studies

2.4.1 Data and Resources

To establish a reliable complex disaster network, it is essential to analyze the contributing factors and causal chains associated with failures in urban transit operations. Given the critical role of metro systems in urban transportation, this study focuses on metro-related disasters. Drawing on real accident cases and credible sources, a wide range of studies have investigated various incidents, including explosions, fires, and stampedes. To identify causal relationships among these disaster events, three primary sources of data are utilized: historical records, engineering reports, and literature, and expert interviews.

Historical accident data provides a foundational basis for identifying disaster nodes and edges. This research collects information on past operational incidents in urban rail transit, which serves to determine key events (nodes) and their causal linkages (edges). For example, in a documented accident on August 19, 2019, a passenger fell from a New Jersey Transit train due to a door opening on the wrong side. The investigation revealed that the incident was caused by an operational error. From this case, the primary disaster node is identified as "passenger falls on the track," the direct causal node is "door opened on the wrong side," the indirect causal node is "misoperation," and the resulting node is "human injuries." In addition to historical data, engineering reports, and academic literature are valuable for understanding the mechanisms underlying transit system failures. These sources support the identification of causal factors based on experimental studies and simulations.

Interviews with domain experts further complement the data collection process by offering qualitative insights grounded in practical experience. As a flexible and widely used method in qualitative research, interviews help capture expert perspectives on disaster evolution in metro systems. This study employs semi-structured interviews, which balance open-ended discussion with a focused framework. The interview process consists of three main stages. First, participants are presented with preliminary disaster chains derived from historical data and asked to provide comments based on their knowledge. Next, they are invited to propose additional events, causes, or consequences that might not have been captured in existing records. Finally, participants are asked to construct new or revised disaster chains based on their professional judgment. The resulting data allow for the comprehensive identification of disaster events and the formation of detailed disaster chains within the urban transit system.

2.4.2 Disaster Chains Identification

Based on the collected data, disaster chains associated with operational failures and accidents in UTS can be identified. From the perspective of source origins, these disaster chains can be classified into four main categories: human behavior, equipment failures, environmental hazards, and management deficiencies. Details of each type of disaster chain are elaborated below.

2.4.2.1 Human Behavior Chains

Human behavior is a particularly critical factor in the safety performance of urban rail transit systems, encompassing both passenger-related and staff-related risks.

As shown in Fig. 2.5, irregular passenger behavior includes a range of unsafe actions such as dangerous boarding or alighting, dropping personal belongings onto the tracks, falling onto the tracks, initiating fires, engaging in terrorist activities, or releasing toxic gases within the station. These behaviors can result in severe consequences and should be carefully considered in the development of disaster response strategies. For instance, rushing onto or off trains while doors are closing or obstructing door closure may cause property damage and lead to delays or injuries due to door clamping. If a passenger's belongings fall onto the tracks, timely intervention can minimize the consequences, though train delays are often unavoidable. When a passenger falls onto the track, injury is likely, and the associated rescue efforts typically cause further delays. In extreme cases, such events can result in fatalities, particularly when trains are operating at high speeds. Fires initiated by passengers may lead to property loss, stampedes, and casualties. Acts of terrorism can delay train operations, incite panic, cause property damage and injuries, or, in the case of toxic gas release, result in poisoning-related casualties and widespread chaos in the station.

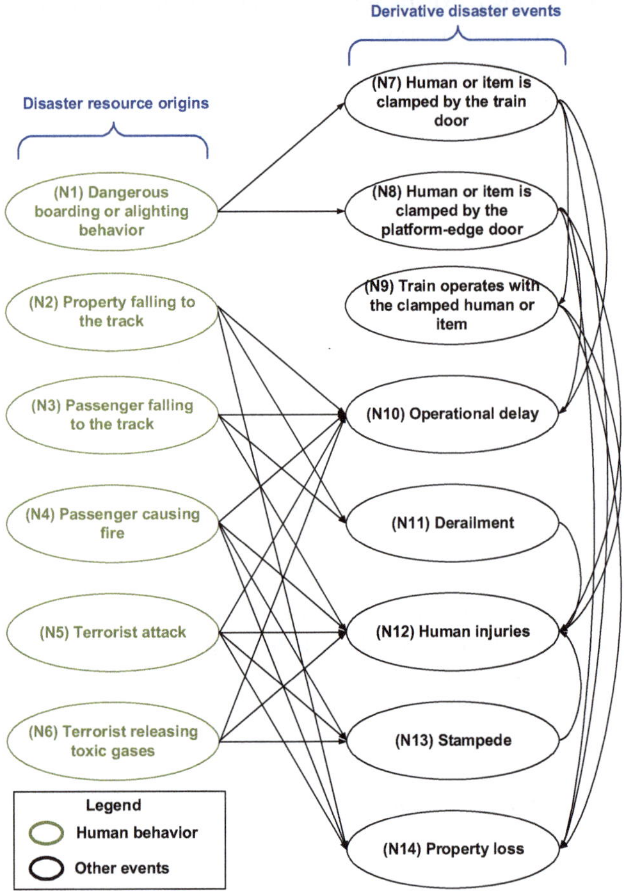

Fig. 2.5 Disaster chains with irregular passenger behaviors as risk source origins

In addition to passenger behavior, irregular staff behavior also poses significant risks, as shown in Fig. 2.6. Three common forms of staff-related errors are identified: misoperations by train stewards, dispatchers, and maintenance technicians. Misoperations by stewards, such as improper door handling, may lead to passengers or objects being caught in doors, contributing to stampedes or delays. Failure to maintain safe passageways between train and platform doors can also lead to accidents. Dispatchers, responsible for monitoring and controlling train operations, must coordinate train movements effectively to prevent delays and rear-end collisions. Errors in this role may compromise the entire system's safety. Maintenance technicians are charged with ensuring the functionality of rolling stock and infrastructure. However, if technicians lack sufficient professional training, their substandard maintenance work may cause equipment malfunctions or failures, which can jeopardize operational safety.

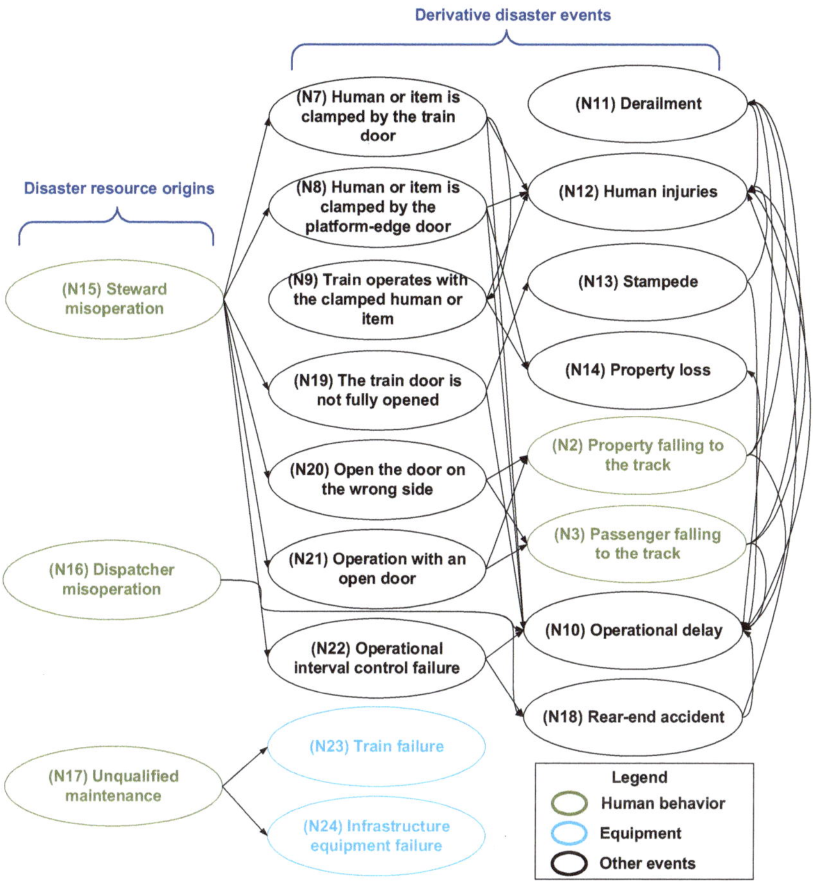

Fig. 2.6 Disaster chains with irregular staff behaviors as risk resource origins

2.4.2.2 Equipment Based Disaster Chains

Equipment-related disaster chains involve failures in both onboard systems and fixed infrastructure within the urban rail transit system. Mechanical wear and aging during operation are the primary causes of equipment failure, while system overloading can further degrade performance. This type of disaster chain analysis focuses on two main categories: train equipment failures and infrastructure equipment failures. As shown in Fig. 2.7, common train equipment failures include malfunctions in the door control system and the train operation control system. Door control failures can result in door malfunctions, such as improper opening or closing, leading to safety hazards. Failures in the train operation control system may cause loss of speed regulation, compromising safe train operation. Infrastructure equipment includes tracks, roadbeds, and platform edge doors. Failures in these

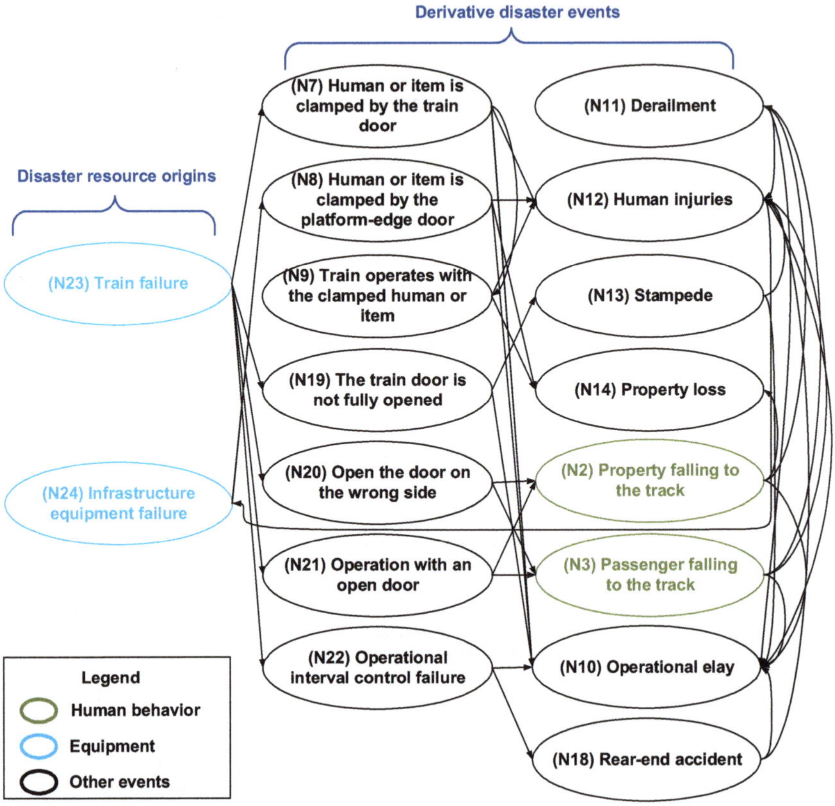

Fig. 2.7 Disaster chains with equipment failures as risk resource origins

components can disrupt normal operations, potentially causing train delays or even passenger injuries.

2.4.2.3 Environmental Disaster Chains

Environmental hazards such as snowfall, earthquakes, and foreign object intrusions pose significant risks to the safe operation of urban rail transit systems. Figure 2.8 presents disaster chains originating from these environmental factors. Snowfall is among the most critical hazards, as snow accumulation on tracks can lead to ice formation, increasing the likelihood of operational delays or derailments. Earthquakes also present substantial threats by damaging infrastructure, disrupting operations, and causing severe casualties. In addition, foreign objects entering the track area can result in serious accidents. For example, items such as balloons or long textile fabrics may obstruct train movement or damage rail infrastructure, leading to service delays or derailments.

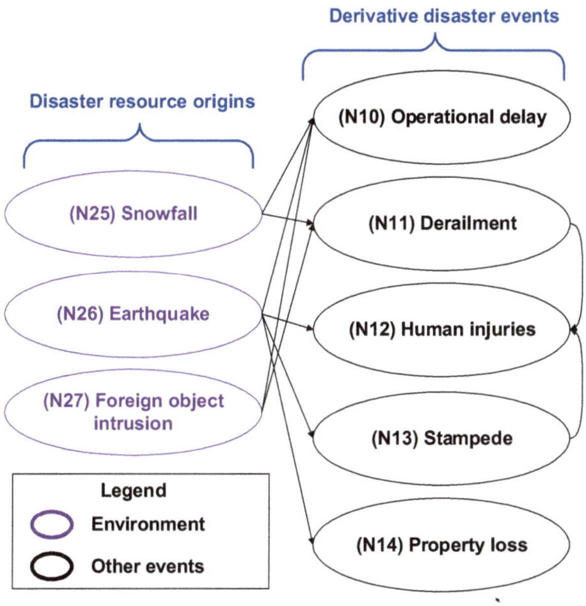

Fig. 2.8 Disaster chains with environment hazards as risk resource origins

2.4.2.4 Management-Based Disaster Chain

Management-related disaster chains highlight the critical role of safety management in urban rail transit operations, where safety management serves as a key component in disaster prevention and control. An effective safety management system can significantly reduce the likelihood of disasters. Therefore, the quality of safety management should be considered when assessing the vulnerability of the transit system within the disaster transmission network. Figure 2.9 presents disaster chains triggered by deficiencies in management practices. This study categorizes management-related factors into two dimensions: maintenance management and behavior-based safety management. Inadequate maintenance systems can lead to substandard maintenance procedures, increasing the risk of equipment and infrastructure failures and ultimately triggering potential disaster events. Behavior-based safety management aims to address human factors through organizational oversight. Promoting staff safety awareness and enforcing standardized operational procedures helps reduce the occurrence of human error. However, the absence or inadequacy of safety education, training, or regulatory enforcement can result in staff misoperations and insufficient supervision of passenger behavior, both of which may contribute to further disaster escalation.

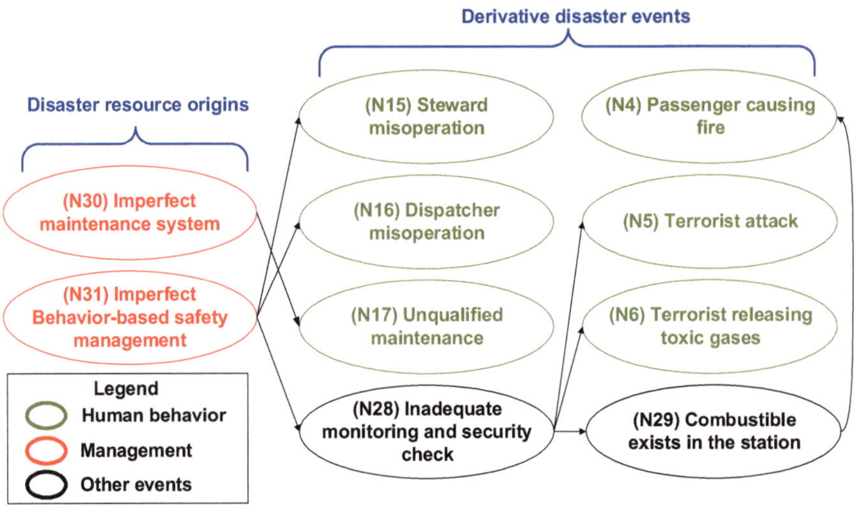

Fig. 2.9 Disaster chains with imperfect management as risk resource origins

2.4.3 Disaster Chain Network Development

Based on the identified disaster chains originating from irregular human behaviors, equipment failures, environmental hazards, and management deficiencies (see Figs. 2.5, 2.6, 2.7, 2.8, and 2.9), a complex network of urban transit disasters is constructed. In this network, disaster events are represented as nodes, and their causal relationships are modeled as directed edges. Figure 2.10 illustrates the integrated disaster network that incorporates various types of disaster chains within the urban transit system. Following the procedures outlined in Sect. 2.3, a weighted complex network is established to simulate disaster transmission, as shown in Fig. 2.11. The network consists of 31 nodes, each representing a specific disaster event, and directed edges indicating transmission paths between them. For example, the event "unqualified maintenance" (Node 17) leads to "train failure" (Node 23) and "infrastructure equipment failure" (Node 24), forming two directed links from Node 17 to Nodes 23 and 24 in the network. A detailed description of each node is provided in Table 2.1.

2.4.4 Analysis of Results

The disaster network of the urban transit system is complex, as the disaster events have a dependence on each other. Both micro- and macro-level analyses are performed in this section to model and assess the vulnerability of the complex disaster network.

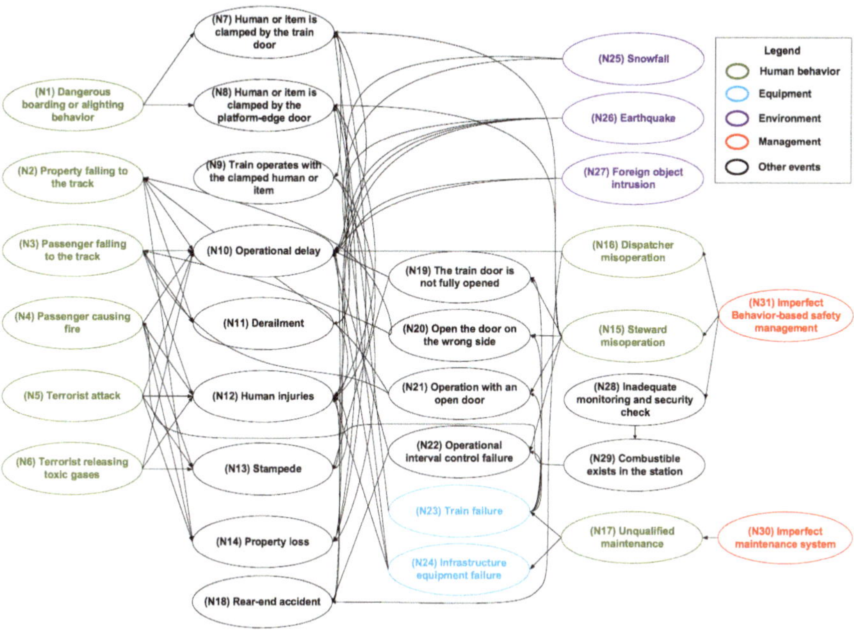

Fig. 2.10 The integrated disaster network considers various types of disaster chains in the urban transit system

2.4.4.1 Microcosmic Analysis

The microcosmic analysis aims to reveal the vulnerability of nodes and edges in the built complex disaster network of the urban transit system. Based on the weighted complex network as shown in Fig. 2.11, this study uses Eqs. (2.3)–(2.5) to obtain the node degree, eigenvector centrality, and node betweenness in the disaster network, respectively. Equations (2.6) and (2.8) are used to obtain the edge betweenness and edge vulnerability, respectively. Figures 2.12 and 2.13 illustrate the vulnerability results of the 31 nodes and 73 edges in the built complex network in terms of different vulnerability measures. The results are analyzed as follows.

1. Node 15 (Steward misoperation), Node 17 (Unqualified maintenance), and Node 23 (Train failure) are identified as the most important nodes in transferring disasters to other nodes, while Node 12 (Human injuries) and Node 10 (Operational delay) are considered the most important nodes in terms of strength centrality. As shown in Fig. 2.12a, c, Nodes 15, 17, and 23 are identified as the top three nodes in both node degree and betweenness, indicating that these three disaster events are very likely to act as the transmission medi in disaster chains or networks. Indeed, a considerable number of disaster events tend to have causal relationships with these three events. As shown in Fig. 2.12b, Node 12 turns out to be the most important node, while Node 10 is the second most important node in terms of the eigenvector centrality. This indicates that these two events display a

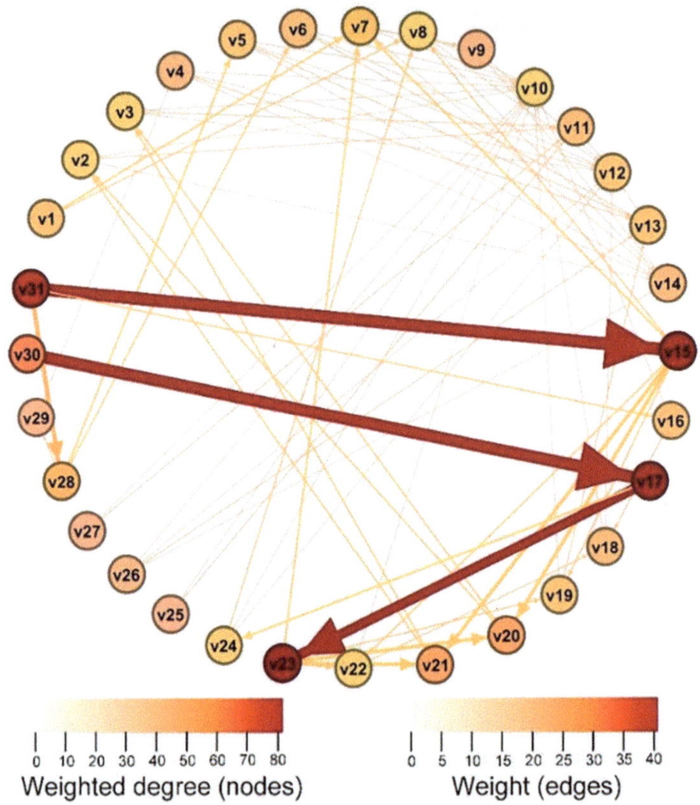

Fig. 2.11 Map of the discovered weighted complex network for modeling the disaster transmission in the urban transit system. (Note: The color of the node reveals the weighted node degree using the above Eq. (2.3). The color of the edge reveals the edge weight, which is also visualized in terms of the edge's thickness)

strong centrality or importance in disaster occurrence. Therefore, these five disaster events should be paid special attention to in disaster prevention and mitigation.

2. The edge with the highest value of vulnerability is E38 (i.e., the edge from Node 17 to Node 23), which should be paid high attention to in urban transit safety management. The betweenness of the edge is one of the most used measures for evaluating the vulnerability of the edge. As shown in Fig. 2.13b, the edge betweenness of most edges is in the interval [1, 7], while several edges have outstanding values representing their relatively high vulnerability. E38 has the highest betweenness with a value of 25 in the built-weighted complex network since E38 is the only path that connects Node 17 with all the consequences of Node 23. That is to say, the management of E38 would contribute significantly to enhancing the operational safety of the urban transit system. Therefore, the operator and regulator of the urban transit system should pay more attention to

Table 2.1 Description of disaster events in the established complex network for the transit system

Node	Description of disaster events	Node	Description of disaster events
N1	Dangerous boarding or alighting behavior	N17	Unqualified maintenance
N2	Property falling onto the track	N18	Rear-end accident
N3	Passengers falling onto the track	N19	The train door is not fully open
N4	Passenger causing fire	N20	Open the door on the wrong side
N5	Terrorist attack	N21	Operation with an open door
N6	Terrorist releasing toxic gases	N22	Operational interval control failure
N7	Human or item is clamped by the train door	N23	Train failure
N8	Human or item clamped by the platform-edge door	N24	Infrastructure equipment failure
N9	Human or item clamped by the platform-edge door	N25	Snowfall
N10	Operational delay	N26	Earthquake
N11	Derailment	N27	Foreign object intrusion
N12	Human injuries	N28	Inadequate monitoring and security check
N13	Stampede	N29	Combustible exists in the station
N14	Property loss	N30	Imperfect maintenance system
N15	Steward misoperation	N31	Imperfect behavior-based safety management
N16	Dispatcher misoperation		

minimizing the possibility of train failure (Node 23) caused by unqualified maintenance (Node 17), which would be efficient in reducing the disaster risk.

3. The results of the proposed edge vulnerability measurement show a strong correlation with traditional edge betweenness while demonstrating greater sensitivity in identifying high-vulnerability edges. As illustrated in Fig. 2.13a, a clear linear relationship is observed between the two methods across the 73 edges, with an adjusted R^2 value of 82.74%. This consistency supports the validity of the proposed approach. However, traditional edge betweenness has certain limitations. In particular, the differences among high betweenness values are often minimal, making them difficult to distinguish. Moreover, this method typically overlooks the edge's impact on the overall dispersion and integrity of the network. To address these issues, Eq. (2.8) is proposed to evaluate edge vulnerability by incorporating average path length and the number of reachable node pairs, thereby accounting for both network dispersion and structural connectivity. As shown in Fig. 2.13b, the proposed method enables clearer differentiation among edge vulnerabilities, especially for edges with high risk. This demonstrates that the proposed metric offers a more precise and comprehensive evaluation of edge vulnerability compared to traditional betweenness-based approaches.

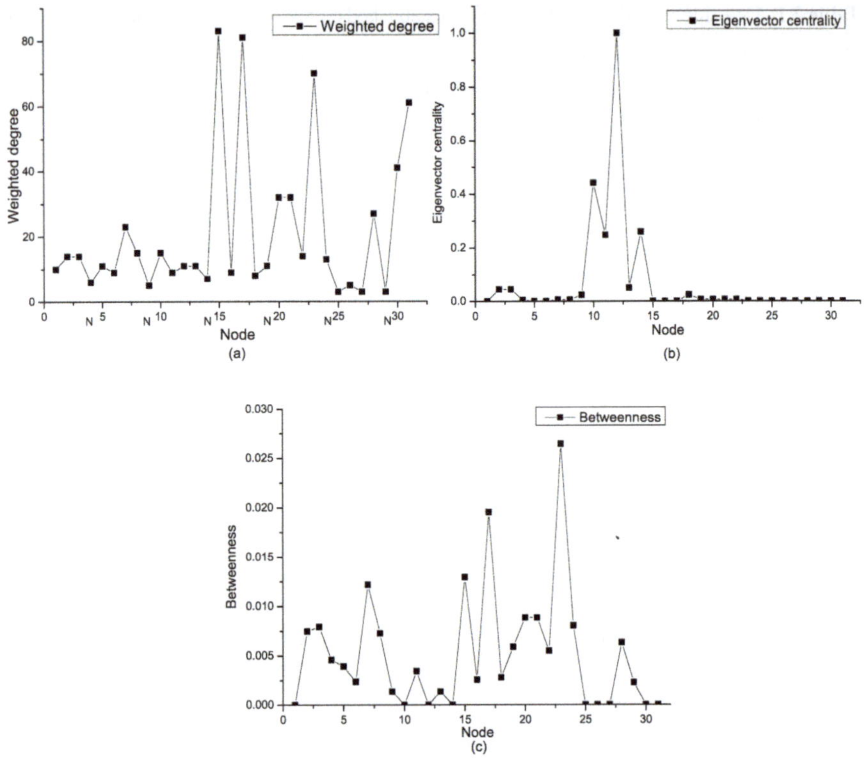

Fig. 2.12 Vulnerability results of the 31 nodes in the built complex network: (**a**) Node degree; (**b**) eigenvector centrality; (**c**) Node betweenness

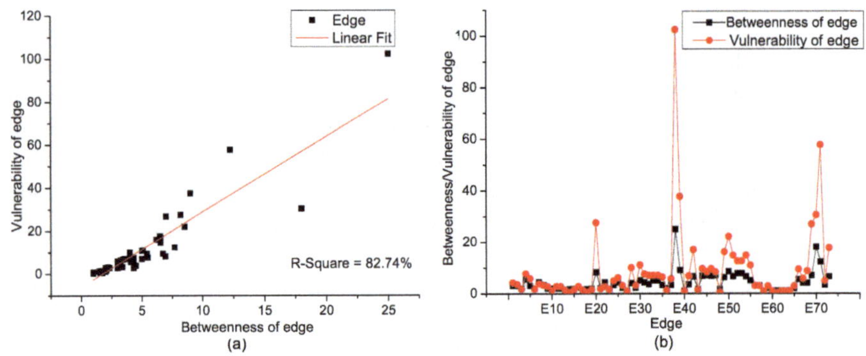

Fig. 2.13 Vulnerability results of the 73 edges in the built complex network: (**a**) Linear fit between the edge betweenness and the edge vulnerability; (**b**) Comparison between the edge betweenness and edge vulnerability

2.4.4.2 Macroscopic Analysis

The macroscopic analysis aims to reveal the vulnerability of the total urban transit network. This section analyzes the network vulnerability by using the network diameter (D) and the efficiency of the network (E). In the real world, there is a high demand to reduce the disaster risk for safety improvement in the transit system, aiming to enhance the value of network efficiency. To improve the operational safety of the transit system, three different strategies (i.e., Strategies 1, 2, and 3) are developed for benchmark purposes. Specifically, Strategy 1 focuses on monitoring and guiding the passenger behavior to minimize the potentially relevant risks (i.e., the objective is to remove the disaster nodes 1–4), Strategy 2 pays more attention to the staff behavior (i.e., the objective is to remove the disaster nodes 15–17), while Strategy 3 highlights the improvement of the terrorist regulation for eliminating the risk sources (i.e., the objective is to remove the disaster nodes 5–6). Figure 2.14 illustrates the distribution of the shortest lengths between nodes in the built complex network (original network, see Fig. 2.11) and the modified complex networks with three different strategies. Table 2.2 reveals the efficiency of the original complex network and the modified complex networks with three different strategies. The results are investigated as follows.

1. The efficiency of the built disaster complex network (see Fig. 2.11) is measured to be 0.1244, which would be the anchor for further improvement. As shown in Fig. 2.14a, the diameter of the built disaster complex network in the urban transit system is 5, indicating there are no reachable routes with more than 5 edges in the shortest path in the built complex network. As shown in Table 2.2, the number of reachable routes with the shortest length of 1, 2, 3, 4, and 5 is 73, 65, 24, 8, and 1, respectively, indicating a descending trend. By using Eq. (2.10), the network efficiency is calculated to be 0.1244. Due to the nature of the disaster network, the network efficiency is expected to be as low as possible. That is because cutting down disaster transmission in the disaster network is helpful in

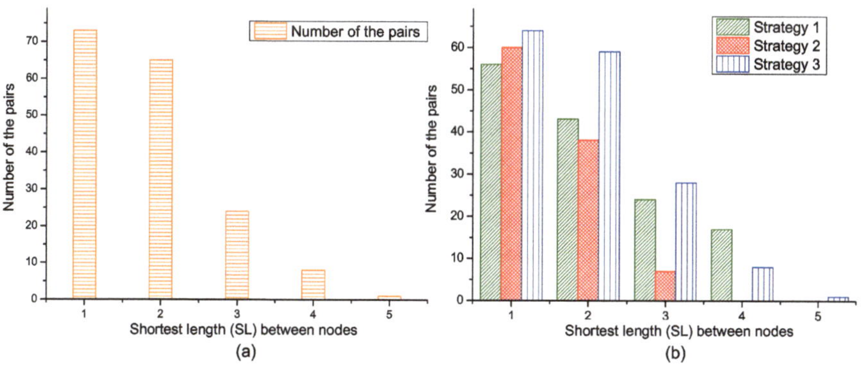

Fig. 2.14 Distribution of the shortest length between nodes in (**a**) the built complex network (original network) and (**b**) the modified complex networks with three different strategies

Table 2.2 Comparison of the efficiency results between the original complex network and the modified complex networks with three different strategies

Item	Original network	Strategy 1 Modified network	Strategy 1 Change	Strategy 2 Modified network	Strategy 2 Change	Strategy 3 Modified network	Strategy 3 Change
SL = 1	73	56	−17	60	−13	64	−9
SL = 2	65	43	−22	38	−27	59	−6
SL = 3	24	24	0	7	−17	28	4
SL = 4	8	17	9	0	−8	8	0
SL = 5	1	0	−1	0	−1	1	0
Total	171	140	−31	105	−66	160	−11
Efficiency	0.1244	0.0965	−2.79%	0.0875	−3.69%	0.1129	−1.15%

preventing the occurrence of further severe accidents or failures. As a result, any strategy leading to the reduction of network efficiency is effective for reducing disaster risk and enhancing operational safety in the urban transit system.

2. Strategy 2 turns out to be the most efficient solution among the three strategies, given that the cost of those three strategies is the same. The implementation of new strategies would modify the original complex network through the removal and/or addition of nodes/edges. That is to say, the proposed three strategies will result in three different modified networks. In the same way, network efficiency can be recalculated by using Eq. (2.10). As shown in Table 2.2, the network efficiency of the modified complex networks with Strategies 1, 2, and 3 is calculated to be 0.0965, 0.0875, and 0.1129, respectively. In comparison to the original complex network with an efficiency of 0.1244, Strategy 2 can achieve the most significant reduction in terms of the network efficiency, which is −3.69% (i.e., (0.0875–0.1244)/0.1244). Given that the requested resources or costs for those three strategies are the same, Strategy 2 becomes the best cost-effective solution for improving the operational safety of the urban transit system.

2.5 Merits Analysis

To further validate the reliability of the proposed approach and offer insights into disaster risk reduction, a robust analysis is conducted on the constructed network using different attack strategies. Attack analysis is a widely used method to assess how network performance is affected by specific disruptions. The most common approach involves removing nodes and/or edges from the network and recalculating efficiency or other performance indicators. In the context of the developed urban transit disaster network, node removal represents the proactive elimination of specific disaster events through control measures, while edge removal simulates the interruption of transmission pathways to block disaster propagation. Network efficiency is adopted as the primary indicator to quantify how different attack strategies affect the network's robustness.

Seven node attack strategies are examined in total: initial degree removal (ID), initial betweenness removal (IB), recalculated degree removal (RD), recalculated betweenness removal (RB), initial eigenvector centrality removal (IEI), recalculated eigenvector centrality removal (REI), and random removal. The first three (ID, IB, IEI) are based on the node measures calculated from the initial network structure. For example, in ID removal, nodes are removed sequentially in descending order of initial node degree. Similarly, IB and IEI removal strategies remove nodes based on their initial betweenness and eigenvector centrality, respectively. In contrast, RD, RB, and REI strategies recalculate node measures after each removal step, allowing for dynamic adjustment of removal priorities. These six methods are intentional strategies, with ID, IB, and IEI considered local attacks and RD, RB, and REI categorized as global attacks. The seventh strategy, random removal, serves as a baseline for comparison.

Regarding the edge attack analysis, there are five attacking strategies in total, including initial edge vulnerability removal (IEV removal), initial edge betweenness removal (IEB removal), recalculated edge vulnerability removal (REV removal), recalculated edge betweenness removal (REB removal), and random removal. Specifically, IEV (or IEB) removal indicates the edges will be ranked in the descending order of the edge vulnerability (or betweenness) in the original complex network and then be removed one by one, starting from the edge with the highest vulnerability (or betweenness). REV (or REB) removal uses information from the dynamic complex network, where the edges will be removed one by one, starting from the edge with the highest vulnerability (or betweenness), with these values recalculated after every step. Obviously, the first four edge attack strategies (i.e., IEV, IEB, REV, and REB removals) are intentional, while the last one is random. Particularly, IEV and IEB removals are local strategies, while REV and REB removals are global ones.

In this research, we analyze attacks on the top nine nodes or edges which are considered the topmost important nodes or edges. Figure 2.15 illustrates the tendency of network efficiency in the node and edge attack analysis. Table 2.3 reveals

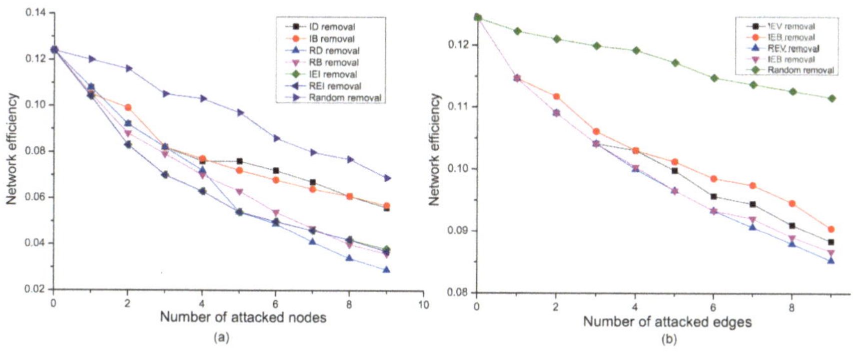

Fig. 2.15 The tendency of the efficiency of the network in different attack strategies: (**a**) Node attack analysis; (**b**) Edge attack analysis

Table 2.3 Impact of attacked nodes on the network efficiency in different node attack strategies

	ID removal		IB removal		RD removal		RB removal		IEI removal		REI removal		Random removal	
Run	Node	*E*	Node	*E*	Node	*E*	Node	*E*	Node	*E*	Node	*E*	Node	*E*
0	Null	0.1	Null	0.12	Null	0.12	Null	0.12	Null	0.124	Null	0.1	Null	0.1
1	15	0.1	23	0.11	15	0.11	23	0.11	12	0.104	12	0.1	16	0.1
2	17	0.1	17	0.1	17	0.09	15	0.09	10	0.083	10	0.1	26	0.1
3	23	0.1	15	0.08	23	0.08	24	0.08	14	0.07	14	0.1	7	0.1
4	31	0.1	7	0.08	28	0.07	28	0.07	11	0.063	11	0.1	9	0.1
5	30	0.1	20	0.07	10	0.05	2	0.06	13	0.054	13	0.1	29	0.1
6	20	0.1	21	0.07	3	0.05	3	0.05	2	0.05	2	0.1	8	0.1
7	21	0.1	24	0.06	2	0.04	4	0.05	3	0.046	3	0	5	0.1
8	28	0.1	3	0.06	13	0.03	18	0.04	18	0.042	18	0	27	0.1
9	7	0.1	2	0.06	7	0.03	11	0.04	9	0.038	30	0	3	0.1

Table 2.4 Impact of attacked edges on the network efficiency in different edge attack strategies

	IEV removal		IEB removal		REV removal		REB removal		Random removal	
Run	Edge	*E*	Edge	*E*	Edge	*E*	Edge	*E*	Edge	*E*
0	Null	0.124	Null	0.124	Null	0.124	Null	0.124	Null	0.124
1	17–23	0.115	17–23	0.115	17–23	0.115	17–23	0.115	2–14	0.122
2	31–15	0.109	30–17	0.112	15–31	0.109	15–31	0.109	1–8	0.121
3	17–24	0.104	31–15	0.106	17–24	0.104	17–24	0.104	26–13	0.120
4	30–17	0.103	17–24	0.103	29–4	0.100	31–28	0.100	4–12	0.119
5	7–9	0.100	7–23	0.101	31–28	0.097	4–29	0.097	23–21	0.117
6	29–4	0.096	7–9	0.099	7–9	0.093	7–9	0.093	20–3	0.115
7	23–7	0.094	23–20	0.098	19–13	0.091	20–3	0.092	5–10	0.114
8	31–28	0.091	23–21	0.095	22–18	0.088	20–2	0.089	6–13	0.113
9	19–13	0.088	4–29	0.091	11–12	0.085	21–3	0.087	19–10	0.112

the impacts of the attacked nodes on the network efficiency of different node attack strategies. Table 2.4 reveals the impacts of the attacked edges on the network efficiency in different edge attack strategies. The results are discussed as follows.

1. All the intentional attack strategies display better performance for reducing the network efficiency than random attack strategies, indicating that the discovery of the important nodes and edges by the developed approach can be meaningful. As shown in Fig. 2.15a and Table 2.2, after nine steps of node attack, the six intentional attack strategies, including ID, IB, IEI, RD, RB, and REI removals, result in the final network efficiency $E = 0.056, 0.057, 0.029, 0.036, 0.038$, and 0.037, respectively, which are all lower than that in the random attack with a final network efficiency $E = 0.069$. That is to say, the recognition of the topmost important nodes helps to reduce network efficiency in a faster manner. As shown in Fig. 2.15b and Table 2.4, after nine steps of edge attack, the four intentional attack strategies, including IEV, IEB, REV, and REB removals, result in the final

network efficiency $E = 0.088$, 0.090, 0.085, and 0.087, respectively, which are lower than that in the random attack with a final network efficiency $E = 0.112$. The results of the random removal strategies display lower impacts on network efficiency than the intentional strategies, indicating that the discovered topmost vulnerable nodes/edges are verified to be truly the most important in the built complex network.

2. Global attack strategies are more effective than local strategies in both node and edge analyses, suggesting that dynamically recalculating node and edge metrics is essential for accurately assessing network vulnerability. As illustrated in Fig. 2.15a and Table 2.3, after nine rounds of node removal, the global strategy RD yields a final network efficiency of 0.029, compared to 0.056 for the local strategy ID. Similarly, RB results in an efficiency of 0.036, while IB achieves 0.057. In the edge attack analysis (Fig. 2.15b and Table 2.4), the global strategies REV and REB result in efficiencies of 0.085 and 0.087, respectively, which are lower than those from the local strategies IEV (0.088) and IEB (0.090). These findings highlight that each attack alters the network structure and affects the relative importance of nodes or edges, thereby reshaping the ranking used in targeted attacks. Consequently, incorporating dynamic updates through global strategies provides a more accurate vulnerability assessment. Furthermore, the REV strategy achieves lower efficiency than REB, indicating that the proposed edge vulnerability measure more effectively identifies critical edges than traditional edge betweenness. This further confirms the robustness of the proposed metric.

2.6 Chapter Summary

To systematically assess the vulnerability of disaster events in UTS and support improved decision-making, this chapter develops a hybrid approach integrating disaster chain analysis and complex network modeling. Based on identified disaster chains originating from irregular human behaviors, equipment failures, environmental hazards, and management deficiencies, a weighted complex network comprising 31 nodes and 73 edges is constructed to model disaster transmission pathways. The proposed method enables the evaluation of vulnerabilities at the node, edge, and network levels, offering a deeper understanding of disaster propagation dynamics. This approach provides a practical decision-support tool for safety engineers to identify bottlenecks and weak links in transit system operations, thereby informing proactive control measures to enhance operational safety.

The empirical findings of this chapter are summarized as follows: (1) Node degree and betweenness centrality effectively identify highly vulnerable nodes, which should be prioritized in safety management strategies; (2) The calculated network efficiency is 0.1244, serving as a baseline for assessing and improving system resilience; (3) The method identifies optimal strategies for improving operational safety, using network efficiency as a performance benchmark; (4) All intentional attack strategies significantly reduce network efficiency compared to random attacks, validating the method's effectiveness in identifying critical vulnerabilities.

The novelty of this work lies in its hybrid integration of disaster chain theory with complex network modeling to assess the vulnerability of UTS. By constructing a weighted complex disaster network where disaster events and their transmission paths are represented as nodes and edges, weighted by real-world frequencies, the proposed model offers a systematic method to pinpoint weak links in UTS and support more effective planning of disaster prevention and mitigation.

References

1. He Y, Zhao, Yang, K.-L., Tsui (2023) Short-term forecasting of origin-destination matrix in transit system via a deep learning approach. Transportmetrica A Transp Sci 19:2033348. https://doi.org/10.1080/23249935.2022.2033348
2. Yin X, Wu J (2023) Research on the performance recovery strategy model of Hangzhou metro network based on complex network and tenacity theory. Sustainability 15:6613. https://doi.org/10.3390/su15086613
3. Chen H, Zhang L, Ran L (2021) Vulnerability modeling and assessment in urban transit systems considering disaster chains: a weighted complex network approach. Int J Disaster Risk Reduct 54:102033. https://doi.org/10.1016/j.ijdrr.2020.102033
4. Meng Y (2024) Vulnerability comparisons of various complex urban metro networks under multiple failure scenarios. Sustainability 16:9603. https://doi.org/10.3390/su16219603
5. Qi Q, Meng Y, Zhao X, Liu J (2022) Resilience assessment of an urban metro complex network: a case study of the Zhengzhou metro. Sustainability 14:11555. https://doi.org/10.3390/su141811555
6. Meng Y, Qi Q, Liu J, Zhou W (2022) Dynamic evolution analysis of complex topology and node importance in Shenzhen metro network from 2004 to 2021. Sustainability 14:7234. https://doi.org/10.3390/su14127234
7. Zhang J, Wang S, Wang X (2018) Comparison analysis on vulnerability of metro networks based on complex network. Phys A Stat Mech Appl 496:72–78. https://doi.org/10.1016/j.physa.2017.12.094
8. Xing Y, Lu J, Chen S, Dissanayake S (2017) Vulnerability analysis of urban rail transit based on complex network theory: a case study of Shanghai metro. Public Transp 9:501–525. https://doi.org/10.1007/s12469-017-0170-2
9. M'cleod L, Vecsler R, Shi Y, Levitskaya E, Kulkarni S, Malinchik S, Sobolevsky S (2017) Vulnerability of transportation networks: the New York City subway system under simultaneous disruptive events. Procedia Comput Sci 119:42–50. https://doi.org/10.1016/j.procs.2017.11.158
10. Sun L, Huang Y, Chen Y, Yao L (2018) Vulnerability assessment of urban rail transit based on multi-static weighted method in Beijing, China. Transp Res A Policy Pract 108:12–24. https://doi.org/10.1016/j.tra.2017.12.008
11. Lu Y, Qiao S, Yao Y (2025) Risk assessment of typhoon disaster chain based on knowledge graph and Bayesian network. Sustainability 17:331. https://doi.org/10.3390/su17010331
12. Su Y, Rong G, Ma Y, Chi J, Liu X, Zhang J, Li T (2022) Hazard assessment of earthquake disaster chains based on deep learning—a case study of Mao County, Sichuan Province. Front Earth Sci 9. https://doi.org/10.3389/feart.2021.683903

13. Huang L, Chen T, Deng Q, Zhou Y (2023) Reasoning disaster chains with Bayesian network estimated under expert prior knowledge. Int J Disaster Risk Sci 14:1011–1028. https://doi.org/10.1007/s13753-023-00530-w
14. Shangguan D, Li D, Ding Y, Liu J, Anjum MN, Li Y, Guo W (2021) Determining the events in a glacial disaster chain at Badswat glacier in the Karakoram range using remote sensing. Remote Sens 13:1165. https://doi.org/10.3390/rs13061165
15. Guo H, Huang C, Zhang C, Shao Q (2024) A novel comprehensive system for analyzing and evaluating storm surge disaster chains based on complex networks. Front Mar Sci 11. https://doi.org/10.3389/fmars.2024.1510791
16. Jiao L, Luo Q, Lu H, Huo X, Zhang Y, Wu Y (2024) Research on the urban rail transit disaster chain: critical nodes, edge vulnerability and breaking strategy. Int J Disaster Risk Reduct 102:104258. https://doi.org/10.1016/j.ijdrr.2024.104258

Chapter 3
Cascading Failure Modeling

3.1 Introduction

Urban transit systems (UTS) have become increasingly critical components of urban infrastructure due to their high convenience and efficiency in public transportation. Mature transit networks feature extensive service coverage and complex interdependencies to meet growing demands for efficient mobility [1]. However, such complexity also poses significant challenges for operational management. Hence, vulnerability assessment is essential for ensuring the safety and security of urban transit operations. In highly interdependent systems, the failure of a single component can have far-reaching effects on other system elements, emphasizing the need for comprehensive vulnerability evaluation [2]. For instance, the disruption of one station can redistribute service loads and trigger cascading failures across the network [3]. Therefore, effective methods are required to assess system vulnerability, particularly under scenarios involving cascade failures.

In response to this need, complex network theory has emerged as a powerful tool for vulnerability assessment in UTS. By representing stations as nodes and tunnel sections as edges, this approach captures real-world structural dependencies [4]. Complex networks often exhibit small-world and scale-free properties, under the assumption of homogeneous transport capabilities. They have been extensively applied to analyze metro system robustness and component-level vulnerabilities [5, 6]. However, studies that primarily focus on static structural analysis often fail to account for changes in network performance following disruptions. Specifically, they tend to overlook how network performance evolves when nodes or edges fail. In reality, disruptions can propagate through the system, causing cascading failures and compounding impacts on operational capacity [7]. To address this limitation, cascade failure theory offers a promising direction for simulating failure mechanisms in complex systems. This chapter builds upon this perspective by modeling

L. Zhang et al., *Resilience in Urban Transit Systems*, Advances in 21st Century Human Settlements, https://doi.org/10.1007/978-981-95-4322-9_3

vulnerability evolution under cascade failures, incorporating dynamic changes in node significance during sequential attacks.

Cascade failure refers to a dynamic process within a network in which the failure of certain nodes or edges triggers subsequent failures in other components, potentially leading to overall network collapse [8]. In cascade failure theory, the load of a node is typically represented by its degree or other vulnerability indicators, while the node's capacity is defined as a function of its load and an adjustable tolerance parameter. When a node fails, its load is redistributed among adjacent nodes. If the redistributed load exceeds the capacity of a node, it also fails, which initiates a new round of load redistribution [9]. This mechanism has been widely applied in various domains, including transportation, power grids, water supply networks, and other complex systems [10].

Building on this foundation, the main objective of this research is to propose a hybrid approach to assess the vulnerability of the urban transit network under cascade failures. Therefore, a novel holistic approach is developed to assess the vulnerability of the urban transit system toward sustainable development. The cascade failure mechanism is incorporated into the established complex network to simulate the process in a dependent system, where the failure of one or a few system components can trigger the failure of other components. Several attack strategies are used to test the system vulnerability, and the impacts of the tolerance parameter and passenger transfer rate on the loss of transport capacity are explored. The developed approach can be used as a decision tool to assess the vulnerability in the existing transit system and optimize the planning of the new transit system.

3.2 Complex Network Simulation

Complex network-based simulation offers valuable insights into how metro systems respond to failure and disruption scenarios. By representing metro networks as graph structures, researchers can systematically analyze how structural configurations influence performance, robustness, and the propagation of vulnerabilities. Such simulations enable the quantitative assessment of systemic risks and provide guidance for designing more resilient transit infrastructures. Recognizing the intricate and interconnected nature of metro systems, researchers have proposed various simulation methods grounded in complex network theory. Among these, static network models and dynamic network models have been widely adopted to address different analytical objectives.

1. **Static network-based simulation methods.** Static network-based simulation methods focus on the intrinsic structural properties of the metro system. Topological analysis is the most common static method, utilizing metrics such as degree centrality, betweenness, clustering coefficient, and average path length to identify critical nodes and evaluate network efficiency [11]. These analyses help reveal structural weaknesses and inform design improvements. Another widely

used static approach is structural robustness analysis, which simulates node or edge removals to assess the connectivity and fault tolerance of the network [12]. Furthermore, redundancy and accessibility evaluations use the topology to measure alternative path availability and service fairness, particularly for edge or peripheral zones within the network [13]. Although these methods provide valuable insights into inherent system vulnerabilities, they are limited to pre-disruption conditions and cannot capture the cascading nature of failures.

2. **Dynamic network-based simulation methods.** Dynamic network-based simulation methods model how disruptions propagate through the metro network over time. Among these, cascade failure simulation is a central approach for capturing systemic vulnerability. This method explores how the failure of a single node or edge, such as a station closure or track failure, can trigger a chain reaction affecting large portions of the system [14]. Cascade models are often based on threshold mechanisms, load redistribution rules, or capacity overflow dynamics, reflecting how network components interdependently respond to local disruptions. Time-varying topology simulation is another important dynamic method, modeling temporary structural changes—such as station closures due to maintenance or localized hazards—and their impacts on connectivity and accessibility [15]. Additionally, data-driven network dynamics leverage historical incident data or real-time monitoring to inform simulation behavior [16]. Methods such as graph neural networks or reinforcement learning are increasingly used to learn risk propagation patterns or optimize intervention strategies based on network topology.

In summary, complex network-based simulation provides a valuable framework for analyzing metro system vulnerabilities. Static models help uncover structural weaknesses and identify robust characteristics, whereas dynamic models offer deeper insights into the time-dependent and systemic nature of disruptions, particularly those caused by cascading failures. Building on these foundations, this research proposes a novel hybrid approach to assessing the vulnerability of urban transit networks under cascade failure scenarios. By incorporating the cascading failure mechanism into the complex network model, the proposed method captures the interdependencies among system components, reflecting real-world operational dynamics more accurately. This enables a more comprehensive assessment of vulnerability and provides practical guidance for optimizing both the operation and planning of sustainable transit systems.

3.3 Cascading Failure Modeling in UTS

To account for real-world cascading effects in UTS, this chapter proposes a hybrid approach that integrates complex network modeling with cascade failure mechanisms. Figure 3.1 presents the overall workflow of the proposed methodology. The process consists of three main stages: (1) complex network construction, (2)

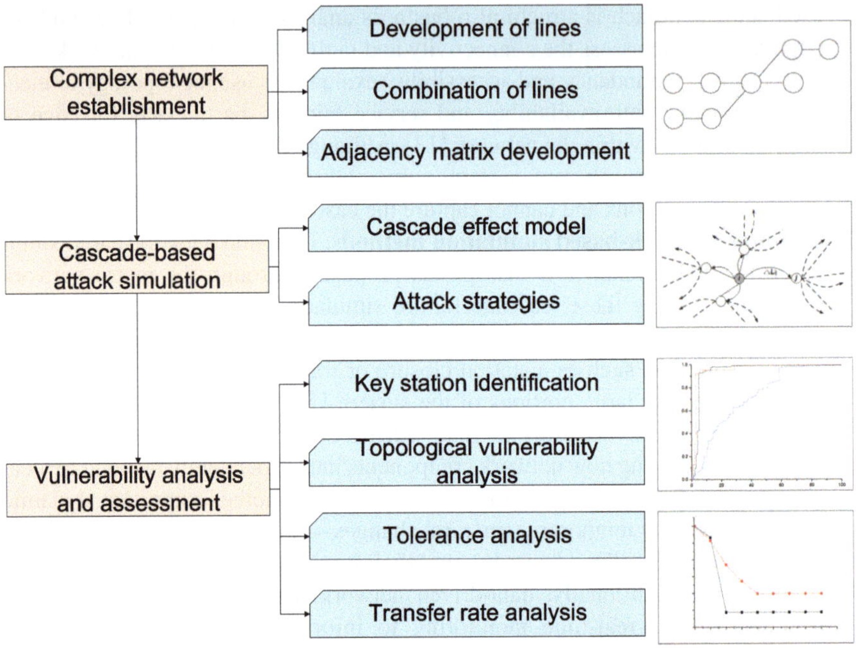

Fig. 3.1 Workflow of the developed approach for vulnerability assessment with cascade failures

cascade-based attack simulation, and (3) vulnerability analysis and assessment. Each of these stages is elaborated in the following sections.

3.3.1 Complex Network Establishment

Complex network theory provides a foundation for modeling real-world systems with scale-free and small-world properties. These characteristics enable the simulation of metro systems that exhibit regular patterns and numerous short paths. In this chapter, the metro complex network is constructed based on a realistic case, following a three-stage process: defining nodes and lines, constructing the network, and developing the adjacency matrix.

The first step involves identifying metro stations as nodes and tunnel sections as edges. A metro system typically consists of several lines, which are interconnected through interchange stations. Each line serves as a structural unit in the network, in which the nodes represent individual stations and the edges represent tunnel connections between them. Since passenger movement between stations is bidirectional, these edges are modeled as undirected. While most station connections are regular, interchange stations create shortcuts across the network, reflecting its

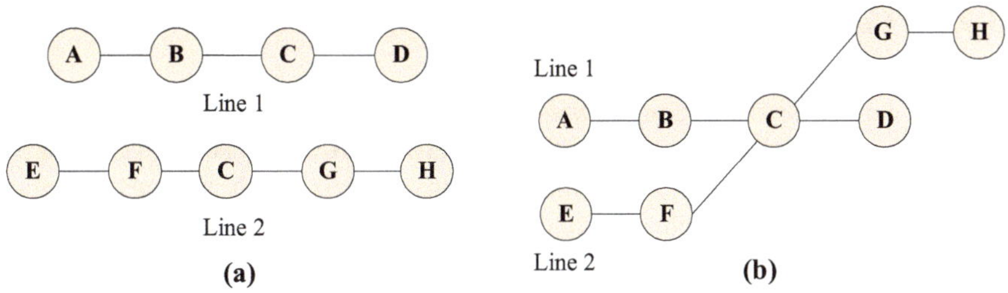

Fig. 3.2 A simple example for constructing a complex network in the transit system: (**a**) Two simple metro lines; (**b**) The established complex network linked by the shared metro station

small-world nature. The varying degrees of these interchange stations also conform to the power-law distribution characteristic of scale-free networks.

After defining nodes and lines, the metro network is assembled by merging stations shared across different lines. Interchange stations, which serve two or more lines, allow passengers to transfer between routes at a single location. Therefore, shared nodes in different lines are consolidated. As illustrated in Fig. 3.2, Line 1 consists of nodes A–B–C–D, and Line 2 includes nodes E–F–C–G–H. Since station C appears in both lines, the two can be merged to form an integrated network. Following this logic, all metro lines are connected to construct the complete complex network.

To enable algorithmic processing, the constructed network is transformed into an adjacency matrix. Based on the defined nodes and edges, the undirected graph $G = (V, U)$ is created, where V is the set of nodes and U is the set of edges. The corresponding $N \times N$ adjacency matrix A is developed based on the graph G, as defined in Eqs. (3.1)–(3.4). Since the network is unweighted, the matrix entry $a_{ij} = 1$, if there is an edge between the node i and node j, and 0 otherwise.

$$V = \{ v_i \mid i = 1,2,3 \ldots N \} \tag{3.1}$$

$$U = \{ u_{ij} \mid v_i, v_j \in V \} \tag{3.2}$$

$$A = \left[a_{ij} \right]_{N \times N} \tag{3.3}$$

$$a_{ij} = \begin{cases} 1, (v_i, v_j) \in \mathrm{U} \\ 0, (v_i, v_j) \notin U \end{cases} \tag{3.4}$$

where V is the node set, U is the edge set, and N is the number of nodes. A is the adjacency matrix, a_{ij} is the matrix element, and N is the number of nodes.

3.3.2 Cascade-Based Attack Simulation

Cascade failure describes a process in which the failure of one or more nodes leads to the successive failure of others within the network. A widely adopted framework for modeling this behavior is the load–capacity model, which assumes that each node or edge bears a certain load under normal conditions. When a node fails, its load will be redistributed to adjacent nodes. Each node has a tolerance capacity, defined by a safety threshold, that determines whether it can sustain both its initial and redistributed loads. This capacity is quantified using a tolerance parameter q, which represents the node's ability to absorb additional load.

In metro systems, node load reflects the volume of passengers passing through a station. In this chapter, the initial load of each node is defined as a function of the product of its degree and the sum of the degrees of its neighboring nodes. A scaling parameter a is introduced to adjust the load intensity, as defined in Eq. (3.5). The node's tolerance capacity, denoted as C_i, is given by Eq. (3.6).

$$L_i = \left[K_i \left(\sum_{j \in \Gamma_i} K_j \right) \right]^a \tag{3.5}$$

$$C_i = (1+q) L_i \tag{3.6}$$

where L_i is the original load of node i, K_i is the value of the degree of node i, a is an adjustable parameter, C_i is the tolerance capacity of the node i, $q \geq 0$ is a tolerance parameter, and Γ_i is the set of nodes that connect with the node i.

Once a node fails, its load is redistributed to its neighboring nodes. If the redistributed load exceeds a neighboring node's capacity, that node will also fail, causing another round of redistribution. As illustrated in Fig. 3.3, when node i becomes invalid, its load is transferred to adjacent nodes. If node j cannot sustain the updated load, it will fail as well, triggering further propagation. Incorporating metro network topology with cascade failure theory, the model assumes that passengers are

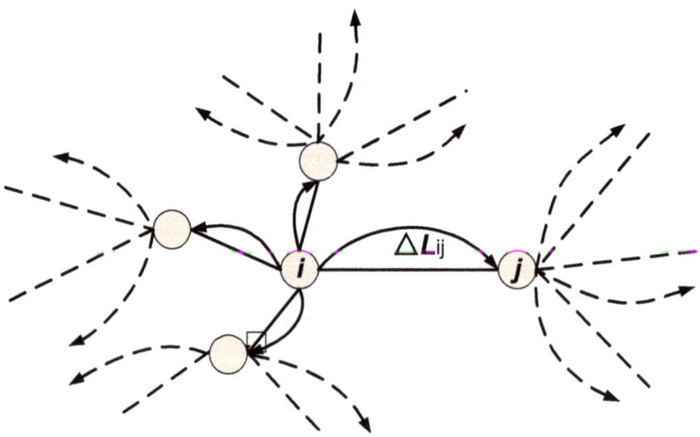

Fig. 3.3 Cascade failure in the transit system

redirected to nearby stations on the same metro line when one station becomes unavailable. Load redistribution is therefore influenced by the transport capacity of adjacent stations. Stations with higher initial loads are likely to receive greater shares of the redistributed burden. However, not all passengers will choose to transfer; some may opt for alternative transportation modes. To reflect this, a passenger transfer rate is introduced to quantify the proportion of load actually reassigned to nearby stations. The redistribution formula is defined in Eq. (3.7) as a function of the relative original loads of neighboring nodes.

$$\Delta L_{ij} = \beta * L_i \frac{L_j}{\sum_{n \in \Gamma_i} L_n} = \beta * L_i \frac{\left[K_j \sum_{m \in \Gamma_j} K_m\right]^{\alpha}}{\sum_{n \in \Gamma_i} \left[K_n \sum_{f \in \Gamma_n} K_f\right]^{\alpha}} \tag{3.7}$$

where L_j is the original load of node j, ΔL_{ij} is the extra load from the attacked node i to the related node j, and β is the passenger transfer rate.

The cascading effect in the metro network is primarily governed by two adjustable parameters: the tolerance parameter and the passenger transfer rate. The tolerance parameter serves as an internal system factor that indicates how much additional load a station can absorb, while the transfer rate represents an external behavioral factor that determines how much passenger flow is redistributed. A higher tolerance parameter increases the system's ability to mitigate cascading failures, whereas a higher transfer rate intensifies pressure on nearby stations. These two parameters are interrelated, and striking an appropriate balance between them is essential to reduce network vulnerability and improve operational resilience. Although the model parameters are initially determined based on a hypothetical normal operating condition, actual system performance may vary due to factors such as changes in passenger volume or train schedules. Several factors influence a station's tolerance capacity, including the number of ticket gates, stairway width, escalator speed, and the train schedules. Therefore, it is necessary to dynamically adjust both the tolerance parameter and the passenger transfer rate to adapt to changing operational conditions and to ensure robust system performance.

The established complex network functions as an integrated system that combines topological structure, passenger transfer behavior, and transport capacity to simulate cascade failure mechanisms, model various attack strategies, and assess network vulnerability. Figure 3.4 presents the flowchart of the cascade-based attack simulation procedure. The simulation begins with all nodes in an unfailed state, and key parameters are initialized based on the actual conditions of the metro system. When selected nodes are attacked according to a predefined strategy, they become unavailable for handling passenger loads. Their loads are then redistributed to adjacent nodes following the network's topology. A recursive loop is initiated to assess whether the redistributed load exceeds each node's tolerance capacity. If the updated load surpasses a node's capacity, the node fails, triggering further redistribution. This process continues iteratively until no additional failures occur. The cumulative loss of transport capacity is recorded as the outcome, representing the extent of

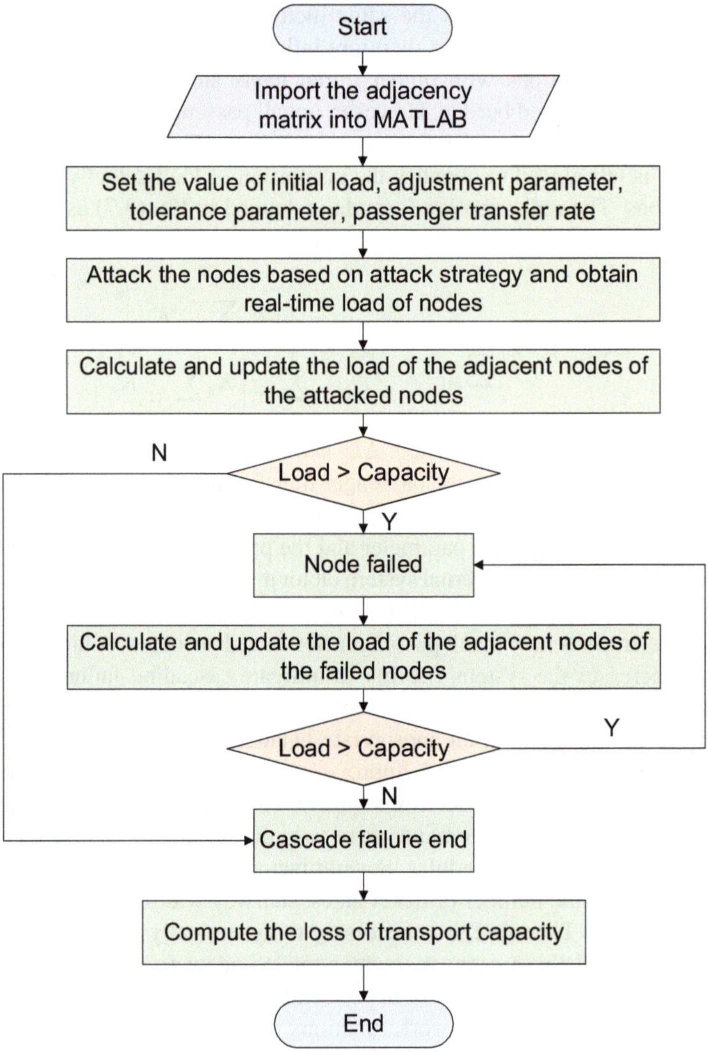

Fig. 3.4 Flowchart of procedures for performing cascade-based attack simulation

network vulnerability. During the simulation, the system dynamically updates the passenger loads of all stations in each iteration. The tolerance capacity and passenger transfer rate are factored into the calculation to determine whether a station can absorb additional load or will contribute to a subsequent cascading failure.

To simulate real-world cascade failures in metro networks and evaluate system vulnerability, it is essential to adopt appropriate attack strategies. In general, network attack strategies fall into two categories: intentional attacks and random attacks. Intentional attacks can be further classified into static and dynamic approaches. These strategies rely on structural information about the network to target nodes ranked by their importance. In static intentional attacks, all nodes are ranked in advance, and attacks proceed sequentially according to this ranking. In dynamic intentional attacks, node rankings are updated after each attack step. In contrast, random attacks target nodes without regard to their structural roles, simulating unexpected or disordered disruptions.

To determine which node should be attacked first in intentional attacks, centrality measures such as node degree and betweenness serve as reference indicators. Nodes with higher values on these metrics are considered more critical and are attacked first. Accordingly, two main types of intentional attack strategies are employed in this chapter. The degree-based strategies include the initial degree (ID) and recalculated degree (RD) attacks, while the betweenness-based strategies include the initial betweenness (IB) and recalculated betweenness (RB) attacks. In the ID or IB strategy, nodes are ranked based on their initial degree or betweenness, and attacks are carried out sequentially without reevaluation. In contrast, the RD or RB strategy recalculates the degree or betweenness of nodes after each attack step, updating the ranking accordingly. In summary, this chapter employs five attack strategies to evaluate network vulnerability: ID, RD, IB, RB, and random attacks.

3.3.3 Vulnerability Analysis and Assessment

Following the establishment of the complex network and the execution of cascade-based attacks, a vulnerability assessment is conducted to evaluate the impact of node failures on the metro network. This assessment employs two key indicators: the rate of connected Origin-Destination (OD) pairs and the Travel Efficiency Rate (TER).

In a complex network, an OD pair refers to two nodes that can be connected via a valid path, representing the origin and destination of a trip. The total number of OD pairs reflects the connectivity level of the network: a higher number indicates broader system accessibility and better structural integrity. In metro systems, full accessibility requires that all stations remain connected. When nodes are attacked and become non-functional, the number of reachable OD pairs decreases, signifying a reduction in network scale and integrity. To quantify this change, the OD pair rate is introduced, defined as the ratio of the number of connected OD pairs in the attacked network to that in the original network, as shown in Eq. (3.8).

$$\mathrm{ODR} = \frac{P_a}{P_O} \tag{3.8}$$

where ODR is the rate of the connected OD pairs, P_a is the number of OD pairs of the attacked network, and P_O is the number of OD pairs of the original network.

Beyond connectivity, transport efficiency is also a critical component of network performance. The Travel Efficiency Rate (TER), derived from the concept of global efficiency, evaluates how efficiently passengers can move through the network. In a metro system, global efficiency is influenced by the number of stations passengers must pass through to reach their destination. Node failures reduce shortcut paths and increase travel distances between reachable pairs, resulting in a decline in overall efficiency. As defined in Eq. (3.9), TER is calculated as the ratio of global efficiency in the attacked network to that in the original network.

$$\mathrm{TER} = \frac{\mathrm{Eff}_a}{\mathrm{Eff}_O} \tag{3.9}$$

where TER is the transport efficiency rate, Eff_a is the global efficiency of the updated network after the attack, and Eff_O is the global efficiency of the original network.

To integrate both connectivity and efficiency impacts, the loss of transport capacity (ΔTC) is used as a comprehensive vulnerability indicator. ΔTC is a normalized metric that combines the weighted OD pair rate and TER, as described in Eq. (3.10). The corresponding weight coefficients can be adjusted according to specific analytical goals (Eq. (3.11)). ΔTC captures the degree of functional degradation caused by node failures and subsequent cascade effects and is positively correlated with overall system vulnerability. In the cascade-based attack simulation, ΔTC serves as the final output and is used to quantify the extent of network vulnerability.

$$\Delta TC = \mu_1 (1 - \mathrm{ODR}) + \mu_2 (1 - \mathrm{TER}) \tag{3.10}$$

$$\mu_1 + \mu_2 = 1 \tag{3.11}$$

where ΔTC is the loss of transport capacity, μ_1 and μ_2 are the weight coefficients which are both set to be 0.5 in this research, ODR is the rate of the connected Origin-Destination pairs, and TER is the transport efficiency rate.

3.4 Experimental Studies

3.4.1 Data and Resources

Wuhan, the largest city in central China, covers an area of 8569.15 square kilometers and had a resident population of approximately 11 million in 2018. The city's extensive territory and large population place considerable pressure on its transportation infrastructure. The Wuhan Metro System was developed to address daily mobility demands through energy-efficient and environmentally friendly means. As of mid-2020, the metro system comprises nine lines and 198 operational stations, including 24 interchange stations (Fig. 3.5). The system accommodates an average daily ridership of over three million, with peak volumes reaching approximately 4.5 million passengers. This accounts for more than 40% of all public transportation trips in the city.

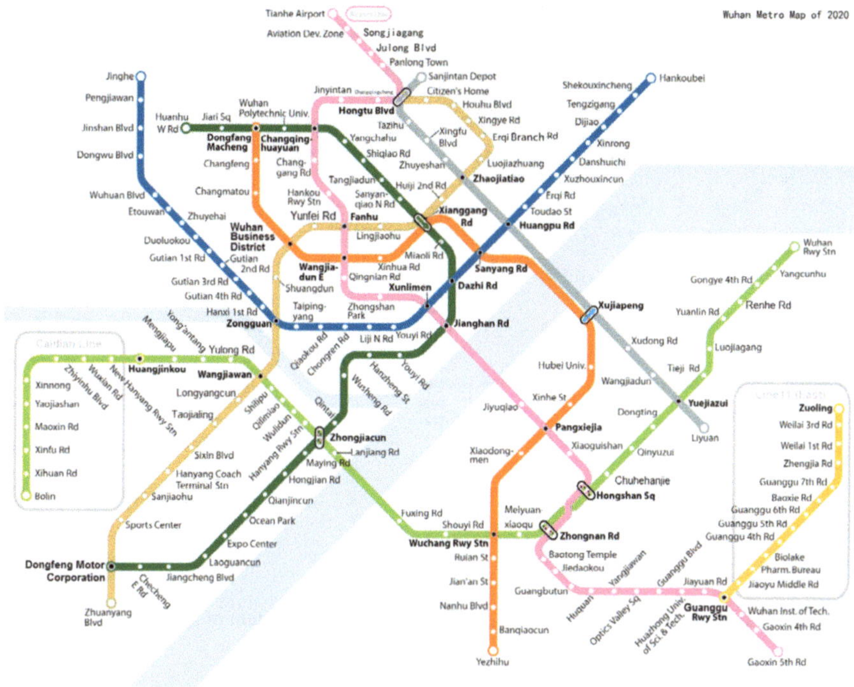

Fig. 3.5 Map of the Wuhan metro system in China by mid-2020

3.4.2 Complex Network Development

To analyze the vulnerability of the metro system under cascade failures, a complex network of the Wuhan Metro is constructed based on the official metro map. In this network, metro stations are modeled as nodes, and the connections between adjacent stations are represented as edges. The resulting network consists of 198 nodes and 215 edges, corresponding to 198 metro stations and their physical connections. The network is constructed line by line. For example, Line 1 includes five interchange stations that connect with other metro lines. These interchange stations are identified during the modeling of Line 1 and reused in the construction of other lines where necessary. For instance, Zongguan station serves as an interchange node between Wangjiawan and Shuangdun stations in the modeling of Line 3. Figure 3.6 presents the final complex network, which forms the basis for the subsequent vulnerability analysis.

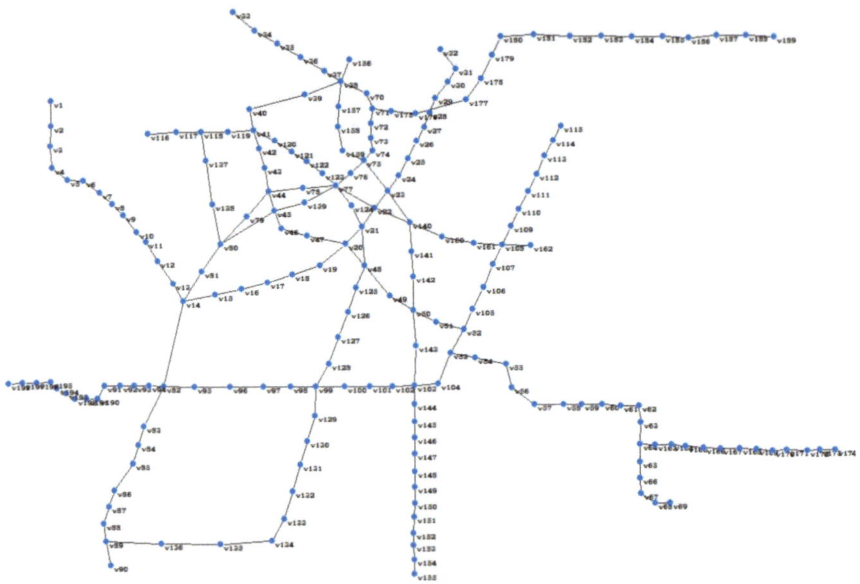

Fig. 3.6 The established complex network of the Wuhan metro system of 2020

3.4.3 Analysis of Results

3.4.3.1 Key Node Discovery

Due to the scale-free and small-world properties of complex networks, nodes exhibit varying levels of degree and betweenness centrality, reflecting differences in their importance and vulnerability. Nodes with high centrality values play a more critical role in overall network vulnerability management. As discussed earlier, four attack strategies (ID, IB, RD, and RB) are employed to assess network vulnerability. In cascade failure modeling, key nodes may fail due to the influence of surrounding nodes, potentially affecting the accuracy of key node identification. In this section, the top ten ranked nodes under each attack strategy are analyzed. Figure 3.7 presents the distribution of node degree and betweenness in the constructed complex network, while Table 3.1 lists the top ten key nodes selected and attacked under different strategies. The findings are summarized as follows:

1. Interchange stations exhibit higher vulnerability and should be prioritized in risk management. As shown in Table 3.1, all key nodes identified by degree-based or betweenness-based rankings are interchange stations. In degree-based strategies, interchange stations have inherently higher degrees due to their connections with multiple metro lines. Although betweenness centrality is not explicitly tied to station type, the results consistently show that interchange stations rank highest under betweenness-based strategies. This confirms that interchange stations are structurally more vulnerable and should be prioritized in safety inspections and

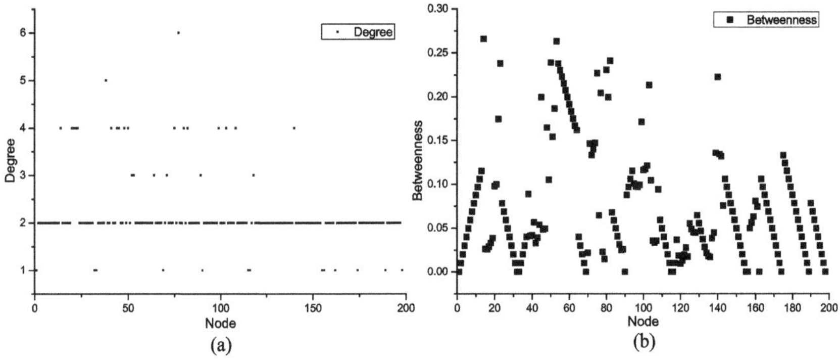

Fig. 3.7 Results of (**a**) node degrees and (**b**) node betweenness in the established complex network

monitoring. This conclusion aligns with real-world operational practices, providing further validation of the network model.

2. Xianggang Road and Hongtu Boulevard stations have the greatest direct influence, while Taipingyang and Zhongnan Road stations are central to transport efficiency. Under both ID and RD attack strategies, Xianggang Road and Hongtu Boulevard stations rank first, indicating their high degree of centrality. As the degree reflects the number of directly connected stations, these nodes are structurally vulnerable due to the widespread impact of their failure. In contrast, Taipingyang and Zhongnan Road stations consistently rank highest under IB and RB strategies, showing they are critical to overall transport efficiency. Their removal results in longer paths and decreased network performance. These four stations should, therefore, receive focused attention and resource allocation in risk prevention and mitigation planning.
3. High-degree stations are located in dense network cores, while high-betweenness stations are found near network boundaries. As shown in Fig. 3.8, Xianggang Road (degree 6) and Hongtu Boulevard (degree 5) are located in the densest parts of the network, where their capacity to manage passenger flow is essential. In contrast, Taipingyang (betweenness 0.265936) and Zhongnan Road (0.263027) are situated near the network edges and serve as crucial connectors between peripheral and central areas. These spatial patterns suggest that node degree and betweenness assess different aspects of node importance. Therefore, both metrics should be considered based on specific management objectives.

3.4.3.2 Topological Vulnerability Analysis

In the topological vulnerability analysis, this research focuses on studying the vulnerability of the total network with the consideration of the cascade failure mechanism under different attack strategies. The loss of transport capacity (ΔTC) acts as an indicator to evaluate the vulnerability of the total network while considering the

Table 3.1 Key nodes under different attack strategies

	ID attack		RD attack		IB attack		RB attack	
Rank	Station	Degree	Station	Degree	Station	Betweenness	Station	Betweenness
1	Xianggang Rd (77)	6	Xianggang Rd (77)	6	Taipingyang (14)	0.265936	Taipingyang (14)	0.26594
2	Hongtu Boulevard (38)	5	Hongtu Boulevard (38)	5	Zhongnan Rd (53)	0.263027	Zhongnan Rd (23)	0.279141
3	Taipingyang (14)	4	Taipingyang (14)	4	Wangjiadun (82)	0.240838	Jianghan Rd (48)	0.302717
4	Xunlimen (20)	4	Xunlimen (20)	4	Xiaoguishan (50)	0.238844	Xujiapeng (140)	0.382832
5	Dazhi Rd (21)	4	Hunagpu Rd (23)	4	Hunagpu Rd (23)	0.237913	Wuchang railway station (103)	0.177677
6	Sanyang Rd (22)	4	Changqinghuayuan (41)	4	Baotong Temple (54)	0.23775	Houhu Boulevard (71)	0.043639
7	Hunagpu Rd (23)	4	Fanhu (44)	4	Wuhan Business District (80)	0.230557	Hongshan Square (52)	0.037696
8	Changqinghuayuan (41)	4	Pangxiejia (50)	4	Jiedaokou (55)	0.230291	Wangjiawan (82)	0.024896
9	Fanhu (44)	4	Wuhan Business District (80)	4	Zhaojiatiao (75)	0.226793	Xianggang Rd (77)	0.024701
10	Wangjiadun East (45)	4	Wuchang railway station (103)	4	Guangbutun (56)	0.222729	Changqinghuayuan (41)	0.037812

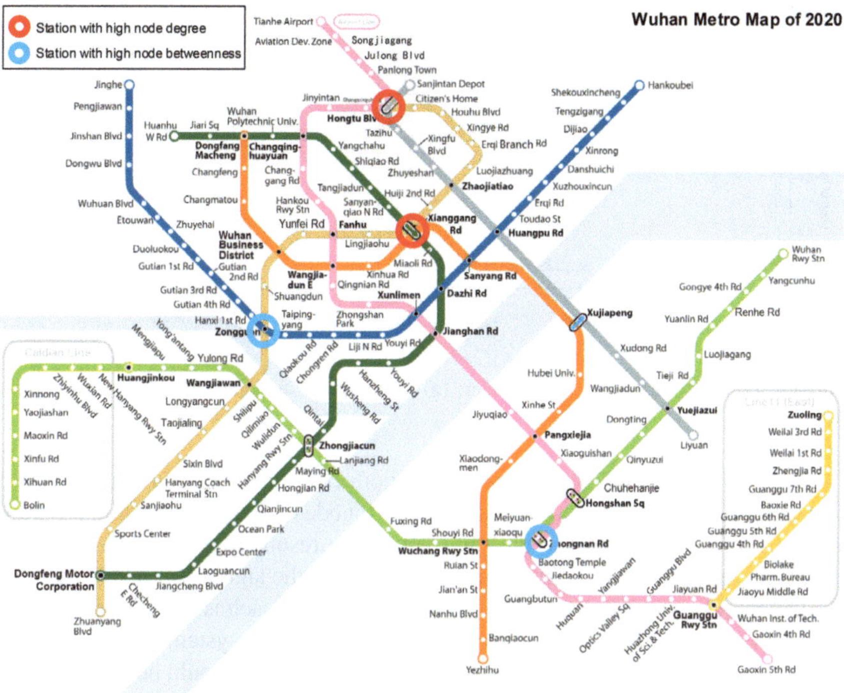

Fig. 3.8 The stations with high degree values and betweenness values (by mid-2020)

damage to the network connectivity and the loss of transport efficiency at the same time, while the degree and betweenness are the two main indices to assess the vulnerability of the nodes. Five attack strategies, including ID, IB, RD, RB, and random attacks, are adopted to testify to the vulnerability of the metro system. In this research, according to the initial design of the urban transit system under the hypothetical normal operation and relevant references [17], the adjustable parameter of the load estimation is 0.8, the tolerance parameter is 0.5, and the passenger parameter is set at 0.6. Based on the established complex network (see Fig. 3.6) and Eq. (3.10), Fig. 3.9 illustrates the loss of transport capacity of the metro complex network under different attack strategies. The results are analyzed as follows.

1. All sorts of intentional attack strategies display superior efficiency compared to the random strategy when attacking the metro network, revealing the significance of discovering the important nodes. As Fig. 3.9 shows, the most intentional attack strategies can almost completely destroy the transport capacity of the network (the loss of transport capacity is more than 0.9) by attacking 10 nodes in the supposed cascade scenario, while the random attack strategy can wreck the network when enough nodes are attacked (about 60 nodes). After 10 nodes are attacked, the values of the loss of transport capacity of ID, IB, RD, RB, and random attack are ΔTC = 0.9491, 0.9531, 0.7468, 0.9568, and 0.2694,

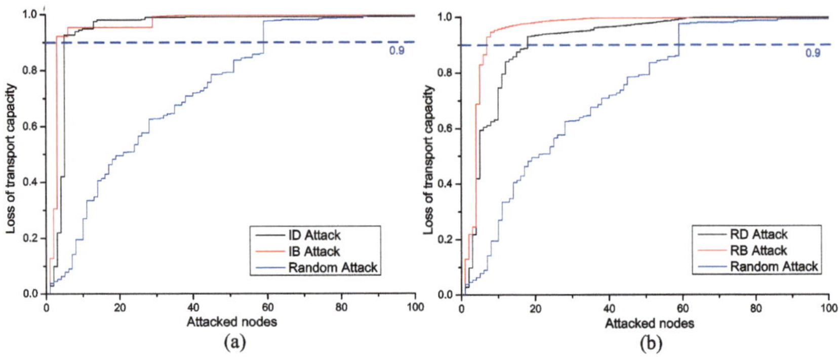

Fig. 3.9 Loss of transport capacity under: (**a**) static attack strategies, and (**b**) dynamic attack strategies

respectively. The RB strategy is the most effective one in causing the loss of the transport capacity to reach 0.9568 when 10 nodes are attacked, while a random attack only gets 0.2694 at the same time. The results illustrate that intentional attack strategies have a better ability to identify the significant nodes than random attack, which indicates that the developed approach is effective in identifying the vulnerability of stations in the urban rail transit system.

2. Using the metric of node betweenness to rank the nodes would be more efficient in identifying the stations with higher vulnerability than using the node degree. The nodes with a higher initial betweenness have more impact on the transport capacity than the nodes with a higher initial degree, while the nodes with a higher recalculated betweenness have more influence than a higher recalculated degree. According to Fig. 3.9a, the curve of the ID attack strategy is below the curve of the IB attack strategy before the loss of the transport capacity reaches up to 0.9, which implies that the IB attack strategy could cause the metro network breakdown by attacking fewer nodes than the ID attack strategy. Besides, Fig. 3.9b presents the loss of the transport capacity in dynamic attack strategies (i.e., RD and RB attacks), which is similar to the static attack strategies. The curve of the RD attack reaches 0.9 when 18 nodes have been attacked, while the RB attack strategy can destroy the network with the same level by attacking only nine nodes. Therefore, for both static and dynamic attacking strategies, the strategies that apply the node betweenness as the criterion of node ranking would be more efficient in impacting the transport capacity of the network.

3.4.3.3 Tolerance-Parameters-Based Analysis

This section evaluates the vulnerability of the metro network by incorporating cascade failure mechanisms under multiple attack strategies. The loss of transport capacity (ΔTC) is adopted as a comprehensive indicator, capturing both the reduction in network connectivity and transport efficiency. Node vulnerability is assessed

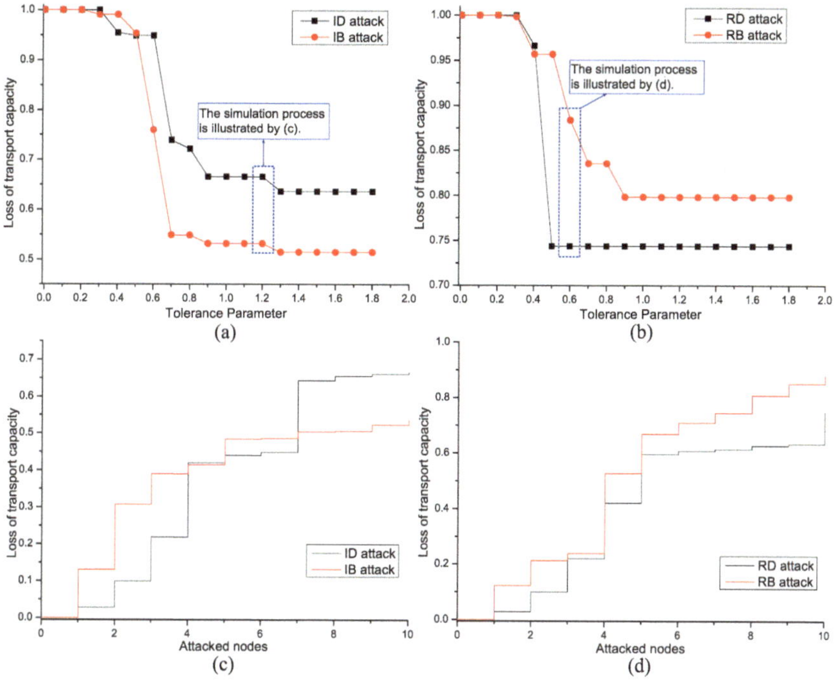

Fig. 3.10 The effect of the tolerance parameter on the loss of transport capacity under (**a**) Static attack strategies; (**b**) Dynamic attack strategies; (**c**) Evolution of the loss of transport capability when the tolerance parameter is set to be 1.2 in the static attack; (**d**) Evolution of the loss of transport capability when the tolerance parameter is set to be 0.6 in the dynamic attack

using two centrality metrics, namely degree and betweenness. Five attack strategies are employed to assess network robustness: initial degree (ID), initial betweenness (IB), recalculated degree (RD), recalculated betweenness (RB), and random attacks. Specifically, the load estimation parameter is set to 0.8, the tolerance parameter to 0.5, and the passenger redistribution parameter to 0.6. Based on the constructed complex network (Fig. 3.6) and Eq. (3.10), Fig. 3.10 illustrates the variations in ΔTC across different attack scenarios. The results are summarized as follows:

1. Intentional attack strategies are markedly more effective than random attacks, highlighting the importance of identifying critical nodes. As shown in Fig. 3.10, intentional attack strategies can cause near-complete network failure ($\Delta TC > 0.9$) after targeting only 10 nodes. In contrast, the random strategy requires the removal of approximately 60 nodes to achieve a similar level of disruption. After 10 nodes are attacked, the observed ΔTC values are 0.9491 for ID, 0.9531 for IB, 0.7468 for RD, 0.9568 for RB, and 0.2694 for the random strategy. Among these, the RB strategy results in the greatest loss, while the random strategy is the least effective. These findings confirm the ability of intentional strategies to identify and disable key nodes, validating the proposed approach for assessing metro network vulnerability.

2. Betweenness-based strategies are more effective than degree-based strategies in identifying highly vulnerable stations. Nodes with higher initial or recalculated betweenness exert greater influence on network performance than those with high degree values. As illustrated in Fig. 3.10a, the IB strategy consistently surpasses the ID strategy before ΔTC reaches 0.9, indicating that fewer nodes need to be removed to destabilize the network. Similarly, Fig. 3.10b shows that in dynamic attacks, the RB strategy achieves $\Delta TC = 0.9$ after attacking only nine nodes, whereas RD requires the removal of 18. These results suggest that using node betweenness as the ranking criterion enables more efficient identification of critical nodes, regardless of whether a static or dynamic strategy is applied.

3.4.3.4 Passenger-Transfer-Based Analysis

The passenger transfer rate is treated as an exogenous factor in network vulnerability modeling. When cascading failures occur, passengers are forced to transfer to nearby stations, thereby increasing their load. A lower transfer rate reduces the risk of cascading failures at adjacent stations but may result in passenger loss and decreased revenue. Moreover, excessive reliance on ground transportation may cause surface systems to become overloaded. Therefore, determining an optimal passenger transfer rate is essential for balancing system reliability and failure risk. This section investigates the effect of the passenger transfer rate on transport capacity under attack scenarios. The findings provide policy insights for managing metro system resilience through targeted adjustments to the passenger transfer rate or related parameters. A what-if scenario analysis is conducted, where the load adjustment parameter is set to 0.8, the tolerance parameter to 1.5, and the transfer rate is varied from 0 to 1.0 in increments of 0.1. A total of 10 nodes are attacked to examine the impact of different transfer rates on transport capacity loss. Based on the established network (see Fig. 3.6) and Eq. (3.10), Fig. 3.11 shows the resulting changes in transport capacity under both static and dynamic attack strategies. The results are summarized below:

1. Higher passenger transfer rates lead to greater transport capacity loss, which is detrimental to network security. As shown in Fig. 3.11, an increase in the transfer rate results in a progressive rise in capacity loss. Under the IB attack strategy, for instance, the capacity loss remains around 0.52 when the transfer rate is below 0.4. However, beyond this threshold, capacity loss sharply increases and approaches 1.0. This indicates that reducing passenger transfer during emergencies and encouraging the use of alternative transportation modes may help alleviate the risk of cascading failures when the metro system becomes overloaded.
2. The range between 0.3 and 0.4 is recommended for minimizing cascading failure risk across all attack strategies. As illustrated in Fig. 3.11, the relationship between transport capacity loss and the passenger transfer rate evolves in four stages. In the range from 0 to 0.3, the transfer rate has minimal effect. Between 0.3 and 0.4, it begins to show a slight influence. A sharp increase occurs between

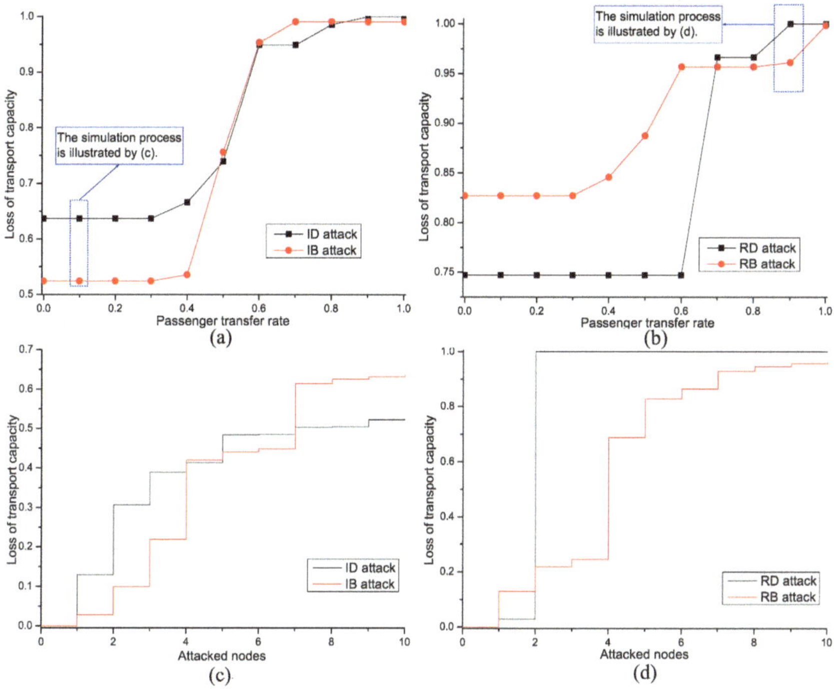

Fig. 3.11 The effect of the passenger transfer rate on the loss of transport capacity under (**a**) Static attack strategies, (**b**) Dynamic attack strategies; (**c**) Evolution of the loss of transport capability when the passenger transfer rate is set to be 0.1 in the static attack; (**d**) Evolution of the loss of transport capability when the passenger transfer rate is set to be 0.9 in the dynamic attack

0.4 and 0.7, after which the effect plateaus and the network approaches complete failure. These patterns suggest that when the transfer rate exceeds 0.4, nearby stations become increasingly overloaded. Once the transfer rate surpasses 0.7, the entire network becomes highly vulnerable. Therefore, identifying the optimal transfer rate under specific conditions can serve as a low-cost strategy for managing cascading failure risks.

3.5 Merits Analysis

This chapter proposes a dynamic approach to assess the vulnerability of urban rail transit networks by incorporating cascade effects. The approach evaluates overall system vulnerability, identifies critical nodes, and determines optimal parameter values to support improved metro system management. It also serves as a decision-making tool to inform the planning and development of future transit lines. The Wuhan metro system is used as a case study to demonstrate the applicability of the proposed approach. The Urban and Rural Planning Bureau of Wuhan is preparing to construct Metro Line 11 to alleviate traffic congestion along the city's east-west corridor. A total of 14 stations have been proposed. The planning is guided by two principles. First, the line must address transportation demands in densely populated areas, with several stations located near subcenters to ease traffic pressure. Second, it should support transit-oriented development by serving new urban expansion zones.

The government aims to enhance accessibility and promote economic development in western Wuhan, which is the main target of the plan. The operation of a new metro line is expected to attract both residents and businesses to this area, thereby facilitating regional growth. Based on the planning principles, two alternative designs for Metro Line 11 were proposed by a team of invited experts. These alternatives, referred to as Plan A and Plan B, are being evaluated across multiple dimensions, including system vulnerability under potential disruptions. The present study contributes to this evaluation by offering recommendations based on vulnerability analysis. The assessment results, together with other expert consultations, were reviewed to support informed decision-making. Figure 3.12 presents the layouts of Plan A and Plan B within the Wuhan metro network. Both plans consist of 14 stations with comparable investment levels and service coverage, making the selection between them particularly challenging. In this context, the study examines how each proposed line affects the vulnerability of the overall transit system to support alternative selection.

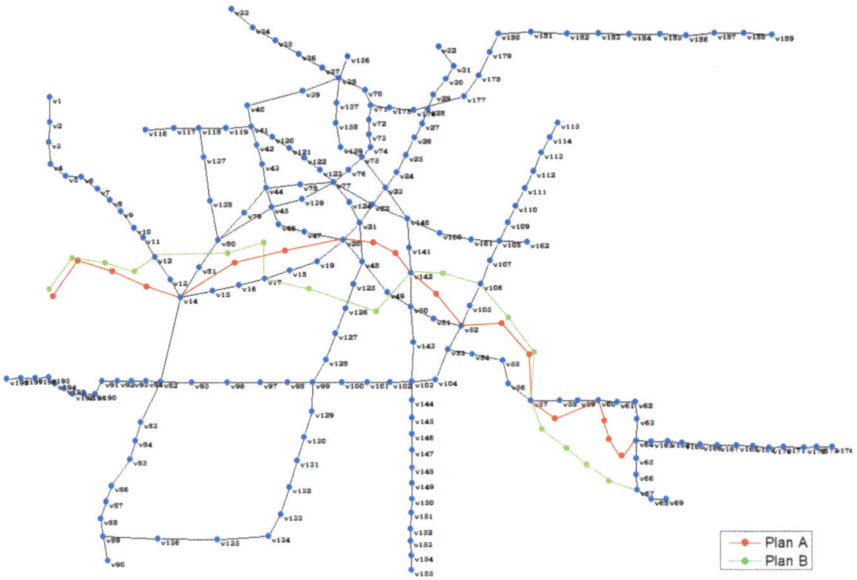

Fig. 3.12 Layout of two alternatives (Plan A and Plan B) for the proposed Line No. 11 in the Wuhan metro system

To compare the two plans, three what-if scenarios are developed using loss of transport capacity as the key indicator of system vulnerability. As established in previous analyses, the tolerance parameter is negatively correlated with capacity loss, while the passenger transfer rate is positively correlated. Scenario I represents a relatively flexible condition with a tolerance parameter of 0.8 and a transfer rate of 0.4. Scenario II is a moderate condition, with a tolerance parameter of 0.5 and a transfer rate of 0.6. Scenario III reflects a more constrained setting, with a tolerance parameter of 0.4 and a transfer rate of 0.7. Figure 3.13 illustrates the resulting transport capacity loss of both plans under a dynamic betweenness-based attack strategy.

Across all three scenarios, Plan B demonstrates lower vulnerability to cascade effects compared to Plan A, particularly under the moderate scenario. As illustrated in Fig. 3.13, the rate of increase in transport capacity loss is slower in Plan B than in Plan A across all scenarios. In Scenario I, the network with Plan A nearly collapses after 17 nodes are attacked ($\Delta TC = 0.90521$), whereas the a comparable threshold is reached in Plan B after 19 nodes ($\Delta TC = 0.90577$). The advantage of Plan B is most evident in Scenario II. In this case, the network with Plan A nearly collapses after nine node attacks ($\Delta TC = 0.93119$), while Plan B requires 16 node failures to reach a similar level ($\Delta TC = 0.91371$). The difference between the two plans is minimal in Scenario III. Although the initial loss of capacity in Plan B is significantly lower than that of Plan A after two attacks ($\Delta TC = 0.0476$ vs. 0.24061), both networks collapse after the third node is attacked. These findings confirm that Plan B is more effective in reducing system vulnerability and improving network reliability. Based on this analysis, Plan B was selected by the Wuhan government as the basis for further design and development.

Scenarios with higher tolerance parameters and lower passenger transfer rates offer greater resilience against transport capacity loss. Figure 3.13d compares the performance of Plan B across the three scenarios under a dynamic betweenness-based attack strategy. In Scenario I, Plan B reaches a capacity loss of 0.90557 after 19 node failures. In Scenario II, this level is reached after 16 nodes ($\Delta TC = 0.91371$), while in Scenario III, the network collapses after only four node failures ($\Delta TC = 1.0000$). These results align with prior findings. A higher tolerance parameter tends to reduce capacity loss, while a higher passenger transfer rate exacerbates vulnerability (see Figs. 3.10 and 3.11). The outcomes presented in Fig. 3.13 validate the reliability of the established complex network model. They also demonstrate the potential of this approach to support network optimization by adjusting tolerance and transfer parameters to enhance transit system resilience.

In conclusion, vulnerability should be a key consideration when planning the expansion of existing metro systems. New metro lines inevitably affect the overall vulnerability of the system. When alternatives have similar resource requirements, the one with lower vulnerability should be prioritized. The approach proposed in this chapter allows for the simulation and evaluation of alternative metro alignments under cascading failure conditions, offering valuable decision support for urban transit planning.

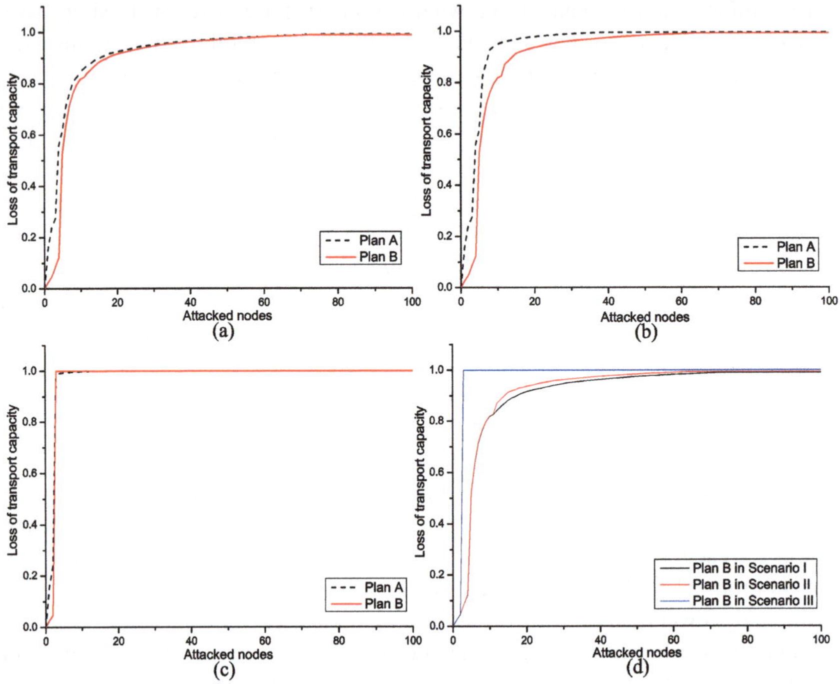

Fig. 3.13 The loss of transport capacity of Plan A and Plan B under the dynamic betweenness-based attack strategy in (**a**) Scenario I; (**b**) Scenario II; (**c**) Scenario III; (**d**) Comparison of the three scenarios in Plan B

3.6 Chapter Summary

This chapter proposes a simulation-based complex network approach to evaluate the vulnerability of urban rail transit systems by incorporating cascade failure mechanisms. The method consists of three primary stages: complex network construction, cascade-based attack simulation, and vulnerability analysis and assessment. A real-world urban transit system is used as a case study to validate the applicability of the proposed framework. The results demonstrate that the method effectively identifies topological vulnerabilities, determines optimal parameter values, and supports the selection of superior design alternatives. It also has potential as a decision-support tool for optimizing metro line alignments within transit planning.

The case study yields several key findings. Interchange stations exhibit higher vulnerability within the network, and four stations (i.e., Xianggang Rd station, Hongtu Boulevard station, Taipingyang station, and Zhongnan Rd) are identified as the most critical. These stations should receive prioritized attention in operational management. Moreover, intentional attack strategies result in significantly greater damage compared to random attacks, indicating that the proposed approach can successfully identify the most vulnerable nodes. Among centrality measures, node betweenness proves more effective in capturing node-level vulnerability under intentional attacks. The analysis also reveals that a tolerance parameter in the range of 0.8–0.9 and a passenger transfer rate between 0.3 and 0.4 can effectively mitigate cascade failures with minimal resource input.

Overall, the novelty of this research lies in the development of a complex network model incorporating a built-in cascade failure mechanism, enabling a better representation of real-world operational dynamics in UTS. The proposed approach supports strategic decision-making related to node criticality, network vulnerability assessment, and system configuration. In addition, it serves as a decision-support tool for optimizing operational strategies, offering practical guidance for enhancing the resilience and sustainability of urban transit networks.

References

1. Liang M, Wang W, Ye C, Dong C (2024) Layout planning of a basic public transit network considering expected travel times and transportation efficiency. Systems 12:550. https://doi.org/10.3390/systems12120550
2. Chen H, Shen Q, Feng Z, Liu Y (2023) Vulnerability assessment in urban metro systems based on an improved cloud model and a Bayesian network. Sustain Cities Soc 98:104823. https://doi.org/10.1016/j.scs.2023.104823
3. Lu Q-C, Zhang L, Xu P-C, Cui X, Li J (2022) Modeling network vulnerability of urban rail transit under cascading failures: a coupled map lattices approach. Reliab Eng Syst Saf 221:108320. https://doi.org/10.1016/j.ress.2022.108320
4. Xu Z, Chopra SS, Lee H (2022) Resilient urban public transportation infrastructure: a comparison of five flow-weighted metro networks in terms of the resilience cycle framework. IEEE Trans Intell Transp Syst 23:12688–12699. https://doi.org/10.1109/TITS.2021.3116667
5. Ouyang M, Pan Z, Hong L, He Y (2015) Vulnerability analysis of complementary transportation systems with applications to railway and airline systems in China. Reliab Eng Syst Saf 142:248–257. https://doi.org/10.1016/j.ress.2015.05.013
6. Wang X, Koç Y, Derrible S, Ahmad SN, Pino WJA, Kooij RE (2017) Multi-criteria robustness analysis of metro networks. Phys A Stat Mech Appl 474:19–31. https://doi.org/10.1016/j.physa.2017.01.072
7. Yin D, Huang W, Shuai B, Liu H, Zhang Y (2022) Structural characteristics analysis and cascading failure impact analysis of urban rail transit network: from the perspective of multi-layer network. Reliab Eng Syst Saf 218:108161. https://doi.org/10.1016/j.ress.2021.108161
8. Xing L (2021) Cascading failures in internet of things: review and perspectives on reliability and resilience. IEEE Internet Things J 8:44–64. https://doi.org/10.1109/JIOT.2020.3018687
9. Ghanbari R, Jalili M, Yu X (2018) Correlation of cascade failures and centrality measures in complex networks. Futur Gener Comput Syst 83:390–400. https://doi.org/10.1016/j.future.2017.09.007
10. Sun J, Tang J, Fu W, Chen Z, Niu Y (2020) Construction of a multi-echelon supply chain complex network evolution model and robustness analysis of cascading failure. Comput Ind Eng 144:106457. https://doi.org/10.1016/j.cie.2020.106457
11. Chen T, Ma J, Zhu Z, Guo X (2023) Evaluation method for node importance of urban rail network considering traffic characteristics. Sustainability 15:3582. https://doi.org/10.3390/su15043582

12. Qi Q, Meng Y, Zhao X, Liu J (2022) Resilience assessment of an urban metro complex network: a case study of the Zhengzhou metro. Sustainability 14:11555. https://doi.org/10.3390/su141811555
13. Wang Z, Jia L, Ma X, Sun X, Tang Q, Qian S (2022) Accessibility-oriented performance evaluation of high-speed railways using a three-layer network model. Reliab Eng Syst Saf 222:108411. https://doi.org/10.1016/j.ress.2022.108411
14. Huang W, Zhou B, Yu Y, Sun H, Xu P (2021) RETRACTED: using the disaster spreading theory to analyze the cascading failure of urban rail transit network. Reliab Eng Syst Saf 215:107825. https://doi.org/10.1016/j.ress.2021.107825
15. Li X, Teng M, Jiang S, Han Z, Gao C, Nekorkin V, Radeva P (2025) A dynamic station-line centrality for identifying critical stations in bus-metro networks. Chaos, Solitons Fractals 194:116102. https://doi.org/10.1016/j.chaos.2025.116102
16. Guan X, Yu F, Xu H, Li C, Guan Y (2024) Flood risk assessment of urban metro system using random forest algorithm and triangular fuzzy number based analytical hierarchy process approach. Sustain Cities Soc 109:105546. https://doi.org/10.1016/j.scs.2024.105546
17. Dui H, Meng X, Xiao H, Guo J (2020) Analysis of the cascading failure for scale-free networks based on a multi-strategy evolutionary game. Reliability Engineering & System Safety, 106919.

Chapter 4
Evolutionary Dynamics Modeling

4.1 Introduction

The rapid development of urban transit systems (UTS) has significantly contributed to economic and social advancement across various sectors [1]. However, the concentration of large crowds within the confined underground spaces of metro networks increases the potential severity of accidents. Such incidents can escalate rapidly, leading to serious consequences [2]. In recent years, failures and accidents have occurred frequently, which have been attributed to diverse causes, including human error, inadequate management, aging infrastructure, and unsafe passenger behavior [3]. These risks underscore the complexity and vulnerability of UTS. The interdependent nature of their components means that errors in any subsystem can trigger widespread failures. Therefore, modeling and assessing system vulnerability is essential for enhancing the safety and reliability of metro operations.

System vulnerability refers to the extent to which a system can be affected or disrupted. It reflects how easily a system and its components can be damaged, as well as their limited capacity to resist or recover from external disturbances [4]. System dynamics (SD), since its introduction by Jay Forrester in the 1950s, has gained widespread attention as a powerful method for analyzing the behavior of dynamic and complex systems over time [5]. A key strength of SD lies in its ability to model circular causality among multiple interacting variables. Unlike linear causal models such as regression analysis, SD captures nonlinear relationships and complex feedback mechanisms. By incorporating feedback loops, stocks, and flows, SD enables the analysis of time delays and interdependent dynamics, allowing for a more accurate representation of long-term system evolution [6].

Given these advantages, SD has been widely adopted in modeling system vulnerability, particularly in domains characterized by multifactorial complexity and nonlinear interactions [7]. SD has also been employed to assess and improve vulnerability in areas such as urban ecological zones [8], rangeland ecosystems [9], and food

L. Zhang et al., *Resilience in Urban Transit Systems*, Advances in 21st Century Human Settlements, https://doi.org/10.1007/978-981-95-4322-9_4

supply chains [10]. These studies demonstrate that SD can be effectively used to construct integrated models linking multiple interrelated variables and governing equations. Such models facilitate the simulation of system evolution and the prediction of future states. The core procedures of SD modeling typically include (1) identifying vulnerability factors, (2) constructing a causal feedback diagram, (3) developing a stock-and-flow model with governing equations, and (4) validating the model and analyzing simulation outcomes.

An operating urban metro network is a complex system composed of multiple interrelated subsystems, with its vulnerability influenced by diverse factors such as personnel, equipment, environment, and management. It is evident that metro system reliability is subject to various changes over time, including evolving human safety awareness and aging infrastructure. These factors often shift gradually and may not be readily detectable in the short term. Therefore, long-term vulnerability modeling is more appropriate for capturing the uncertainty and complexity of metro systems over extended periods. Proactive strategies can be formulated in advance to address scenarios in which system vulnerability is projected to increase, helping to balance short- and long-term resilience needs [11]. Modeling long-term vulnerability requires addressing three key aspects: (1) circular causal relationships among internal and external variables, (2) the influence of long-term dynamics, and (3) uncertainty and probabilistic variation over time. As discussed earlier, SD is well-suited to capturing the first two dimensions. To address the third, the Monte Carlo (MC) simulation offers a complementary approach. Widely used in probabilistic risk analysis, the MC method enables global analysis of system performance under uncertainty, making it a valuable tool for long-term vulnerability assessment [12].

In summary, most existing modeling and simulation studies based on SD adopt a deterministic framework, which limits the ability to account for global uncertainty, conduct comprehensive sensitivity analysis, or enhance model robustness. While SD simulations can predict system behavior and evaluate the effectiveness of different strategies, they typically focus on deterministic outputs. Traditional SD approaches often overlook the global interactions among influential factors and the propagation of uncertainty across the system. To address these limitations, this chapter proposes a probabilistic SD approach by integrating MC simulation with SD modeling. This combined method enables global modeling and assessment of long-term vulnerability in UTS, incorporating different levels of uncertainty to support more robust decision-making.

4.2 Vulnerability Evolution Modelling in UTS

A deep understanding of the evolving vulnerability in UTS is fundamental for advancing resilient operations and enabling risk-informed planning. Vulnerability modeling provides a structured means to uncover hidden risk factors, reveal interdependencies, and anticipate potential failure pathways, thereby supporting strategic decision-making for long-term system maintenance and emergency preparedness.

Given the inherent structural complexity and dynamic nature of metro operations, researchers have developed a range of analytical techniques. These approaches are typically categorized into empirical methods, network-based methods, multi-criteria decision analysis (MCDA), and SD methods.

1. **Empirical methods.** Empirical and heuristic-based approaches focus on observed incidents and expert-derived risk factors. Case-Based Reasoning (CBR) identifies vulnerability patterns by comparing past accident data [13], while Performance Shaping Factors (PSF) frameworks categorize causes of operational risk into human, technical, and organizational domains [14]. These approaches offer practical insights and can be valuable for risk diagnosis and management training. However, they lack the capacity to simulate the dynamic interactions between system components, limiting their utility for predicting future failure scenarios or evaluating response strategies.
2. **Network-based methods.** Network-based methods approach metro systems as interconnected components, enabling vulnerability assessment through structural analysis. Complex network theory represents metro stations and connections as nodes and links, respectively, to evaluate topological robustness, identify critical points of failure, and simulate cascading effects [15]. FMECA (Failure Mode Effects and Criticality Analysis), when combined with network analysis, supports vulnerability prioritization based on failure likelihood and impact [16]. These theories are well-suited to evaluate structural fragility but do not inherently account for temporal changes or dynamic feedback.
3. **MCDA methods.** MCDA techniques, such as the Technique for Order Preference by Similarity to an Ideal Solution (TOPSIS), are used to rank vulnerability levels by integrating diverse indicators under fuzzy or uncertain conditions [17]. By incorporating expert knowledge and quantifiable data, MCDA approaches facilitate prioritization among competing risks. However, they typically provide static assessments and do not simulate evolving system behavior.
4. **SD methods.** Dynamic simulation offers a long-term perspective by modeling the feedback-rich and nonlinear nature of metro systems. SD has been widely used to simulate how changes in workforce capacity, equipment reliability, organizational policies, and environmental factors influence system vulnerability over time. SD enables scenario testing and policy evaluation, helping stakeholders anticipate future trends and potential tipping points. However, traditional SD models are typically deterministic and assume fixed relationships, which can limit their ability to fully capture the uncertainties arising from sensor inaccuracy, environmental variability, and human behavior.

In conclusion, existing approaches offer valuable tools for identifying risk factors and assessing system vulnerability in operating metro systems. Nonetheless, most traditional models are limited by static assumptions and insufficient representation of time-dependent processes. To address these gaps, this chapter introduces a hybrid modeling framework combining SD and MC simulation. The SD model captures causal feedback among four subsystems (workforce, equipment, organization, and environment), while MC simulation introduces probabilistic variability. Together, they enable long-term, uncertainty-informed assessments of system vulnerability, offering actionable insights for sustainable metro resilience planning.

4.3 System Vulnerability Assessment Framework

To conduct long-term vulnerability modeling and assessment of UTS, this chapter proposes a probabilistic system dynamics (SD) approach that integrates Monte Carlo (MC) simulation with SD modeling. This hybrid framework addresses the three key requirements identified earlier: circular causality, long-term dynamic behavior, and uncertainty evaluation. Figure 4.1 presents the workflow of the proposed approach, which consists of four main stages: (1) variable identification, (2) model development, (3) model validation, and (4) model analysis. Each step is described in detail below. It is worth noting that the development of an SD model is often time-consuming and inherently involves subjectivity and uncertainty.

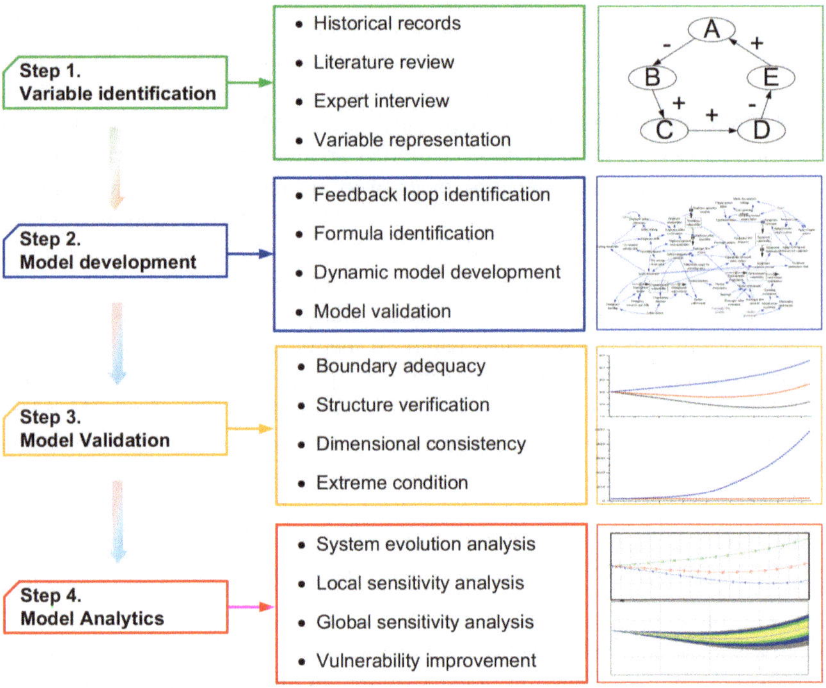

Fig. 4.1 Workflow of the proposed approach for long-term vulnerability modeling and assessment

4.3.1 System Variables Identification

SD is employed to simulate real-world systems by capturing the internal relationships among system variables. Variable identification serves as the foundational step in model development, as it provides the structural basis for constructing and analyzing the system. In this approach, all relevant variables and their interconnections must be clearly defined.

4.3.1.1 Data Collection and Knowledge Resources

This study relies on historical accident records and an extensive literature review to identify the key variables influencing metro system vulnerability. In addition, expert consultations contribute to the development of a robust model framework. UTS consist of multiple interdependent subsystems and operate within dynamic external environments. Both internal and external factors can lead to operational failures, ultimately triggering accidents. Based on accident case analysis and prior studies, these factors can be broadly categorized into four main dimensions: workforce, organization, equipment, and environment.

Workforce vulnerability arises from uncertainties related to employee performance, including both frontline staff and management. Factors such as professional skills, safety awareness, physical and mental health, and compliance with safety protocols all influence this dimension. Organizational vulnerability reflects the effectiveness of institutional structures and emergency preparedness, and is further shaped by elements such as ticket pricing strategies, organizational hierarchy, regulatory enforcement, and the quality of emergency management systems. Equipment vulnerability pertains to the reliability and functionality of the core infrastructure. To facilitate analysis, equipment is classified into four subsystems: electrical, signal, vehicle, and electromechanical systems. Environmental vulnerability is associated with external conditions that influence system performance. It is evaluated across three domains: natural (e.g., weather events), social (e.g., passenger behavior and density), and operational environments (e.g., surface traffic and multimodal integration). These dimensions collectively shape the overall resilience and safety of metro systems and serve as the foundation for variable selection in the proposed system dynamics model. Accordingly, the SD model developed in this chapter integrates the following four components.

4.3.1.2 Variable Representation

In SD modeling, three primary types of variables are used in flow diagrams: stocks, flows, and converters. Stock variables, represented by rectangles, indicate the state of a system at a specific point in time. Flows, typically placed between cloud symbols and stock variables, capture the rate of change in stocks. Converter variables define the internal mechanisms of the system and can be further categorized as constants or auxiliaries. Auxiliaries are informational variables that enrich the model's logic, while constants are fixed values that remain unchanged throughout the simulation. Table 4.1 outlines the types of variables defined in the metro system vulnerability model.

SD models also require a boundary test, which distinguishes between exogenous and endogenous variables. Exogenous variables are defined externally and are not influenced by the model's internal feedback, whereas endogenous variables are determined by the model's structure and interactions. In this study, variables influenced by external conditions or directly assigned by decision-makers are treated as exogenous. In contrast, endogenous variables are computed within the model framework and are considered simulation outputs. For instance, constants such as "Safety training", "Safety system", "Equipment maintenance frequency", and "Passenger safety awareness" are externally assigned and remain unaffected by model evolution and are thus classified as exogenous variables in Table 4.1.

Additionally, SD includes shadow variables, which are denoted using angle brackets (e.g., "< >"). These special variables are used to reference another variable, often across subsystems, or to link model elements with time functions. Shadow variables help reduce diagram complexity and enhance model readability. As illustrated in Fig. 4.2, "Workforce vulnerability" is modeled as a stock variable influenced by a flow and cloud input, while the remaining variables act as converters that define internal system dynamics.

Table 4.1 Types of the defined variables in the vulnerability model of the overall metro system

Types	Defined variables
Stock variables	Workforce vulnerability, organizational vulnerability, equipment vulnerability, environmental vulnerability
Flow variables	Employee safety operation, employee operation accident, management improvement, management decline, equipment maintenance, equipment recession, environment optimization, environmental degradation
Auxiliary variables	Passenger flow volume, operating income, safety management pressure, safety investment, training investment, employee safety awareness, employee skills, employee physical and mental state, employee safety performance, employee mistake, emergency resources and drill, emergency handling, organization structure, system enforcement, metro line network mileage, Total operating mileage, vehicle system failure, electromechanical system failure, equipment failure, aging of the power supply system, aging of the signal system, aging of strong and weak electric equipment, equipment renewal, equipment maintenance, natural environment, passenger flow pressure, passenger attitude, infrastructure conditions, surrounding environment, operating environment, vulnerability of operating subway
Constants (Exogenous variables	The average ticket price, government subsidy ratio, vulnerability target for operating metro, natural disasters, sabotage, recession rate, safety training, safety system, equipment maintenance frequency, passenger safety awareness

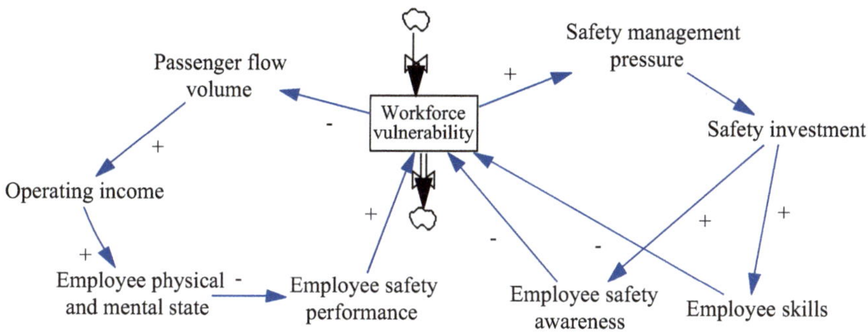

Fig. 4.2 A simple example of the SD flow diagram

4.3.2 *SD Model Development*

After the factor identification, the relationships of the identified factors would be analyzed to establish the model. In this part, the model development consists of three procedures, including feedback loop identification, formula determination, and model updates with feedback.

4.3.2.1 Feedback Loop Identification

The foundation of SD modeling lies in representing variable interdependencies through feedback loops, which reflect circular causality within the system. Feedback loops are categorized into two types: positive (reinforcing) and negative (balancing). In the causal loop diagram, a positive arrow denotes a reinforcing relationship, while a negative arrow indicates an inverse effect.

Semi-structured expert interviews and literature reviews were conducted to define the causal structure of metro system vulnerability. These informed the development of the system's causal loop and stock-flow diagrams. The resulting model includes 16 feedback loops: seven positive and nine negative. Figure 4.3 presents the complete feedback loop diagram for the metro system vulnerability.

In Fig. 4.3, red, green, black, and blue circles denote stock, flow, auxiliary, and constant variables, respectively. Each stock variable represents a subsystem's vulnerability, influenced by corresponding flow variables. Arrows indicate the direction and nature of variable relationships. The polarity of each loop is determined by the number of negative causal links it contains. For example, the loop "Employee safety performance-Employee safety operation-Workforce vulnerability-vulnerability of overall metro system-Passenger attitude-Passenger flow volume-Operation income-Employee physical and mental health-Employee safety performance" forms a positive feedback loop, as it includes six reinforcing and two opposing effects.

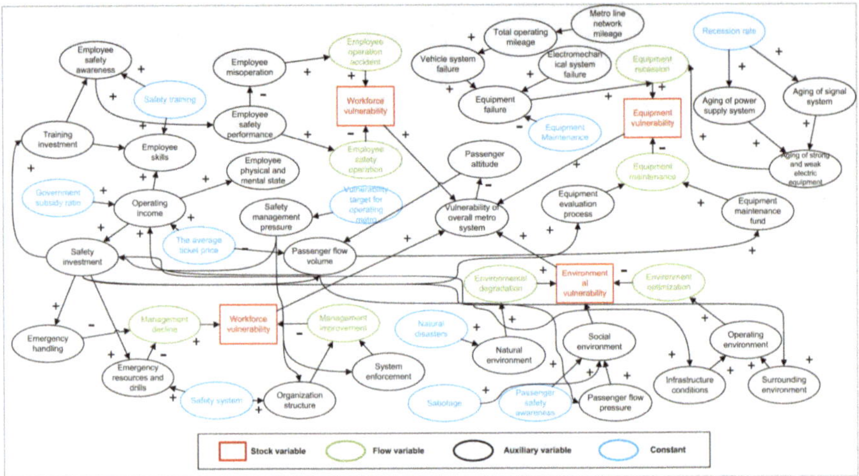

Fig. 4.3 Feedback loop diagram for revealing the vulnerability of the overall metro system

4.3.2.2 Formula Determination

The variables in the model are set according to the results of the questionnaire and interviews with experts to obtain objective and reasonable formulas. Taking the employee subsystem as an example, the main variable is the workforce vulnerability, and the corresponding formula is given as follows:

$$\begin{aligned}\text{Workforce vulnerability}(t) &= \text{Employee safety peroformance}(t_0)\\&+\int_t^{t0}(\text{Employee operation accident} - \text{Employee safety operation})dt \qquad (4.1)\end{aligned}$$

where t_0 is the starting time of the SD simulation, t is the current time, and the workforce vulnerability at time t is the initial value of employee safety performance plus the changes during the simulation.

Heinrich's law suggests that one major safety accident is accompanied by 29 minor accidents and 300 potential accidents. Out of a total of 330 times (i.e., 1 + 29 + 300) of employee mistakes and/or misoperation, there are very likely to result in 30 accidents (i.e., one major and 29 minor accidents). Therefore, as a flow variable, the formula for employee operation accidents is given by Eq. (4.2).

$$\text{Employee operation accident} = \text{Employee mistake}/11 \qquad (4.2)$$

where 11 represents that one accident is likely to happen within 11 (330/30) times of employee mistakes and/or misoperation on average.

Other variables in the model are derived following similar logic, drawing on established theoretical principles and expert knowledge. Once all equations are defined, a preliminary SD model for the metro system vulnerability is established.

4.3.2.3 Model Updates with Feedback

The reasonability of the established SD model is vital to perform the simulation of the actual system. To ensure the reliability of the SD model for simulating metro system vulnerability, expert feedback was collected and incorporated into the model structure. Several key refinements were made based on insights related to workforce, organizational, equipment, and environmental subsystems.

Workforce vulnerability is influenced by employee safety awareness and skills, both of which can be improved through safety training, with effectiveness determined by training investment. Organizational vulnerability is linked to metro revenue, which depends on ticket pricing and passenger volume. Equipment vulnerability reflects differences across subsystems: while power and signal systems degrade gradually due to aging, the reliability of electromechanical and vehicle systems is strongly affected by maintenance frequency. Environmental vulnerability, though relatively stable in terms of physical infrastructure, is significantly impacted by unpredictable passenger behavior and safety awareness.

Following expert recommendations, the model was revised to strengthen causal links among key variables. The final SD model includes 4 stock variables, 8 flow variables, 10 constants, and 30 auxiliary variables, as illustrated in Fig. 4.4.

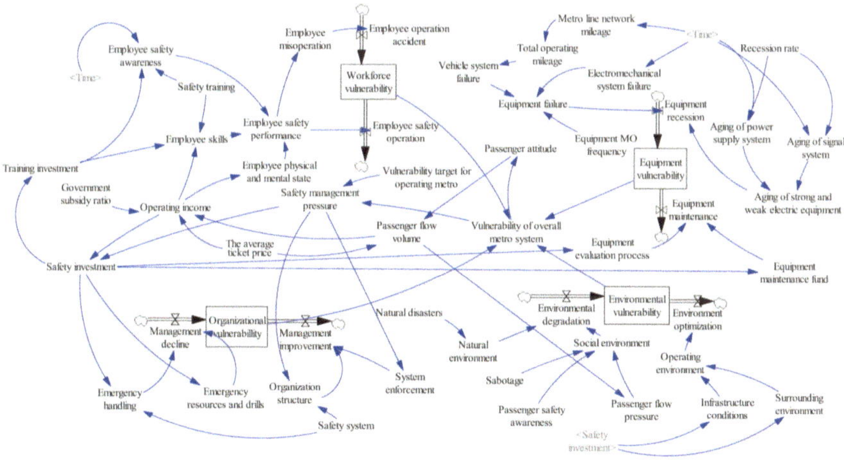

Fig. 4.4 Finalized SD model revealing the long-term vulnerability of the metro system in operation

4.3.3 SD Model Validation

To ensure the applicability and robustness of the developed SD model, structural validation was conducted through four standard tests: boundary adequacy, structure verification, dimensional consistency, and extreme condition testing.

Boundary adequacy evaluates whether key concepts influencing system behavior are included within the model boundaries. Most variables in the model are endogenous, which are identified through literature review and expert consultation. Exogenous variables are categorized into two types: (1) national-level, such as average ticket price, subsidy ratio, natural disasters, and macroeconomic recession rate, and (2) metro-specific, such as safety training, maintenance frequency, and passenger safety awareness. The former requires adjustment when applying the model to other regions, while the latter can be directly optimized to improve local system performance.

Structure verification assesses whether the logic and relationships in the model reflect real-world operations. Previous studies and expert feedback were used to verify the appropriateness of causal links and the structure of subsystems such as workforce, organization, equipment, and environment. Dimensional consistency ensures that equations and variables are dimensionally coherent. All model variables were assigned consistent units, and Vensim's built-in "Units Check" tool confirmed that the model passed this test without errors. Extreme condition testing verifies the model's ability to behave logically under extreme input values. For instance, the variable "safety training" (range: 0–100) was tested at values of 1 and 99. As shown in Fig. 4.5, under low training investment, system vulnerability

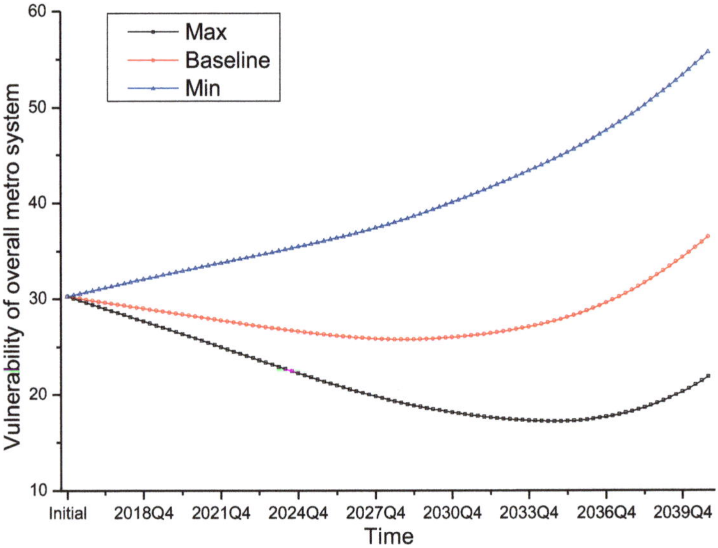

Fig. 4.5 Simulation results of the vulnerability of the overall metro system under extreme conditions of the variable of safety training

increases continuously over time. Under high investment, vulnerability first decreases and remains low over the long term. This behavior aligns with expert expectations, demonstrating that the model reasonably simulates metro system dynamics under boundary conditions.

Overall, the results of the structural validation tests confirm the internal logic, reliability, and applicability of the developed SD model for long-term vulnerability analysis of UTS.

4.3.4 System Vulnerability Analytics

The well-developed SD model is used to predict the long-term vulnerability of the urban metro system from various perspectives. The impacts of important variables and subsystems on the vulnerability of the overall metro system are investigated. The model analytics consist of system evolution analysis, local sensitivity analysis, and global sensitivity analysis.

System dynamics simulation is used to explore dynamic relationships within the model and forecast long-term trends in both individual factors and overall system behavior. The simulation results support model refinement and provide insights into variable performance. In this study, two aspects are analyzed: the evolution and vulnerability of subsystems, revealing their individual contributions to overall vulnerability, and the evolution of key factors, which shed light on behavioral patterns and long-term vulnerability trends.

Local sensitivity analysis evaluates the extent to which changes in individual factors affect model outputs. In the SD model, this is typically done by adjusting constant values and comparing results. Selected critical variables in metro system operations are modified (e.g., increased or decreased by 10% or 20%) to observe their impact on system vulnerability. Variables that induce greater output variation are considered more sensitive.

Global sensitivity analysis examines the combined effects of simultaneous variations in multiple variables, accounting for their interactions. Monte Carlo (MC) simulation, implemented via the built-in Vensim application, is used for this purpose. Also referred to as multivariate sensitivity simulation (MVSS), this method samples constant variables over a defined range and performs numerous simulations to capture output variability. Latin Hypercube sampling is applied to ensure efficient coverage of the input space. Input variables are assumed to vary within their possible range (e.g., 10% or 20% of their initial values) of their initial values, reflecting measurement uncertainty. Hundreds of iterations are run to assess model output under uncertainty, and the resulting vulnerability is reported at 50%, 75%, 95%, and 100% confidence levels from a long-term perspective.

4.4 Experimental Studies

4.4.1 Data and Resources

Wuhan, the capital of Hubei Province in the People's Republic of China, is the largest city in the province and the most populous city in Central China, with a population of 12.32 million. Its vast territory and dense population create substantial demand for public transportation. Since the launch of the Wuhan metro system in 2004, both its service coverage and passenger volume have grown rapidly. According to the Ministry of Transport (2021), by the end of 2020, the Wuhan metro system included 321 stations and spanned a total operating length of 409.1 km. Figure 4.6 presents the Wuhan metro map alongside real-world images of the system. In this chapter, an SD model incorporating various types of variables is developed to predict the long-term vulnerability of metro system development over a 25-year period (i.e., 100 quarters), with a simulation time step of one quarter. Most variables in the model are dimensionless.

4.4.2 System Evolution Analysis

The SD simulation results are used to analyze the long-term vulnerability of the metro system and its subsystems, providing insights for improved decision-making. Figure 4.7 illustrates the simulated vulnerabilities of various subsystems and their associated variables, while Fig. 4.8 presents the relationship between overall metro system vulnerability and that of its subsystems. The analysis is summarized as follows.

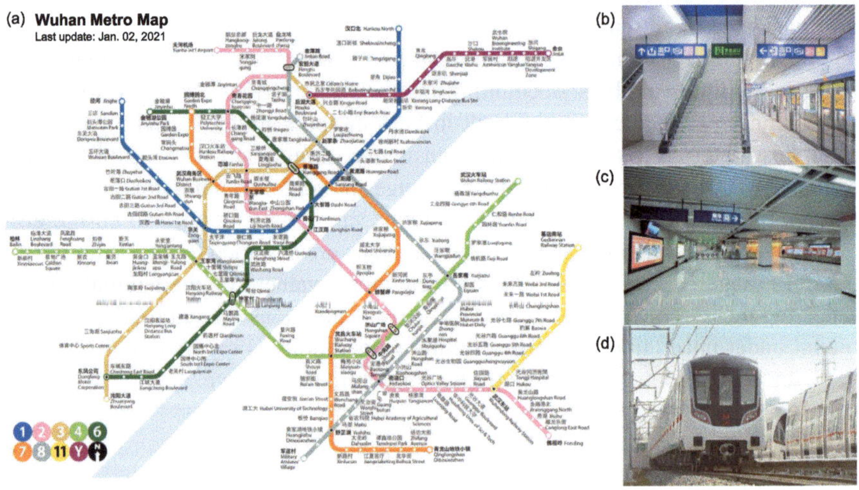

Fig. 4.6 Case profile: (**a**) Map of the Wuhan metro system (by Jan 2021), and (**b–d**) Field scenes of passenger guides and the power supply system

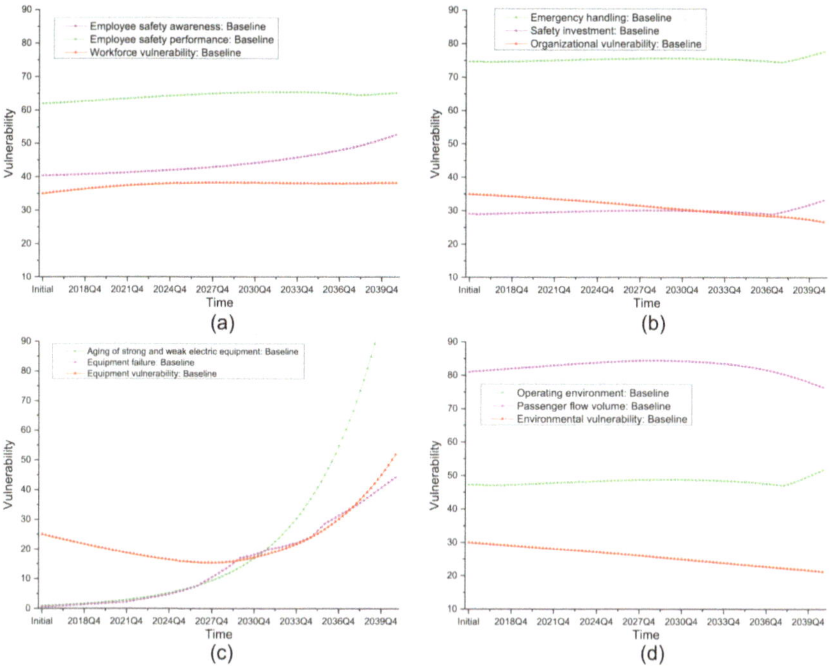

Fig. 4.7 Simulation results of the subsystems and the related significant variables over time: (**a**) Workforce subsystem; (**b**) Organizational subsystem; (**c**) Equipment subsystem; (**d**) Environmental subsystem

1. Most subsystems remain relatively stable and in a moderate state of vulnerability over time, with the equipment subsystem exhibiting the most dynamic changes. As shown in Fig. 4.7, the vulnerability of the workforce subsystem remains relatively constant in the range of [35, 40], and the organizational and environmental subsystems display a steady decline in vulnerability. In contrast, the equipment subsystem shows significant fluctuations. Specifically, its vulnerability decreases during the first 48 quarters (12 years), then begins to rise, eventually reaching a peak of 51.97, substantially higher than its initial value (Fig. 4.7c).
2. The overall system vulnerability initially declines but exhibits an inflection point in the mid-term, primarily driven by changes in equipment vulnerability. As seen in Fig. 4.8, the vulnerabilities of the workforce, organizational, and environmental subsystems all show long-term reductions (red curves), contributing to the initial decrease in overall system vulnerability. This trend reflects improvements in safety culture and resilience within these subsystems over time. However, aging equipment and infrastructure lead to a sharp rise in the vulnerability of the equipment subsystem after approximately 12 years, reversing the overall trend (blue curves in Fig. 4.8). This suggests that equipment renewal should receive increased attention after extended periods of operation. Preventive and predictive maintenance are essential not only for ensuring equipment reliability but also for sustaining the resilience of the metro system as a whole.

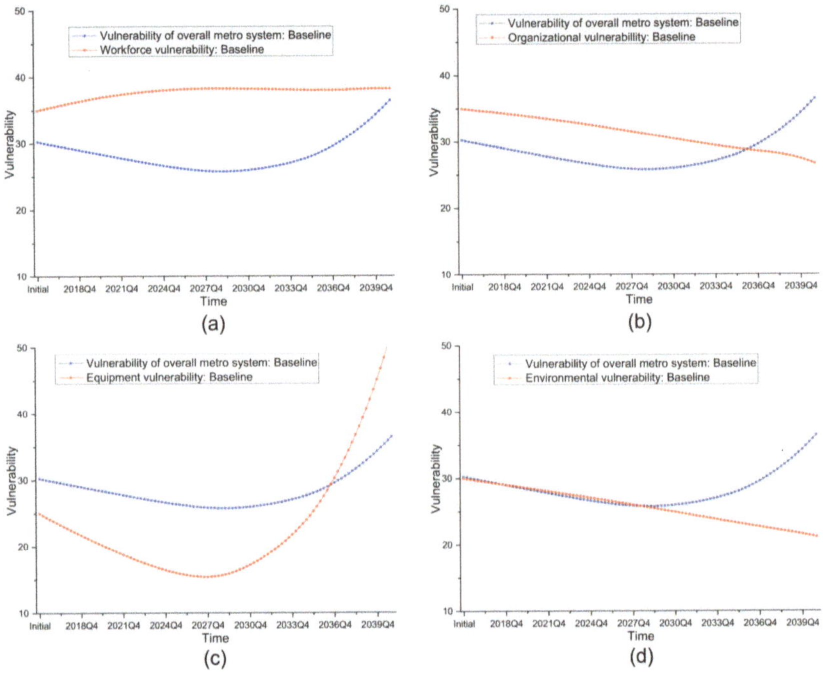

Fig. 4.8 Simulation results of the vulnerability of the overall metro system together with its four subsequent systems: (**a**) Workforce subsystem; (**b**) Organizational subsystem; (**c**) Equipment subsystem; (**d**) Environmental subsystem

4.4.3 Local Sensitivity Analysis

Local sensitivity refers to the extent to which changes in individual factors influence system outcomes. In SD models, this analysis is typically conducted by varying constant values and comparing the resulting outputs. The input factors are selected from constants defined in the developed SD model. As shown in Table 4.1, these constants include average ticket price, government subsidy ratio, vulnerability target for metro operations, natural disasters, sabotage, recession rate, safety training, safety system, equipment maintenance frequency, and passenger safety awareness. To identify strategies for enhancing metro system robustness, the chosen input variables must be both adjustable and variable. Variables such as the government subsidy ratio and vulnerability target are determined by government policy and regulation, making them relatively stable and difficult to modify. Similarly, natural disasters, sabotage, and recession rates are largely beyond the control of metro system managers. The "safety system" variable represents mandated safety infrastructure, regulations, and emergency programs, which also exhibit limited variability. Consequently, four variables (safety training, average ticket price, equipment maintenance frequency, and passenger safety awareness) are selected for the sensitivity analysis. These variables are more susceptible to managerial interventions and fluctuations, offering meaningful insights into how their variation influences system vulnerability and informs resilience strategies.

During the analysis, each selected constant is individually increased or decreased by 10% or 20% from its baseline value. The corresponding changes in metro system vulnerability are recorded. The effects of these variations on subsystems and overall system vulnerability are illustrated in Figs. 4.9 and 4.10. Results are analyzed as follows.

1. Variables typically exert the greatest influence on the specific subsystems to which they belong, although some, like average ticket price, affect multiple domains, including employees, organizations, and equipment. As shown in Fig. 4.9a–d, safety training significantly influences workforce vulnerability and moderately affects organizational vulnerability, with negligible impact on the remaining subsystems. In Fig. 4.9e–h, the average ticket price is shown to substantially affect the first three subsystems but exhibits limited sensitivity to environmental vulnerability. Figure 4.9i–l presents the effects of equipment maintenance frequency on each subsystem. Among them, equipment vulnerability shows the most pronounced response to changes in maintenance frequency, while the other three subsystems exhibit minimal variation. This suggests that equipment maintenance frequency is a key determinant for the equipment subsystem but has limited relevance elsewhere. As illustrated in Fig. 4.9m–p, passenger safety awareness significantly impacts environmental vulnerability and slightly influences workforce vulnerability, but does not notably affect the other subsystems. These findings indicate that variables display varying levels of sensitivity across subsystems and should be treated accordingly in vulnerability management. Furthermore, the simulations suggest interdependencies among subsystems, as changes in one variable may influence components beyond its immediate domain.
2. The identified sensitive variables include average ticket price, safety training, equipment maintenance frequency, and passenger safety awareness. Among these, average ticket price and safety training exhibit the highest sensitivity with respect to overall metro system vulnerability, whereas equipment maintenance frequency and passenger safety awareness are less influential. Average ticket price correlates positively with overall system vulnerability, whereas the other three variables exhibit negative correlations. As shown in Fig. 4.10a, b, a rise in safety training reduces overall vulnerability from +14.80% to −19.01% (a range of 33.81%), while an increase in average ticket price raises vulnerability from −11.08% to +27.46% (a range of 38.54%). This demonstrates that the average ticket price is the most sensitive factor affecting overall system vulnerability. According to Fig. 4.10c, d, a ±20% change in equipment maintenance frequency shifts overall vulnerability from +8.19% to −5.83% (a range of 14.02%). Enhancing passenger safety awareness adjusts vulnerability from +4.25% to −3.05% (a range of 7.30%). These findings support the practical adjustment of these variables to reduce overall vulnerability and justify their inclusion in further vulnerability analyses.

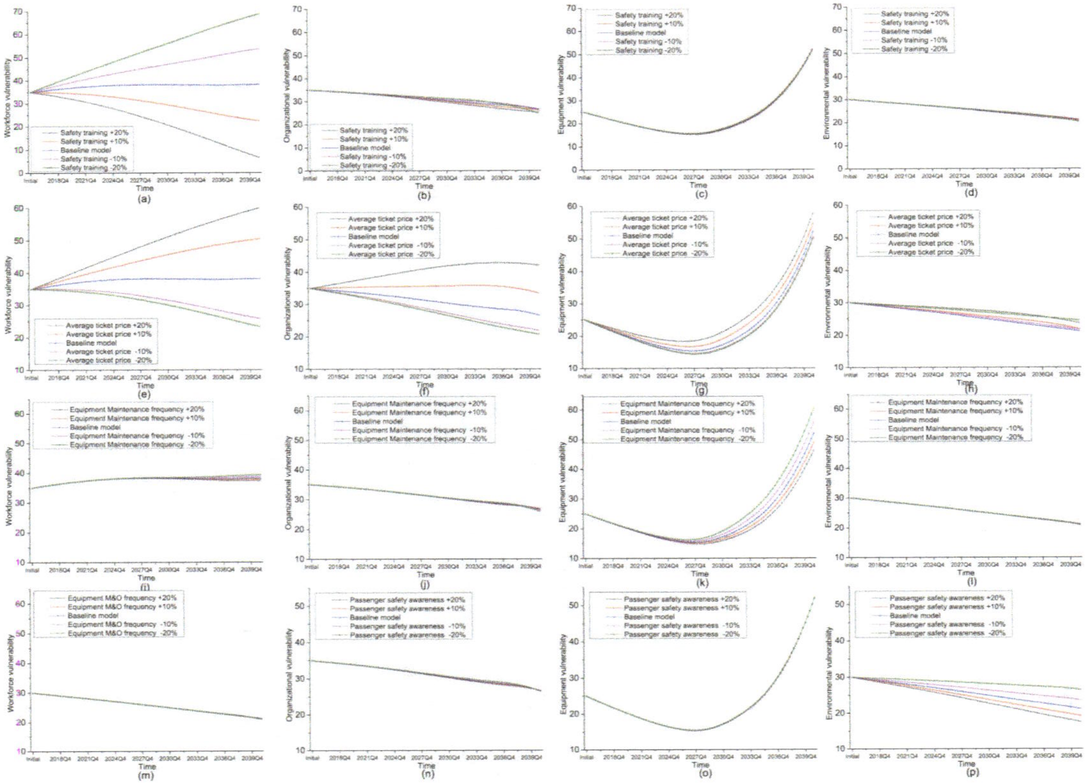

Fig. 4.9 Impacts of the variations of significant variables on the vulnerability of the subsystems: **(a–d)** Safety training; **(e–h)** Average ticket price; **(i–l)** Equipment maintenance frequency; **(m–p)** Passenger safety awareness. Note: Column at the left end represents the impacts of the four variables (i.e., Safety training, Average ticket price, Equipment maintenance frequency, and Passenger safety awareness) on the workforce vulnerability; Column at the middle-left represents the impacts of the four variables on the organizational vulnerability; Column at the middle-right represents the impacts of the four variables on the equipment vulnerability; Column at the right end represents the impacts of the four variables on the environmental vulnerability

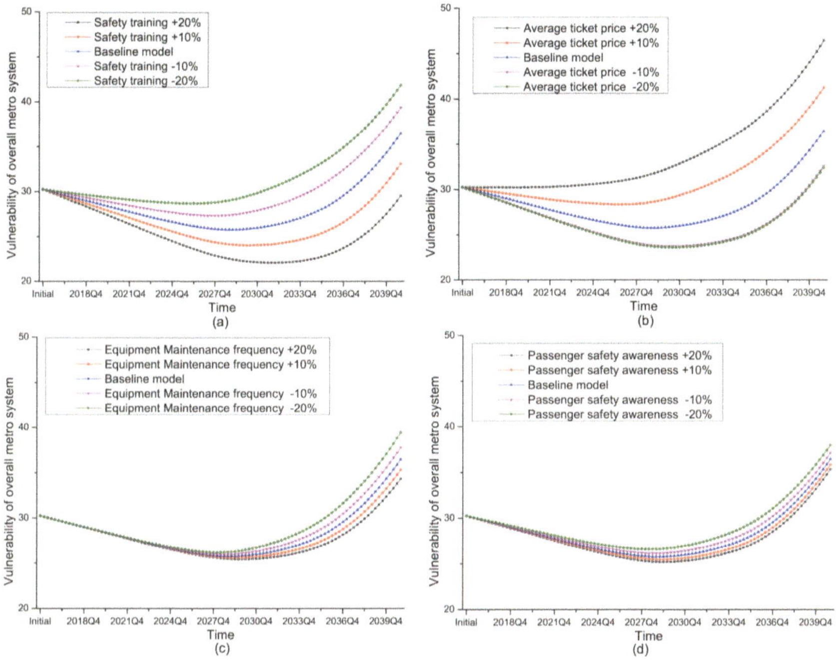

Fig. 4.10 Impacts of the variations of significant variables on the vulnerability of the overall metro system: (**a**) Safety training; (**b**) Average ticket price; (**c**) Equipment maintenance frequency; (**d**) Passenger safety awareness

4.4.4 Global Sensitivity Analysis

In metro operation, many uncertain factors could jointly affect the vulnerability of the metro system in operation. The MVSS functions built in the Venism software are used to perform the global sensitivity analysis. The selected four variables (i.e., safety training, average ticket price, equipment maintenance frequency, and passenger safety awareness) are set to follow the uniform distributions, where the maximum and minimum values are 110% and 90% of the variable values in the baseline model, respectively, in the 10% fluctuation scenario (or 120% and 80% in the 20% fluctuation situation). Based on the Latin Hypercube sampling technique, hundreds of simulations are produced to obtain vulnerability changes of the overall metro system and its subsystems. Figures 4.11 and 4.12 show the global sensitivity results of the subsystems under two fluctuations (i.e., 10% and 20% fluctuation) conditions, respectively. Figure 4.13 illustrates the global sensitivity of the overall metro system. Results are analyzed as follows.

1. Among the subsystems, workforce vulnerability exhibits the highest global sensitivity, followed by the organizational subsystem. In contrast, equipment and environmental vulnerabilities demonstrate substantially lower sensitivity to vari-

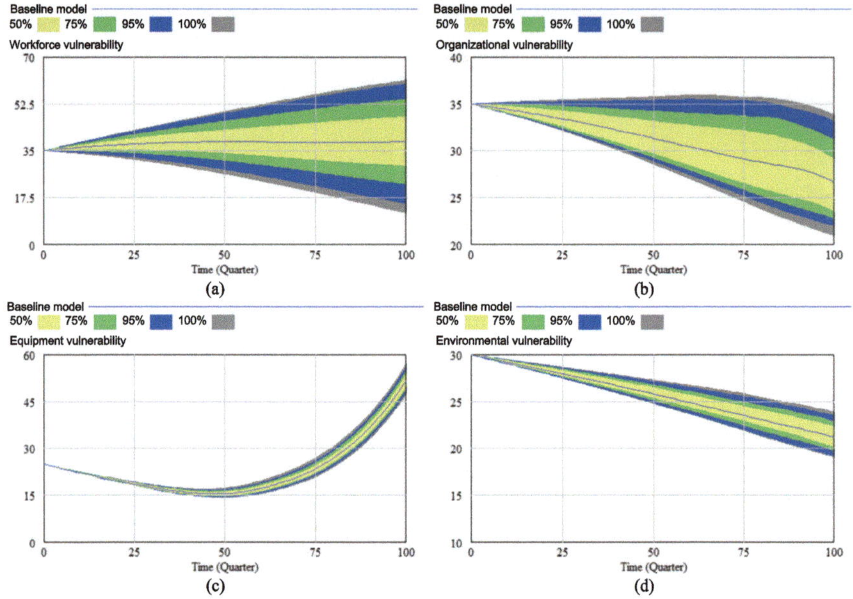

Fig. 4.11 Global sensitivity analysis results of subsystems with 10% fluctuation: (**a**) Workforce subsystem; (**b**) Organizational subsystem; (**c**) Equipment subsystem; (**d**) Environmental subsystem

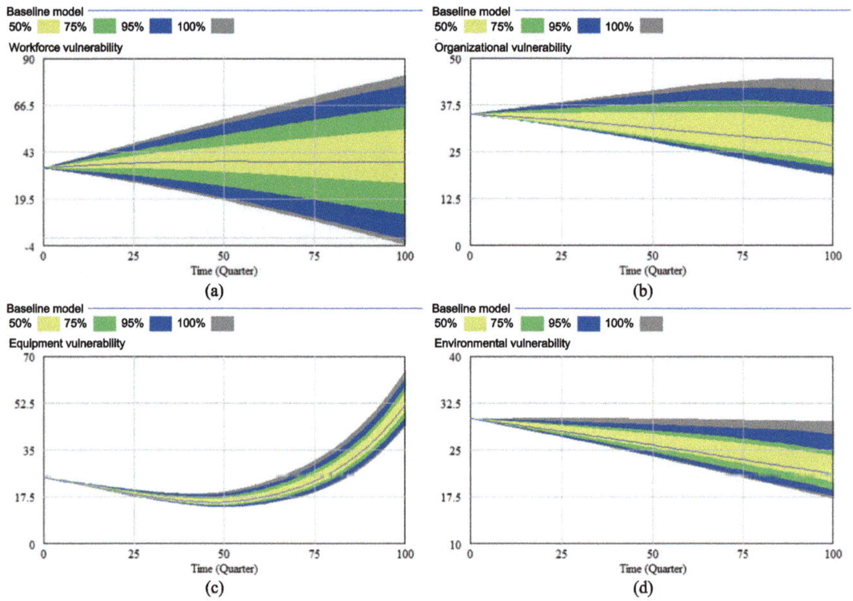

Fig. 4.12 Global sensitivity analysis results of subsystems with 20% fluctuation: (**a**) Workforce subsystem; (**b**) Organizational subsystem; (**c**) Equipment subsystem; (**d**) Environmental subsystem

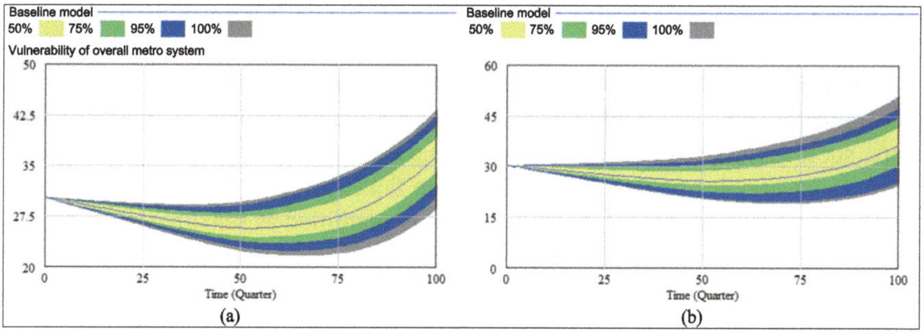

Fig. 4.13 The results of global sensitivity analysis of operating metro with: (**a**) 10% fluctuation; (**b**) 20% fluctuation

able fluctuations. As shown in Fig. 4.11, the vulnerability value of the subsystems fluctuates in different ranges after 100 quarters of operation when the input values have a 10% fluctuation. At the 100th quarter, the workforce vulnerability has the largest interval of about 49, and the interval of organizational vulnerability is about 13. The equipment and environmental subsystems display greater stability, with fluctuation ranges of approximately 9 and 5 units, respectively. Figure 4.12 shows that increasing the input fluctuation from 10% to 20% does not alter the relative ranking of subsystem sensitivities. Subsystems with high sensitivity under the 10% scenario display even greater sensitivity under the 20% scenario. This consistent ranking supports the robustness and rationality of the proposed method.

2. The overall vulnerability of the metro system follows a similar trend across both fluctuation scenarios. In general, greater input variability is associated with greater output variability. As shown in Fig. 4.13, under the 10% fluctuation scenario, the model output ranges from 29 to 43 (a spread of 14 units) after 100 quarters. In the 20% fluctuation scenario, the range expands from 25 to 50 (a spread of 25 units). These results confirm the reliability of the model. Moreover, Fig. 4.13 indicates that under lower fluctuation conditions, the system can produce both extremely high and extremely low vulnerability values. In contrast, under higher fluctuation, the system tends toward high vulnerability and struggles to achieve low vulnerability values to the same extent. Therefore, managing input uncertainty is essential for improving the robustness of the overall metro system.

4.4.5 System Vulnerability Improvement

To further reduce the vulnerability of the metro system and promote sustainable transit development, several strategies are proposed through what-if scenario analysis. This section focuses on evaluating how these strategies can enhance system robustness by applying the proposed modeling approach. The scenario analysis is conducted using the established SD model (see Fig. 4.4) as a baseline to assess and compare the performance of alternative strategies for vulnerability reduction. As shown in Fig. 4.4, the SD model includes 10 constant variables, most of which are constrained by physical conditions and cannot be directly influenced by metro operators, such as natural disasters, sabotage, and recession rates. As discussed previously, four variables (safety training, average ticket price, equipment maintenance frequency, and passenger safety awareness) are identified as sensitive inputs affecting system vulnerability. Notably, among these, only the average ticket price exhibits a negative correlation with system vulnerability, while the remaining variables show positive correlations. These variables can be adjusted through increased safety investment, changes in fare pricing, enhanced maintenance efforts, and improved public awareness campaigns. In consultation with the Wuhan Metro operations team, five strategic plans were developed. Table 4.2 presents the specific values assigned to the four sensitive variables under each strategy.

Each strategy emphasizes different areas of improvement. Strategy 1 enhances safety training and equipment maintenance frequency to reduce workforce and organizational vulnerability. Strategy 2 reduces average ticket prices and promotes passenger safety awareness, targeting management and environmental subsystems. Strategy 3 strengthens safety training and passenger awareness, while Strategy 4 improves all three variables except ticket pricing. Strategy 5 focuses on safety training and equipment maintenance to reduce vulnerabilities in the workforce and equipment subsystems. It is assumed that a 1% improvement in each variable requires an equal number of resources. Since the overall magnitude of improvement is consistent across strategies (as shown in Table 4.2), the resource cost of each strategy is considered equal. The long-term vulnerability of the metro system is then simulated under each strategy. The evolution of system vulnerability with and without variable adjustments is shown in Figs. 4.14 and 4.15. Table 4.3 summarizes the predicted vulnerability levels of the metro system at selected years under each strategy. The results are discussed below.

1. All five strategies effectively reduce the vulnerability of the metro system and delay the onset of the inflection point in vulnerability over time. As shown in

Table 4.2 Values of variables in the initial setup and different strategies

Strategy	Safety training	Average ticket price	Equipment maintenance frequency	Passenger safety awareness
Initial setup	70	2.7	80	70
Strategy 1	77 (+10%)	2.565 (−5%)	80	70
Strategy 2	70	2.43 (−10%)	80	73.5 (+5%)
Strategy 3	77 (+10%)	2.7	80	73.5 (+5%)
Strategy 4	70	2.565 (−5%)	84 (+5%)	73.5 (+5%)
Strategy 5	73.5 (+5%)	2.7	88 (+10%)	70

Note: The value in the bracket indicates the variation magnitude of the variables compared to their initial setup values

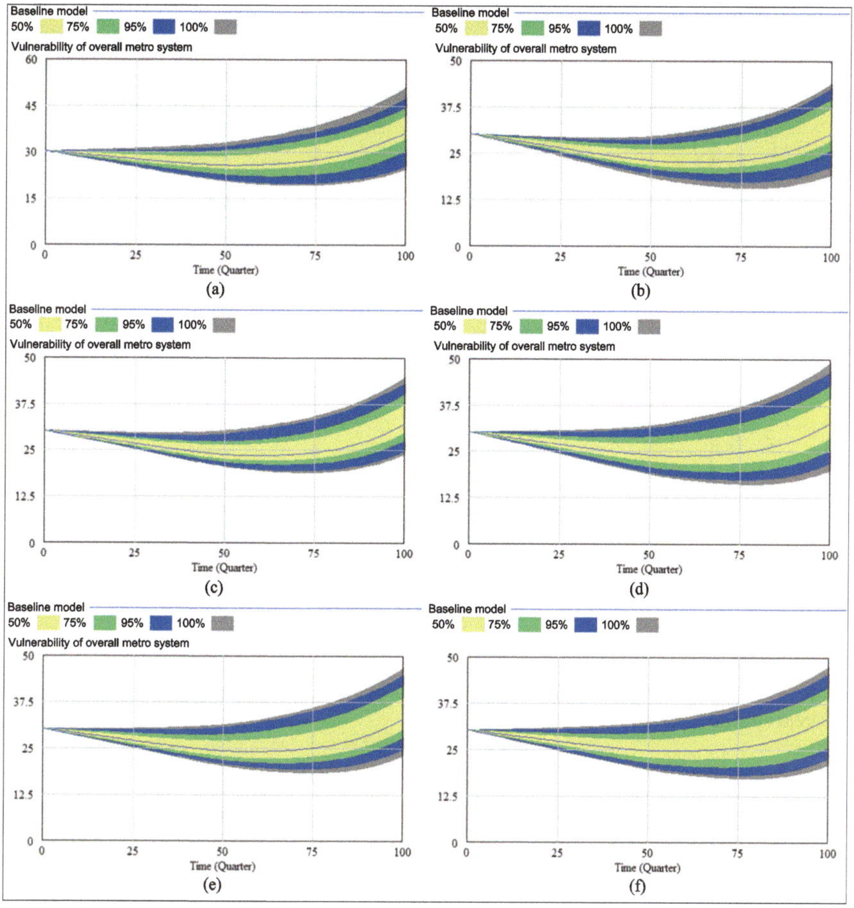

Fig. 4.14 Evolution of the long-term vulnerability of the overall metro system with 20% variations under different strategies: (**a**) Initial setup; (**b**) Strategy 1; (**c**) Strategy 2; (**d**) Strategy 3; (**e**) Strategy 4; (**f**) Strategy 5

Fig. 4.15, the vulnerability trajectories under all strategies lie significantly below those of the baseline model (i.e., the scenario without intervention). Specifically, the lowest vulnerability value in the baseline model is 25.77, which is reduced to 22.70, 23.56, 23.86, 24.22, and 24.71 under Strategies 1 through 5, respectively. Additionally, the inflection point, which occurs at the 54th quarter in the baseline scenario, is delayed to the 63rd, 60th, 59th, 58th, and 57th quarters under the respective strategies. These findings indicate that adjusting the selected variables significantly enhances the robustness of the metro system. In other words, the proposed approach not only predicts long-term vulnerability but also guides practical improvements in system resilience.

2. Among all strategies, Strategy 1 is the most effective in reducing overall metro system vulnerability, and its impact increases over time. As illustrated in

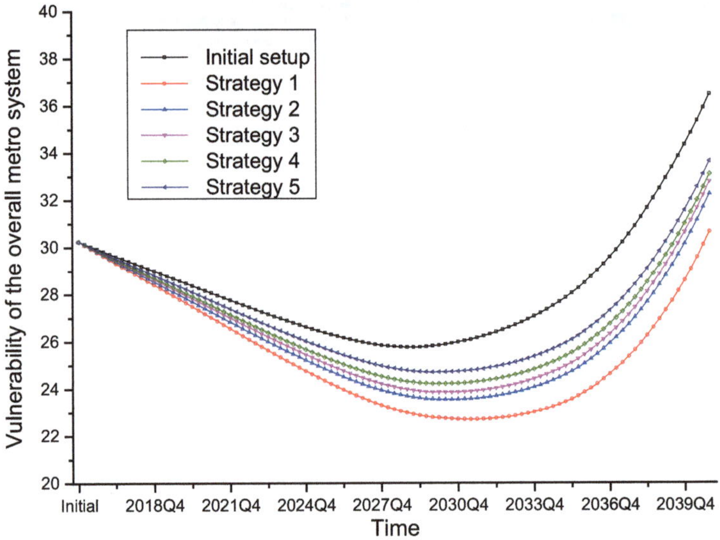

Fig. 4.15 Evolution of the long-term vulnerability of the overall metro system without variations under different strategies

Table 4.3 Predicted vulnerability values of the overall metro system at different years under different strategies

	10 years		15 years		20 years		25 years	
Strategy	0[a]	20%[b]	0	20%	0	20%	0	20%
Initial setup	26.31	[21.82, 31.82]	25.98	[19.55, 34.55]	28.49	[19.55, 40]	36.46	[24.55, 50.45]
Strategy 1	24.23	[20.45, 29.17]	22.71	[17.05, 31.06]	23.85	[15.53, 35.23]	30.64	[19.32, 43.56]
Strategy 2	24.74	[21.97, 29.54]	23.56	[19.32, 31.44]	25.08	[19.32, 35.61]	32.25	[23.48, 44.7]
Strategy 3	25.00	[20.45, 31.06]	23.87	[17.05, 33.71]	25.45	[15.9, 38.64]	32.77	[19.7, 48.48]
Strategy 4	25.24	[21.97, 30.68]	24.24	[19.32, 32.58]	25.88	[18.56, 37.12]	33.08	[22.73, 46.59]
Strategy 5	25.63	[21.59, 31.44]	24.74	[18.18, 33.71]	26.43	[17.05, 38.26]	33.62	[20.83, 47.35]

[a]represents that there are no variations for the input variables in the SD modeling
[b]indicates that there are 20% variations for input variables in the SD modeling

Fig. 4.15 and Table 4.3, Strategy 1 consistently yields the lowest vulnerability values, reaching 24.23, 22.71, 23.85, and 30.64 at the end of the 10th, 15th, 20th, and 25th years, respectively. Compared to the baseline model, Strategy 1 reduces the minimum vulnerability by 11.91% and the 25-year endpoint by 13.01%. Furthermore, it produces the longest delay in the inflection point, extending the declining phase of vulnerability by 16.98%. Therefore, improving safety training and increasing equipment maintenance frequency are recommended as an effective real-world strategy for enhancing metro system resilience.

4.5 Merits Analysis

The local sensitivity analysis reflects the system's vulnerability response to changes in individual variables, whereas the global analysis evaluates vulnerability under simultaneous fluctuations of multiple variables. A comparison of these two analyses reveals certain consistent trends despite differences in absolute values. As illustrated in Figs. 4.9, 4.11, and 4.12, some subsystems exhibit similar developmental patterns in both analyses. For example, equipment vulnerability remains stable initially but increases sharply in the mid-to-late stages, while environmental vulnerability steadily declines in both scenarios. Scenario simulation results indicate that Strategy 1 offers the most substantial improvement in system robustness, highlighting the critical role of employee safety quality. This factor should be prioritized in the design and implementation of vulnerability mitigation strategies and related decision-making processes.

From a theoretical standpoint, this chapter enhances the understanding of causal relationships among key factors influencing metro system vulnerability. Effective long-term safety assessment requires recognizing the complexity and dynamism of metro operations, which are shaped by interrelated subsystems such as workforce, management, equipment, and the environment. Predicting vulnerability trends is challenging due to these interactions. This chapter develops an SD model that integrates relevant risk factors and captures their interdependencies through defined variable relationships, forming dynamic feedback loops. The SD approach enables the identification of critical risk factors and supports long-term vulnerability prediction. Scenario simulation and sensitivity analysis can be achieved by changing the values of constants in the SD model, and the MC function in Vensim can also be used to conduct global sensitivity analyses.

From a practical perspective, the SD model assists metro operators in forecasting vulnerability trends and anticipating risks. By simulating long-term system behavior, it allows operators to evaluate vulnerability across different periods and formulate tailored responses. In the case of the Wuhan metro system, the simulation suggests initial robustness, followed by rising vulnerability in later stages due to equipment degradation. This finding aligns with real-world observations in many cities, where aging infrastructure contributes to increased accident rates despite operational maturity. Developing effective, targeted strategies for long-term vulnerability management is thus a common challenge. The SD model is intuitive and adaptable, making it suitable for real-world applications. Operators can modify the model structure, add variables, or adjust parameters to better reflect actual conditions, supporting its broader feasibility and practical implementation.

4.6 Chapter Summary

Motivated by the need to assess metro system vulnerability from a long-term perspective, this chapter develops a hybrid approach integrating system dynamics (SD) and Monte Carlo (MC) simulation. The methodology comprises three key steps: variable identification, model development, and model analysis. An SD model is constructed, consisting of four stock variables (workforce, organizational, equipment, and environmental vulnerabilities), eight flow variables, ten constants, and thirty auxiliary variables. This model captures the complex interrelationships among influencing factors and represents the evolving vulnerability of the metro system in Wuhan, China. Model validity is confirmed through structural validation and extreme condition testing. Various sorts of analyses, including system evolution analysis and local/global sensitivity analysis, have been conducted to demonstrate the capabilities of the developed approach.

In view of the case study, some conclusions are summarized as follows. (1) The development of the overall metro system vulnerability decreases at first and meets an inflection in the middle stage. (2) The average ticket price and safety training are identified as the two most sensitive variables to the vulnerability of the overall metro system. (3) The workforce subsystem displays the highest global sensitivity to the overall system vulnerability, followed by the organizational subsystem. (4) Strategy 1 is identified as the most effective strategy to reduce the vulnerability of the overall metro system, which contributes to reducing the bottom value of vulnerability by 11.91%, while the final value after 25 years is reduced by 13.01%. In brief, the results indicate that the developed approach in this chapter is able to not only predict long-term vulnerability but also improve the baseline strategy for system vulnerability reduction.

The novelty of this chapter lies in the simultaneous consideration of the interactions among influential factors and the measurement uncertainties in the global modeling and robustness enhancement of metro system vulnerability. The developed approach not only facilitates the identification of critical bottlenecks within the system but also supports the formulation of effective strategies for vulnerability reduction and long-term system improvement. By capturing complex system dynamics and uncertainty, this method offers a practical decision-support tool for enhancing the resilience and sustainability of urban transit networks.

References

1. Lin D, Nelson JD, Beecroft M, Cui J (2021) An overview of recent developments in China's metro systems. Tunn Undergr Space Technol 111:103783. https://doi.org/10.1016/j.tust.2020.103783
2. Hong L, Ouyang M, Xu M, Hu P (2020) Time-varied accessibility and vulnerability analysis of integrated metro and high-speed rail systems. Reliab Eng Syst Saf 193:106622. https://doi.org/10.1016/j.ress.2019.106622

3. Chen H, Chen B, Zhang L, Li HX (2021) Vulnerability modeling, assessment, and improvement in urban metro systems: a probabilistic system dynamics approach. Sustain Cities Soc 75:103329. https://doi.org/10.1016/j.scs.2021.103329
4. Gjorgiev B, Sansavini G (2022) Identifying and assessing power system vulnerabilities to transmission asset outages via cascading failure analysis. Reliab Eng Syst Saf 217:108085. https://doi.org/10.1016/j.ress.2021.108085
5. Gong S, Chen L, Zhou Q, Gao X, Shen F (2025) Vulnerability evolution of critical infrastructures: a multidimensional environment-integrated system dynamics analysis. Reliab Eng Syst Saf 256:110719. https://doi.org/10.1016/j.ress.2024.110719
6. Zarghami SA, Dumrak J (2021) A system dynamics model for social vulnerability to natural disasters: disaster risk assessment of an Australian city. Int J Disaster Risk Reduct 60:102258. https://doi.org/10.1016/j.ijdrr.2021.102258
7. Hu W, Dong J, Hwang B, Ren R, Chen Y, Chen Z (2020) Using system dynamics to analyze the development of urban freight transportation system based on rail transit: a case study of Beijing. Sustain Cities Soc 53:101923. https://doi.org/10.1016/j.scs.2019.101923
8. Lu W, Wang F, Bu L, Liu L (2019) Game approach to vulnerability analysis of evacuation highway networks. J Transp Eng Part A: Syst 145:04019044. https://doi.org/10.1061/JTEPBS.0000263
9. Godde C, Dizyee K, Ash A, Thornton P, Sloat L, Roura E, Henderson B, Herrero M (2019) Climate change and variability impacts on grazing herds: insights from a system dynamics approach for semi-arid Australian rangelands. Glob Chang Biol 25:3091–3109. https://doi.org/10.1111/gcb.14669
10. Galli F, Cavicchi A, Brunori G (2019) Food waste reduction and food poverty alleviation: a system dynamics conceptual model. Agric Hum Values 36:289–300. https://doi.org/10.1007/s10460-019-09919-0
11. Finucane ML, Acosta J, Wicker A, Whipkey K (2020) Short-term solutions to a long-term challenge: rethinking disaster recovery planning to reduce vulnerabilities and inequities. Int J Environ Res Public Health 17:482. https://doi.org/10.3390/ijerph17020482
12. Liu X, Lei Z, Duan Z (2024) Assessing metro network vulnerability with turn-back operations: a Monte Carlo method. Phys A Stat Mech Appl 646:129923. https://doi.org/10.1016/j.physa.2024.129923
13. Yu F, Fan B, Qin C, Yao C (2023) A scenario-driven fault-control decision support model for disaster preparedness using case-based reasoning. Nat Hazards Rev 24:04023040. https://doi.org/10.1061/NHREFO.NHENG-1722
14. Kyriakidis M, Majumdar A, Ochieng WY (2015) Data based framework to identify the most significant performance shaping factors in railway operations. Saf Sci 78:60–76. https://doi.org/10.1016/j.ssci.2015.04.010
15. Jing D, Li W, Wen J, Hou W, Wu H, Liu J, Zhang M, Zhang W, Tian T, Ding Z, Guo H (2025) Critical node failure, impact and recovery strategy for metro network under extreme flooding in Shanghai. Int J Disaster Risk Reduct 121:105414. https://doi.org/10.1016/j.ijdrr.2025.105414
16. Kopsidas A, Kepaptsoglou K (2022) Identification of critical stations in a metro system: a substitute complex network analysis. Phys A Stat Mech Appl 596:127123. https://doi.org/10.1016/j.physa.2022.127123
17. Ma Z, Liu J, Chien SI-J, Hu X, Shao Y (2023) Identifying critical stations affecting vulnerability of a metro network considering passenger flow and cascading failure: case of Xi'an metro in China. ASCE-ASME J Risk Uncertain Eng Syst Part A Civ Eng 9:04023014. https://doi.org/10.1061/AJRUA6.RUENG-1013

Part II
System Vulnerability Evaluation

Chapter 5
Flood Hazard Analysis

5.1 Introduction

With climate change and rapid urbanization, urban floods have become one of the critical global challenges [1]. Climate change increases the frequency and intensity of rainfall [2], while the urbanization process greatly enlarges the area of impermeable paving and thus increases the total runoff [3], both of which enhance the urban flood risk and lead to loss of life and severe damage to infrastructures under flood disasters. As a crucial infrastructure of urban transit systems (UTS), the underground metro network is low-lying with closed space, which makes it vulnerable to floods [4, 5]. Therefore, the authorities must evaluate the flood risk of metro networks so that efficient risk mitigation measures can be implemented before disasters to ensure the reliability of the systems.

In the past years, researchers have developed many methods to perform the flood risk assessment for urban areas, including UTS. These methods can be generally classified into two main categories: the numerical-simulation-based methods and the multi-criteria decision analysis (MCDA) methods. The numerical-simulation-based methods [6, 7] require a large amount of data and considerable computational resources; hence, these methods can only be implemented in a small region or some specific metro stations. The MCDA methods establish a unified assessment structure containing multiple factors for all the assessment objects, and an overall risk level can be obtained by fusing all factors' information. However, there are many subjective parameters involved in the MCDA methods that domain experts should determine, such as factor weight, cutoff values between different risk levels for each factor, etc. Naturally, these subjective parameters are uncertain due to various reasons, due to a divergence of subjective parameter assignment in the expert panel. To deal with the uncertainty of subjective parameters, this chapter proposes an optimality approach based on evidential reasoning (ER), putting the uncertainty of all kinds

L. Zhang et al., *Resilience in Urban Transit Systems*, Advances in 21st Century Human Settlements, https://doi.org/10.1007/978-981-95-4322-9_5

of subjective parameters into a unified framework, which could yield a more reliable result of flood risk assessment.

The metro station is one of the most vulnerable infrastructures to urban flood due to its structural characteristics and geographic location; hence, the countermeasures against flood risk for metro stations should be carefully considered by municipal departments. Some researchers investigated flood-prevention measures from the perspective of engineering equipment and construction [8]. However, these studies do not explore the short-term and long-term benefits of the measures. In this chapter, the short-term and long-term benefits are based on the improvement degree of the influential factors rather than the temporal scale. It is necessary to consider both the short-term and long-term benefits when adopting flood risk mitigation measures so that the improvement degrees of the influential factors can be properly determined. In this chapter, a novel indicator of sensitivity analysis will be proposed based on risk assessment results, which can consider both the short-term and long-term benefits of the flood risk mitigation measures, contributing to a rational decision on the flood control problem.

To summarize, this chapter proposes an ER-based optimality approach to address the uncertainty of subjective parameters involved in risk assessment [9]. Specifically, we consider the subjective parameters to vary within certain intervals, and then employ evolutionary optimization to solve the maximum risk expectation, which is adopted as the index to measure the flood risk in the worst case under the uncertainty of subjective parameters. Furthermore, we propose a sensitivity indicator termed the combined gradient to identify critical influential factors and correspondingly provide risk mitigation measures for the high-risk metro stations. A case study is conducted on Shanghai metro stations to verify the applicability and validity of the proposed approach.

5.2 Flood Risk Assessment in UTS

Currently, there is a substantial body of research on flood risk assessment for UTS. From a methodology perspective, these studies can be categorized into risk assessment based on numerical simulations and MCDA:

1. **Numerical-simulation-based risk assessment.** Flood risk assessment based on numerical simulation generally follows two steps: firstly, establishing an appropriate numerical model to simulate the flood processes within the metro systems, and secondly, identifying high-risk areas based on quantitative results obtained from numerical simulations. Numerical-simulation-based risk assessment primarily targets a particular metro station or tunnels within a specific area. For instance, Lin et al. [10] employed VOF to simulate the flood intrusion process in an interchange metro station, finding that the property development floor was the highest risk area. With regard to the flood risk of metro tunnels, Kusano et al. [11] adopted InfoWorks CS to conduct a flood analysis for the tunnel network in

Osaka, Japan, in which the maximum water depth was taken as the risk indicator. Researchers have also employed an alternative approach to flood risk assessment for a metro system. This approach firstly simulates urban surface flooding, then analyzes flood conditions at metro station entrances to determine whether the corresponding station will experience flood ingress. This approach is relatively macroscopic, allowing for the simultaneous assessment and comparison of flood risks across multiple metro stations within the same area. An exemplary work employing this approach was conducted by Lyu et al. [12].

2. **MCDA-based risk assessment.** MCDA is a branch of operations research that systematically considers multiple criteria involved in decision-making and provides a comprehensive assessment of the assessment object. Well-known methods such as the analytic hierarchy process (AHP), evidential reasoning, Bayesian network, etc., can all be applied in MCDA. In the context of flood risk assessment for metro systems, employing MCDA firstly necessitates the establishment of a unified index system that encompasses various influential factors related to flood risk. Subsequently, the MCDA method is implemented to fuse information from all factors and yield flood risk values or risk levels for the assessed entities. In the area of flood risk assessment, the most popular MCDA method is AHP and its variants. By employing the 1–9 scale method in pairwise comparisons conducted by experts, AHP computes the weight of each factor and ultimately generates a normalized flood risk value. Typical works in this regard can be found in Lyu et al. [5] and Lyu et al. [13], in which the flood risk of Guangzhou metro lines was assessed. However, the MCDA methods are susceptible to the subjectivity of parameter assignment, and AHP is no exception. To address this issue, a series of AHP variants that utilize fuzzy numbers have been introduced for metro flood risk assessment, including triangular fuzzy AHP [14, 15] and trapezoidal fuzzy AHP [16].

However, the above AHP-based research can only consider the uncertainty of weights and cannot incorporate other subjective parameters. Therefore, to address this issue, this chapter develops an ER-based optimality approach which puts the uncertainty of all kinds of subjective parameters into a unified framework. The specific methodologies are detailed in the next section.

5.3 Flood Risk Assessment Framework

As shown in Fig. 5.1, a framework of the ER-based optimality approach is developed to assess and mitigate the flood risk of metro stations. The detailed methodology of each step of the framework will be elaborated on in the following sections.

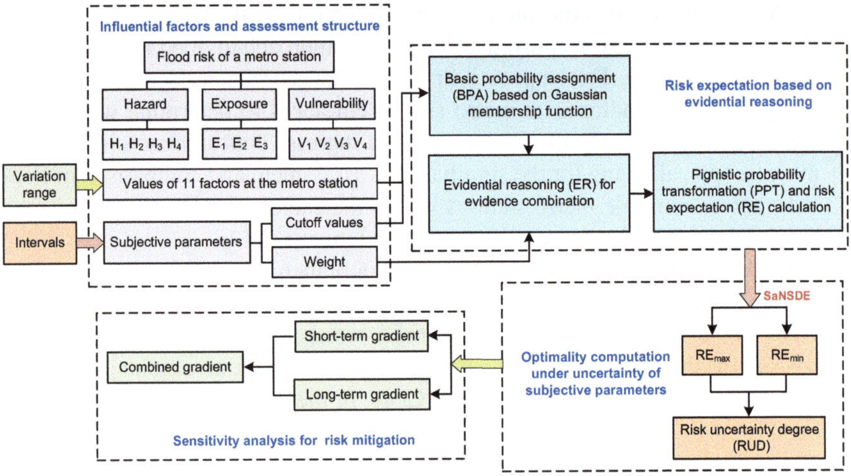

Fig. 5.1 General framework of this chapter

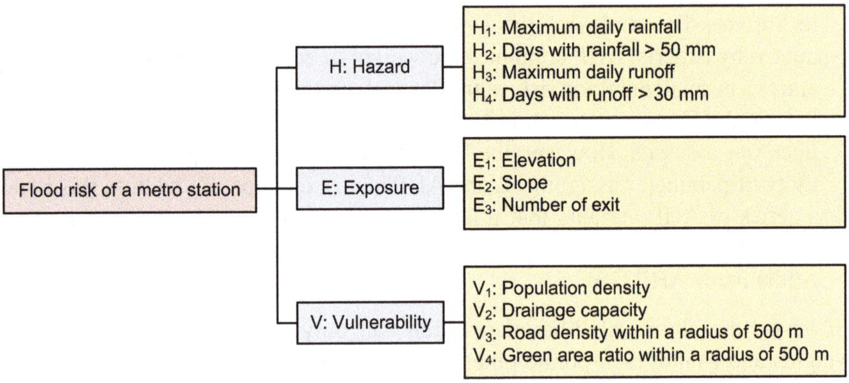

Fig. 5.2 Assessment structure of flood risk of a metro station

5.3.1 *Influential Factors and Assessment Structure*

As a premise of implementing MCDA-based risk assessment, it is necessary to determine the influential factors and assessment structure. Generally, there are three components leading to the overall risk: hazard, exposure, and vulnerability [17]. In the context of flood risk assessment for metro stations, several influential factors under each risk component are selected based on technical documents and domain experts, which constitute the risk assessment structure (see Fig. 5.2). Specifically, hazard is measured by four factors: maximum daily rainfall (H_1), days with rainfall exceeding 50 mm (H_2), maximum daily runoff (H_3), and days with runoff exceeding 30 mm (H_4). As the most critical cause of metro floods, extreme rainfall and runoff should be considered in the assessment structure. Exposure is measured by elevation (E_1) [5, 13], slope (E_2) [5, 13], and the number of exits (E_3) [4]. Vulnerability is measured by population density (V_1) [18], drainage capacity (V_2), road density [14] within a radius of 500 m (V_3), and green area ratio [4] within a radius of 500 m (V_4).

Among the 11 influential factors in the assessment structure, the flood risk will decrease with the increase of the four factors: elevation (E_1), slope (E_2), drainage capacity (V_2), and green area ratio (V_4), which are referred to as negative factors. By contrast, the other seven factors are positive factors, whose values positively correlate with flood risk. In the following calculation procedures of risk assessment and mitigation, there will be some slight differences between the negative and positive factors.

5.3.2 *Risk Expectation Function Based on Evidential Reasoning*

5.3.2.1 BPA Construct

Generally, suppose the metro station to be assessed has the factor values of $x_1, x_2, \ldots, x_L$ for the L influential factors ($L = 11$ in this chapter), and all the L influential factors are assessed using the same set of N risk assessment levels R_n, $n = 1, 2, \ldots, N$, in which R_1 and R_N represent the lowest and highest risk level, respectively. The N assessment levels constitute a frame of discernment $R = \{R_1, R_2, \ldots, R_N\}$ in the D-S theory of evidence [19]. For the lth influential factor, define $c_{i,l}$ ($i = 1, 2, \ldots, N-1$) as the cutoff value between the levels R_i and R_{i+1}. For a positive factor, we have $c_{1,l} < c_{2,l} < \ldots < c_{N-1,l}$; for a negative factor, we have $c_{1,l} > c_{2,l} > \ldots > c_{N-1,l}$; According to the cutoff values, the Gaussian membership function of the risk level R_i ($i = 2, 3, \ldots, N-1$) of the lth influential factor ($l = 1, 2, \ldots, L$), $\mu_{i,l}(x_l)$, can be constructed as Eq. (5.1), in which $a_{i,l}$ and $b_{i,l}$ are calculated as Eqs. (5.2) and (5.3), respectively.

$$\mu_{i,l}\left(x_l\right) = e^{-\frac{\left(x_l - a_{i,l}\right)^2}{2b_{i,l}^2}} \tag{5.1}$$

$$a_{i,l} = \frac{c_{i-1,l} + c_{i,l}}{2} \tag{5.2}$$

$$b_{i,l} = \frac{c_{i,l} - c_{i-1,l}}{2} \tag{5.3}$$

The membership function $\mu_{i,l}(x_l)$ is a fuzzy number reflecting the degree to which the metro station belongs to the risk level R_i according to the lth influential factor. However, the risk assessor may not always be confident enough to provide assessments to an individual risk level only, but at times wishes to assess beliefs to subsets of p adjacent risk levels from i to $(i + p)$, i.e., $R_{i(i+1)\ldots(i+p)}$ (referred to as local ignorance). Similarly, there exists global ignorance, $R_{12...N}$. For the Gaussian membership model, the membership degree of local and global ignorance can be calculated by taking the minimum of the membership degrees of all individual risk levels it covers [20], written as Eqs. (5.4) and (5.5):

$$\mu_{i(i+1)\ldots(i+p),l}\left(x_l\right)=\min\left\{\mu_{i,l}\left(x_l\right),\mu_{i+1,l}\left(x_l\right),\ldots,\mu_{i+p,l}\left(x_l\right)\right\} \tag{5.4}$$

$$\mu_{12\ldots N,l}\left(x_l\right)=\min\left\{\mu_{1,l}\left(x_l\right),\mu_{2,l}\left(x_l\right),\ldots,\mu_{N,l}\left(x_l\right)\right\} \tag{5.5}$$

From the above Gaussian membership model, we can obtain the membership degrees of all individual risk levels and local and global ignorance for a given x_l. Then, the BPA is generated according to the rules put forward by Jiang and Deng [20]. The detailed procedure is demonstrated below:

1. Rank the membership degrees of all individual risk levels in descending order.
2. For the individual risk level with the largest membership degree, we assign the membership degree to it as its belief degree β.
3. For the local ignorance constituted by the top k individual risk levels, we assign it the kth largest membership degree of the individual risk level as the belief degree.
4. Assign the lowest membership degree of the individual risk level to the global ignorance as its belief degree.
5. Assign zero to other subsets of the frame of discernment.
6. If the sum of belief degrees exceeds 1, normalize each belief degree; otherwise, assign the redundant belief degree to global ignorance.

Based on the above rules, we can obtain the BPA regarding the lth factor, i.e., $\beta_{\gamma,l}(\gamma \subseteq R)$. Alternatively, the BPA can be expressed as $\beta_l(\gamma)(\gamma \subseteq R)$, which is a more common way to treat γ as an independent variable.

5.3.2.2 Evidential Reasoning for Evidence Combination

In risk assessment, the importance of the influential factors to the overall risk is different, and such importance can be represented by a weight. Suppose $0 \leq w_l \leq 1$ ($l = 1, 2, \ldots, L$) is the weight of the lth factor, with $w_l = 0$ and 1 standing for "not important at all" and "the most important", respectively. Then, by discounting the BPA with the weight and assigning the redundant belief degree to the power set of R, we construct the weighted belief degree (WBD), written as Eq. (5.6):

$$m_{\gamma,l}=m_l\left(r\right)=\begin{cases}0,\gamma=\varnothing\\ w_l\beta_{\gamma,l},\gamma\subseteq R,\gamma\neq\varnothing\\ 1-w_l,\gamma=P\left(R\right)\end{cases} \tag{5.6}$$

Suppose L pieces of independent evidence have their WBDs represented by Eq. (5.6). Combining L WBDs needs a recursive procedure [21]. To be specific, suppose $e(l)$ denotes the combination of the first l pieces of evidence and $m_{\gamma, e(l)}$ is the WBD to which each γ is supported jointly by $e(l)$, with $m_{\gamma, e(1)} = m_{\gamma, 1}$ and $m_{P(R), e(1)} = m_{P(R), 1}$, and the combination of the first l WBDs is calculated by Eq. (5.7), in which $\hat{m}_{\gamma,e(l)}$ and $\hat{m}_{P(R),e(l)}$ are computed by Eqs. (5.8) and (5.9), respectively. Through the recursive procedure, we can successively solve $m_{\gamma, e(1)}$, $m_{\gamma, e(2)}$, …, and $m_{\gamma, e(L)}$, and the intermediate variable $\hat{m}_{\gamma,e(1)}$, $\hat{m}_{\gamma,e(2)}$, …, and $\hat{m}_{\gamma,e(L)}$ can be also obtained. Finally, the BPA combined from L pieces of independent evidence is given by Eq. (5.10).

$$m_{\gamma,e(l)} = \left[m_1 \oplus m_2 \oplus \ldots \oplus m_l\right](\gamma) = \begin{cases} 0, \gamma = \varnothing \\ \dfrac{\hat{m}_{\gamma,e(l)}}{\sum_{r \subseteq R} \hat{m}_{r,e(l)} + \hat{m}_{P(R),e(l)}}, \gamma \neq \varnothing \end{cases} \tag{5.7}$$

$$\hat{m}_{\gamma,e(l)} = \left[(1 - w_l) m_{\gamma,e(l-1)} + m_{P(R),e(l-1)} m_{\gamma,l}\right] + \sum_{B \cap C = \gamma} m_{B,e(l-1)} m_{C,l}, \forall \gamma \subseteq R \tag{5.8}$$

$$\hat{m}_{P(R),e(l)} = (1 - w_l) m_{P(R),e(l-1)} \tag{5.9}$$

$$\beta_{\gamma,e(L)} = \beta_{e(L)}(\gamma) = \begin{cases} 0, \gamma = \varnothing \\ \dfrac{\hat{m}_{\gamma,e(L)}}{\sum_{r \subseteq R} \hat{m}_{r,e(L)}}, \gamma \neq \varnothing \end{cases} \tag{5.10}$$

where $m_{\gamma, e(l)}$ is the WBD to which each γ is supported jointly by $e(l)$; w_l is the weight of the lth factor; where $\beta_{\gamma, e(L)}$ is the belief degree of γ, which is supported jointly by the L pieces of evidence.

5.3.2.3 Pignistic Probability Transformation

Generally, a BPA will assign belief to local and global ignorance; hence, it is difficult to make decisions or assessments directly from a BPA in most cases. In view of this problem, mapping BPA to probabilities is an effective way to improve the decision or assessment credibility and reduce computational complexity as well. To transform the BPA into a probability mass function, we adopt the PPT approach [22], written as Eq. (5.11):

$$P(R_n) = \sum_{R_n \in S \subseteq R} \frac{\beta_{S,e(L)}}{|S|} = \sum_{S \subseteq R} \beta_{S,e(L)} \frac{|R_n \cap S|}{|S|} \quad n = 1,2,\ldots,N \tag{5.11}$$

where $P(R_n)$ is the probability that the metro station is at the risk level of R_n; S is any subset of the frame of discernment R that contains R_n; $|S|$ is the number of elements belonging to S.

Afterward, inspired by the concept of "utility", we introduce the function written in Eq. (5.12) to transform the risk level R_n to a numerical value within the range of [0,1]. With the probability and quantitative expression of each R_n, RE can be finally acquired by calculating the mathematical expectation of $u(R_n)$, written as Eq. (5.13). RE∈[0,1] is a quantitative measurement for the flood risk of metro stations. A larger RE value closer to 1 represents a higher flood risk, while a lower RE closer to 0 indicates a lower flood risk.

$$u(R_n) = \frac{n-1}{N-1} \quad n = 1,2,\ldots,N \tag{5.12}$$

$$\mathrm{RE} = \sum_{n=1}^{N} u(R_n) P(R_n) \tag{5.13}$$

where $u(R_n)$ is the quantitative expression of the risk level R_n; RE is the risk expectation.

5.3.3 Optimality Computation Under Uncertainty of Subjective Parameters

In Sect. 5.3.2, the subjective parameters involved in RE can be uncertain. One typical uncertain situation is that the subjective parameters, i.e., the cutoff values $c_{i,l}$ ($i = 0, 1, \ldots, N$, $l = 1, 2, \ldots, L$) and weights of factors w_l ($l = 1, 2, \ldots, L$), are determined by a panel of domain experts, and different experts may assign different values for a same subjective parameter. To deal with the uncertainty of the subjective parameters, we use intervals instead of crisp values to describe subjective parameters. For instance, if five experts assign five values, w_{1a}, w_{1b}, w_{1c}, w_{1d}, w_{1e}, for the subjective parameter w_1, and $w_{1a} \le w_{1b} \le w_{1c} \le w_{1d} \le w_{1e}$, we can assume that w_1 is within the interval of $[w_{1b}, w_{1d}]$ if we remove the maximum and minimum values. Under this circumstance, the RE value will fluctuate in a certain range with the subjective parameters varying in their intervals. Defining $\mathcal{F}$ as the functional relationship between RE and the subjective parameters, we can solve the maximum and minimum values of RE given the upper and lower bounds of all the subjective parameters. This problem can be mathematically written as Eq. (5.14):

$$\max/\min \mathrm{RE} = \sum_{n=1}^{N} u(R_n) P(R_n) = \mathcal{F}(c_{n,l}, w_l)$$

$$\text{s.t.} \quad c_{n,l}^{-} < c_{n,l} < c_{n,l}^{+}, \qquad n = 0,1,\ldots,N \text{ and } l = 1,2,\ldots,L$$

$$w_l^{-} < w_l < w_l^{+}, \qquad l = 1,2,\ldots,L \tag{5.14}$$

where $c_{n,l}^{-}$ and $c_{n,l}^{+}$ are the lower and upper bounds of $c_{n,l}$, respectively; w_l^{-} and w_l^{+} are the lower and upper bounds of w_l, respectively.

To obtain the maximum and minimum of RE, a novel evolutionary optimization algorithm, SaNSDE, is employed in this chapter. The detailed computational process and the pseudocode of the SaNSDE algorithm can be referred to in the paper by Hsieh [23]. Using the SaNSDE algorithm, the maximum and minimum RE (abbreviated to RE_{max} and RE_{min}, respectively) can be obtained. Here, RE_{max} is the main index measuring the flood risk, which reflects the worst case under the uncertainty of subjective parameters. A larger RE_{max} closer to 1 indicates a higher flood risk of a metro station. As a supplement to RE_{max}, risk uncertainty degree (RUD), defined as the difference between RE_{max} and RE_{min} (written as Eq. (5.15)), is another index that represents the magnitude of RE's fluctuation range when the subjective parameters vary in their intervals. A larger RUD suggests that the RE value of a metro station is sensitive to the variations of subjective parameters, which means that the experts' opinions are deeply divided on the flood risk of this metro station.

$$\text{RUD} = \text{RE}_{\text{max}} - \text{RE}_{\text{min}} \tag{5.15}$$

5.3.4 Sensitivity Analysis for Risk Mitigation

From the above sections, we know that there is a uniquely determined value of RE_{max} for a metro station when the intervals of subjective parameters are given. Hence, regarding the L factor values as the independent variables and the RE_{max} as the dependent variable, we can obtain a function relationship $\mathcal{G}$ between RE_{max} and factor values, written as:

$$\text{RE}_{\text{max}} = \mathcal{G}\left(x_1, x_2, \ldots, x_L\right) \tag{5.16}$$

For those metro stations with high RE_{max}, it is necessary to take some risk mitigation measures to lower their RE_{max} values. An idea that comes naturally is to adjust the values of the influential factors. However, there is another question: which factor is the most important for the reduction of RE_{max}? To solve the question and suggest the most effective measures for flood risk mitigation, we propose a novel sensitivity indicator termed the combined gradient to quantify the factor importance for a metro station. The combined gradient integrates the short-term and long-term gradients, so it can consider both the short-term and long-term benefits of a risk mitigation measure. The detailed derivation of the combined gradient is given below.

As written in Eq. (5.16), RE_{max} has a uniquely determined value when all factor values are determined. When the lth factor value, x_l, varies in a wide range while other factor values remain unchanged, the RE_{max} value will also change accordingly, and we can then obtain a curve depicting the relationship between RE_{max} and x_l, shown as the black curves in Fig. 5.3 (a) (for positive factor) and (b) (for negative factor). It can be seen from Fig. 5.3 that RE_{max} will significantly change only when x_l varies in a certain range. Suppose that $c_{0,l}^m$ and $c_{N,l}^m$ are the two cutoff values beyond which the RE_{max} value will be basically unchanged. The two cutoff values can be approximately computed by taking the interval mid-values of $c_{0,l}$ and $c_{N,l}$, written as:

$$c_{0,l}^m = \frac{1}{2}\left(c_{0,l}^- + c_{0,l}^+\right) \tag{5.17}$$

$$c_{N,l}^m = \frac{1}{2}\left(c_{N,l}^- + c_{N,l}^+\right) \tag{5.18}$$

where $c_{0,l}^-$ and $c_{0,l}^+$ are the lower and upper bounds of $c_{0,l}$, respectively; $c_{N,l}^-$ and $c_{N,l}^+$ are the lower and upper bounds of $c_{N,l}$, respectively. Suppose that x_l^* is the original value of the lth factor, and $\left(x_l^*, \mathcal{G}\left(x_1^*, x_2^*, \ldots, x_l^*, \ldots, x_L^*\right)\right)$ is the original point. We define two gradients, $k_{st,m}$ and $k_{lt,m}$, as follows:

$$k_{st,m} = \begin{cases} \lim\limits_{\Delta \to 0^-} \dfrac{\dfrac{\mathcal{G}\left(x_1^*, x_2^*, \ldots, x_l^* + \Delta, \ldots, x_L^*\right) - \mathcal{G}\left(x_1^*, x_2^*, \ldots, x_l^*, \ldots, x_L^*\right)}{\Delta}}{c_{N,l}^m - c_{0,l}^m}, & x_l \text{ is the positive factor} \\ \lim\limits_{\Delta \to 0^+} \dfrac{\dfrac{\mathcal{G}\left(x_1^*, x_2^*, \ldots, x_l^* + \Delta, \ldots, x_L^*\right) - \mathcal{G}\left(x_1^*, x_2^*, \ldots, x_l^*, \ldots, x_L^*\right)}{\Delta}}{c_{N,l}^m - c_{0,l}^m}, & x_l \text{ is the negative factor} \end{cases} \tag{5.19}$$

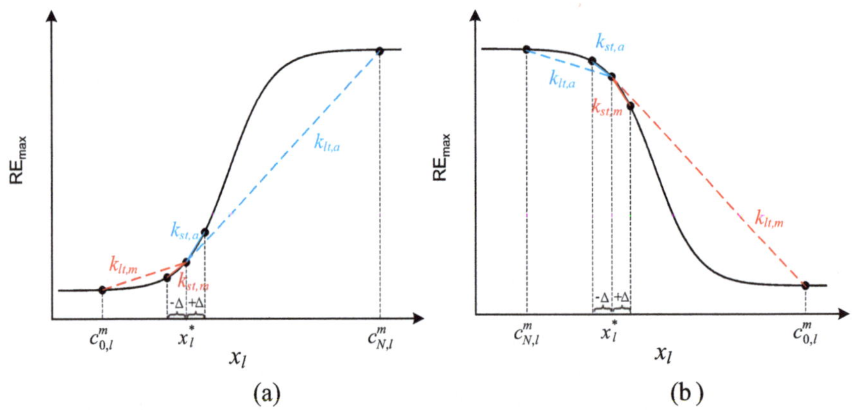

Fig. 5.3 Short-term and long-term gradients for risk mitigation and aggravation: (**a**) positive factor; (**b**) negative factor

$$k_{\mathrm{lt},m} = \frac{\mathcal{G}\left(x_1^*,x_2^*,\ldots,c_{0,l}^m,\ldots,x_L^*\right)-\mathcal{G}\left(x_1^*,x_2^*,\ldots,x_l^*,\ldots,x_L^*\right)}{\dfrac{c_{0,l}^m - x_l^*}{c_{N,l}^m - c_{0,l}^m}} \tag{5.20}$$

where $k_{\mathrm{st},m}$ and $k_{\mathrm{lt},m}$ are the short-term and long-term gradients for risk mitigation, respectively. Note that the x-coordinate, i.e., the factor value, is divided by $c_{N,l}^m - c_{0,l}^m$, which is because we need to nondimensionalize the combined gradient to make the combined gradients of different influential factors comparable.

As shown in Fig. 5.3, $k_{\mathrm{st},m}$ is the left-hand (right-hand) derivative at the original point for a positive (negative) factor, which represents the reduction of $\mathrm{RE}_{\max}$ when the lth factor is improved by a unit in the neighborhood of its original value; $k_{\mathrm{lt},m}$ is the slope of the line connecting the original point and the lowest risk point, i.e., $\left(c_{0,l}^m,\mathcal{G}\left(x_1^*,x_2^*,\ldots,c_{0,l}^m,\ldots,x_L^*\right)\right)$, which represents the reduction of $\mathrm{RE}_{\max}$ when the lth factor is improved to the superior limit. The short-term and long-term gradients reflect the short-term and long-term benefits of a risk mitigation measure, respectively, relying only on one of them will cause bias in decision-making. On this basis, we propose the combined gradient for risk mitigation, k_m, to fully consider the magnitude of both gradients. Written as Eq. (5.21), the combined gradient is the weighted average of the short-term and long-term gradients, i.e.,

$$k_m = \alpha k_{\mathrm{st},m} + (1-\alpha) k_{\mathrm{lt},m} \tag{5.21}$$

where $0 \le \alpha \le 1$ is the weight of $k_{\mathrm{st},m}$. Different values of α indicate different preferences for the short-term gradient. In this study, we suggest $\alpha = 0.5$ to treat both gradients equally. The combined gradient is a comprehensive indicator that measures the sensitivity of $\mathrm{RE}_{\max}$ to an influential factor in terms of risk mitigation. Improving the factors with higher k_m can be suggested for effective flood risk mitigation at a metro station.

Similarly, we can propose the short-term and long-term gradients for risk aggravation, i.e., $k_{\mathrm{st},a}$ and $k_{\mathrm{lt},a}$, written as Eqs. (5.22) and (5.23). Correspondingly, the combined gradient for risk aggravation, k_a, is proposed as the weighted average of $k_{\mathrm{st},a}$ and $k_{\mathrm{lt},a}$, written as Eq. (5.24). Municipal departments should pay special attention to those influential factors with higher k_a, preventing their further deterioration that could cause the aggravation of the flood risk at a metro station.

$$k_{\mathrm{st},a} = \begin{cases} \lim\limits_{\Delta\to 0^+} \dfrac{\mathcal{G}\left(x_1^*,x_2^*,\ldots,x_l^*+\Delta,\ldots,x_L^*\right) - \mathcal{G}\left(x_1^*,x_2^*,\ldots,x_l^*,\ldots,x_L^*\right)}{\dfrac{\Delta}{c_{N,l}^m - c_{0,l}^m}}, & x_l \text{ is the positive factor} \\ \lim\limits_{\Delta\to 0^-} \dfrac{\mathcal{G}\left(x_1^*,x_2^*,\ldots,x_l^*+\Delta,\ldots,x_L^*\right) - \mathcal{G}\left(x_1^*,x_2^*,\ldots,x_l^*,\ldots,x_L^*\right)}{\dfrac{\Delta}{c_{N,l}^m - c_{0,l}^m}}, & x_l \text{ is the negative factor} \end{cases} \tag{5.22}$$

$$k_{\mathrm{lt},a} = \frac{\mathcal{G}\left(x_1^*, x_2^*, \ldots, c_{N,l}^m, \ldots, x_L^*\right) - \mathcal{G}\left(x_1^*, x_2^*, \ldots, x_l^*, \ldots, x_L^*\right)}{\dfrac{c_{N,l}^m - x_l^*}{c_{N,l}^m - c_{0,l}^m}} \tag{5.23}$$

$$k_a = \alpha k_{\mathrm{st},a} + (1-\alpha) k_{\mathrm{lt},a} \tag{5.24}$$

5.4 Experimental Studies

To verify the applicability and validity of the proposed approach, a total of 378 metro stations (as of Aug 2021) in Shanghai, China, are utilized as the case study for demonstration purposes. Detailed elaborations on the case background, data processing, and result analysis are presented in Sects. 5.4.1, 5.4.2, and 5.4.3, respectively.

5.4.1 Studied Area

Shanghai is located between longitude 121°0′0″–122°0′0″ E and latitude 31°0′0″ N on the southern estuary of the Yangtze River in China. As one of the largest and most populous cities in the world, Shanghai possesses the world's biggest metro system by route length (831 km, by Dec 2021), and the second biggest by the number of stations (408 unique stations, by Dec 2021). Figure 5.4 shows the map of the Shanghai metro system. Besides, the Shanghai metro system ranks first in the world in terms of annual ridership, with 3.88 billion rides delivered in 2019 [24]. Therefore, the efficient and safe operation of the Shanghai metro system is closely related to people's lives. However, significant rainfall often occurs in Shanghai in summer due to the plum rain season and typhoon storms, resulting in urban waterlogging and even flooding in the metro system. As reported, the Yishan Road Station and the Shanghai Station suffered from flood events in the early years [4]. Hence, it is necessary to carry out a reliable flood risk assessment and provide some useful risk mitigation measures for Shanghai metro stations to ensure the safety of people's lives and property.

5.4.2 Model Development

To acquire the 11 factor values for a total of 378 metro stations in Shanghai, the GIS platform is employed for raw data preprocessing and result visualization. In this chapter, we collect (1) the daily rainfall and runoff recorded at 11 weather stations in Shanghai in 2021 from the China Meteorological Data Website (http://data.cma.

Fig. 5.4 Map of the Shanghai metro system (as of Dec 2021)

cn); (2) the digital elevation model including the elevation and slope data from Geospatial Data Cloud (http://www.gscloud.cn/) hosted by Chinese Academy of Sciences; (3) population density of Shanghai from the WorldPop research program (https://www.worldpop.org/) based on the University of Southampton; (4) road network and green areas vectorized from the electronic map. Afterward, we utilize the GIS platform to process the raw data and extract the factor values of each metro station in Shanghai. The factor values for Shanghai metro stations are shown in Fig. 5.5.

Then, we ask five experts on urban hydrology and infrastructure construction to assign the values for the subjective parameters. In this study, we set three risk levels for the overall flood risk and each influential factor; hence, there are five subjective parameters (four cutoff values and one factor weight) for each factor. For a specific subjective parameter, e.g., w_1, suppose the five experts assign five values which are sorted in ascending order like $w_{1a} \leq w_{1b} \leq w_{1c} \leq w_{1d} \leq w_{1e}$, we consider that w_1 is

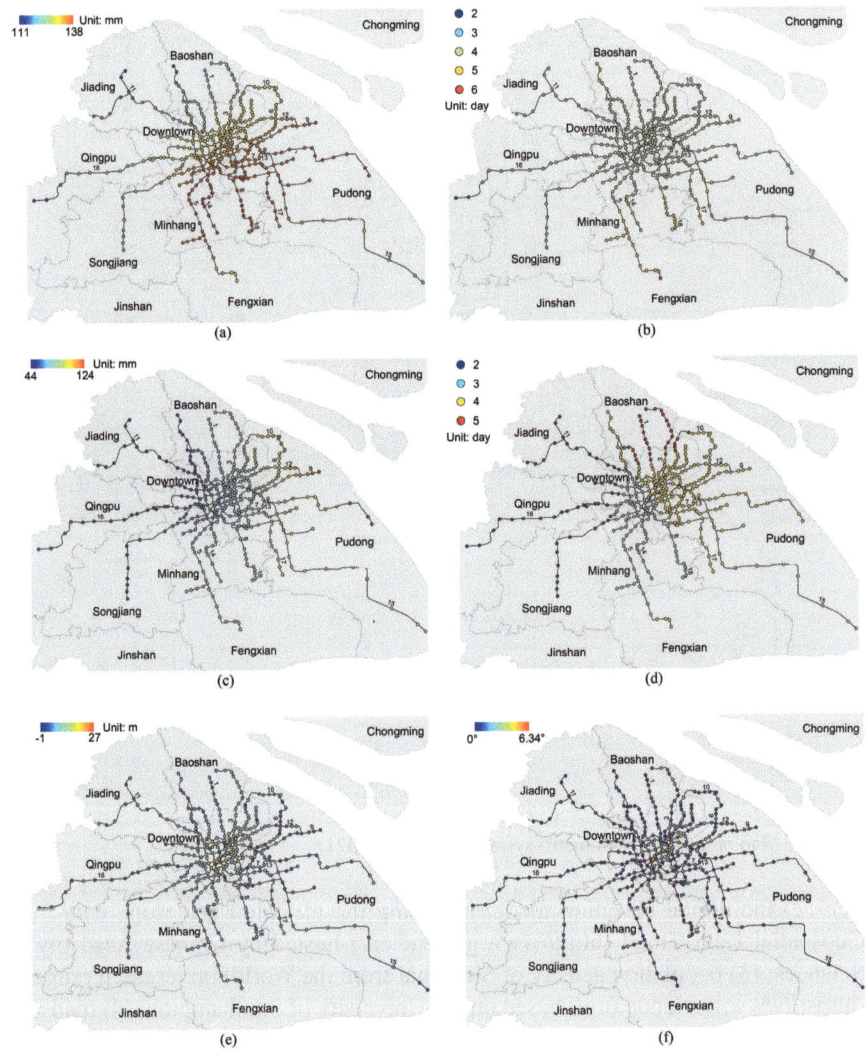

Fig. 5.5 Actors of metro stations in Shanghai: **(a)** H_1; **(b)** H_2; **(c)** H_3; **(d)** H_4; **(e)** E_1; **(f)** E_2; **(g)** E_3; **(h)** V_1; **(i)** V_2; **(j)** V_3; **(k)** V_4

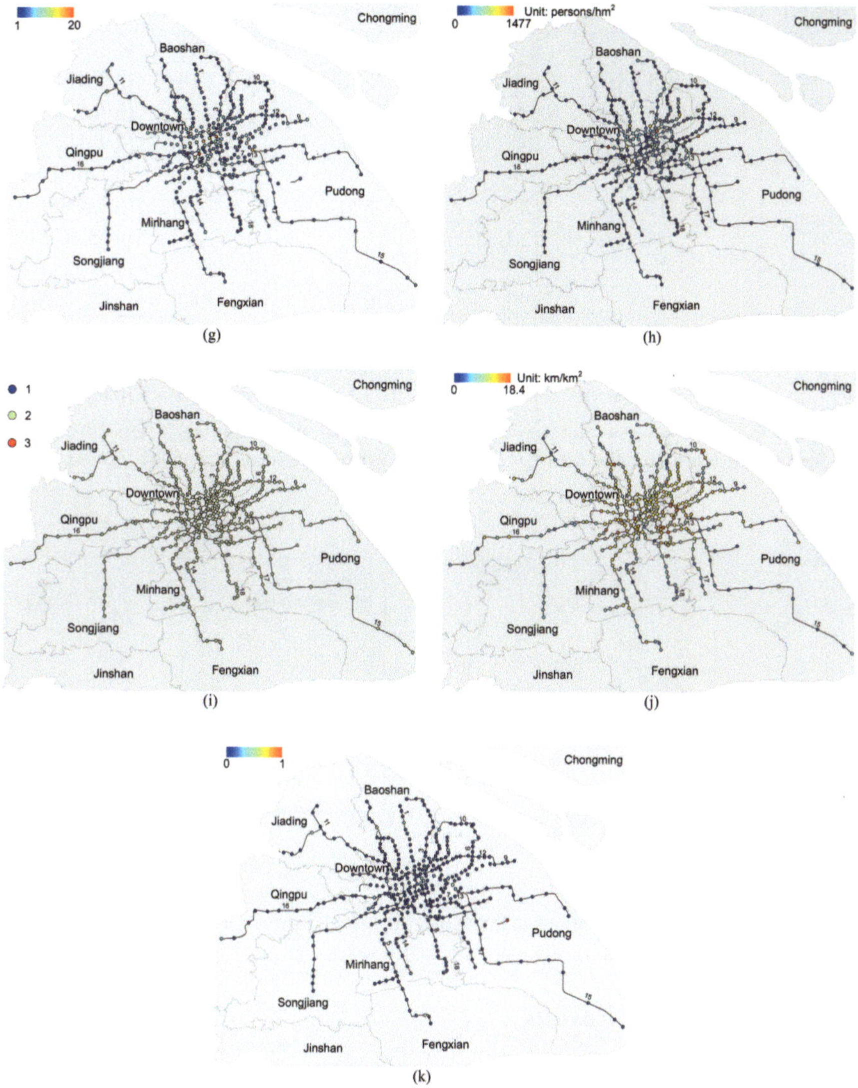

Fig. 5.5 (continued)

within the interval of $[w_{1b}, w_{1d}]$ if the maximum and minimum values are removed. In this way, we finally obtain the intervals of all the subjective parameters, as listed in Table 5.1. Note that some subjective parameters have crisp values rather than intervals because the experts agree on the assignment of those parameters.

Table 5.1 Intervals or crisp values of subjective parameters

Influential factor	c_0	c_1	c_2	c_3	w
H_1	[80, 100]	[115, 125]	[135, 145]	[160, 180]	[0.9, 1]
H_2	3	3.5	4.5	5	[0.7, 0.8]
H_3	[30, 40]	[60, 70]	[80, 100]	[130, 150]	[0.9, 1]
H_4	2	3.5	4.5	6	[0.7, 0.8]
E_1	[10, 20]	[7, 8]	[4, 5.5]	0	[0.8, 0.85]
E_2	[5, 10]	[2.5, 3]	[1, 1.5]	0	[0.65, 0.75]
E_3	2	4.5	8.5	[10, 12]	[0.5, 0.7]
V_1	0	100	[250, 300]	[350, 400]	[0.75, 0.85]
V_2	3	2.5	1.5	1	1
V_3	[2, 3]	5.5	8	[15, 20]	[0.6, 0.75]
V_4	[0.7, 0.8]	[0.3, 0.5]	0.05	0	[0.8, 0.85]

Note: $w = 0$ and 1 represent "not important at all" and "the most important", respectively. It is unnecessary to normalize the factor weight in ER when both local and global ignorance are considered

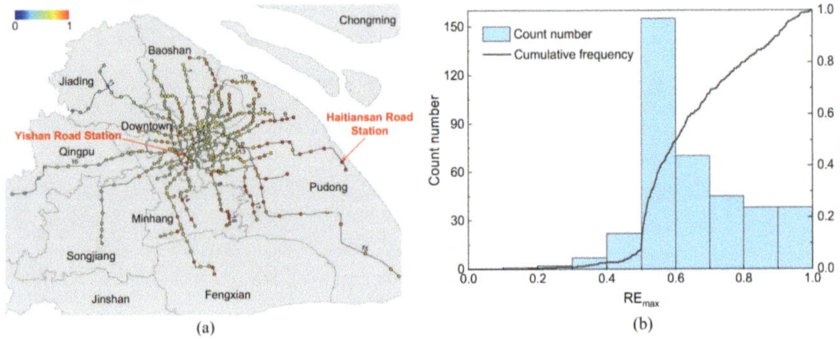

Fig. 5.6 RE$_{max}$ of 378 metro stations in Shanghai: (**a**) spatial distribution map; (**b**) histogram with cumulative frequency

5.4.3 Analysis of Results

The ER-based optimality approach can be implemented after the influential factor values of the 378 metro stations in Shanghai are acquired and the intervals or crisp values of the 55 subjective parameters are determined. As a result, Figs. 5.6 and 5.7 show the RE$_{max}$ and RUD of the 378 metro stations in Shanghai, both of which are demonstrated in a spatial distribution map and a histogram with cumulative frequency. For a clearer demonstration of computation results, Fig. 5.8 shows the interval plot of RE$_{max}$ and RE$_{min}$ values with RUD of all 378 metro stations. Furthermore, taking the Haitiansan Road Station and Yishan Road Station to conduct the sensitivity analysis, we further calculate the combined gradient for risk mitigation and aggravation (see Fig. 5.9) and then propose the corresponding measures to mitigate the flood risk at the two stations. The results are analyzed below.

1. In general, the metro stations in eastern Shanghai have much higher flood risks than those in central and western Shanghai. According to Fig. 5.6a, most eastern metro stations have RE_{max} values approaching 1. Among them, the Haitiansan Road Station (marked out in Fig. 5.6a) is a station with the RE_{max} of 0.996, which is the largest among all Shanghai metro stations. The higher flood risk of eastern metro stations results from the unbalanced rainfall and runoff distributions. By contrast, most metro stations in central Shanghai have a RE_{max} of around 0.5, which shows medium flood risks. However, some specific metro stations in the central area have much higher RE_{max} values, which need to be taken seriously. For example, the RE_{max} of the Yishan Road Station (marked out in Fig. 5.6a) is 0.958, ranking the highest in the downtown district. The reason is that the Yishan Road Station lies at a high-risk level on multiple exposure and vulnerability factors such as slope (E_2), population density (V_1), road density (V_3), green area ratio (V_4), etc. Besides, as illustrated in Fig. 5.6b, nearly half of the metro stations (177/378) have the RE_{max} values within the range of [0.4, 0.6], and about a quarter (95/378) of stations' RE_{max} values are over 0.8, which demonstrates that the Shanghai metro system is generally safe given that the RE_{max} represents the worst case of flood risk under uncertainty.
2. The risk assessment results can be verified by the government flood-prevention decisions and the past flood event. As mentioned, the metro stations in eastern Shanghai have higher flood risks, with the RE_{max} close to 1. At the end of June 2021, Shanghai experienced an extreme rainfall event with a maximum daily rainfall of over 130 mm, during which the Shanghai municipal department once closed 45 out of 113 metro stations in Pudong (the easternmost district in Shanghai), including the Haitiansan Road Station, for safety reasons [25]. To be more specific, among the 45 closed metro stations in Pudong, there are 24 stations with the RE_{max} over 0.8, while 36 stations have the RE_{max} over 0.7. The basic consistency between our risk assessment results and government decisions validates the effectiveness of the proposed approach, although the government would have more information and consider more factors than we do. Besides, we find that the Yishan Road Station is the highest-flood-risk station in the downtown district. According to Lyu [26], heavy rainfall with a return period of 70 years hit Shanghai in 2009 and made the Yishan Road Station suffer from a serious flood event. This can also validate that the proposed approach could identify those metro stations with high flood risks.
3. The RUD values of the metro stations in the peripheral area of Shanghai are basically larger than those in the central area. According to Fig. 5.7a, most metro stations in the downtown district have the RUD below 0.2, while those stations in the outer districts are generally larger than 0.3. The two metro stations with the largest RUD are the Disney Station (RUD = 0.835) and the Yishan Road Station (RUD = 0.790), both of which are marked out in Fig. 5.7a. Hence, the RE values at these stations tend to have substantial uncertainties when the subjective parameters vary in their ranges, which implies big differences in the experts' judgments on the flood risks at these metro stations. Furthermore, it can be seen

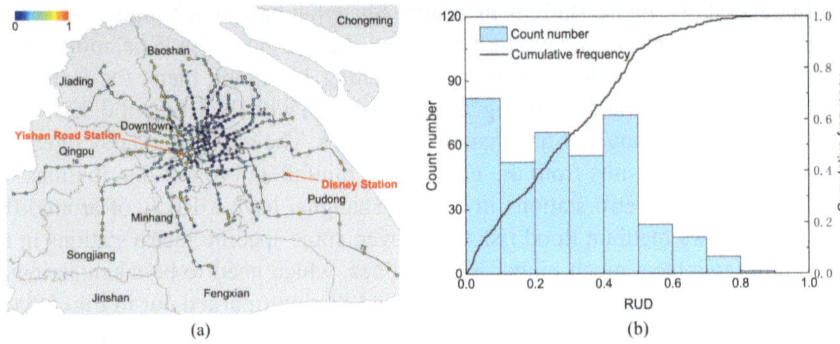

Fig. 5.7 RUD of 378 metro stations in Shanghai: (**a**) spatial distribution map; (**b**) histogram with cumulative frequency

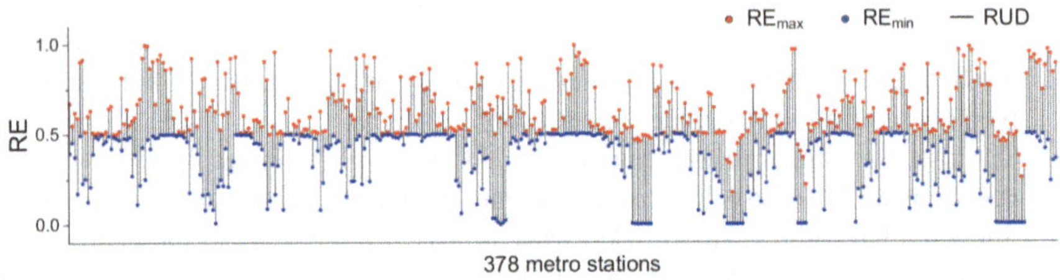

Fig. 5.8 Interval plot of RE_{max} and RE_{min} values with RUD of 378 metro stations in Shanghai

from Fig. 5.8 that there are no obvious correlations between RE_{max} and RE_{min} or RUD. Nevertheless, most RE_{max} are above 0.5 while all RE_{min} are below 0.5, which demonstrates that the flood risk of the whole metro system fluctuates around a medium level.

4. Using the combined gradient as the sensitivity indicator, we are capable of identifying key factors and proposing effective measures for flood risk mitigation of the metro stations. We herein take the Haitiansan Road Station and Yishan Road Station as examples. As shown in Fig. 5.9a, b, the drainage capacity (V_2) is the most important factor for risk mitigation, with the combined gradient of 0.419 at the Haitiansan Road Station and 0.972 at the Yishan Road Station. Hence, enhancing the drainage capacity is of the utmost priority for flood control of metro stations. As a validation, we keep other factor values unchanged, adjust the drainage capacity of the two metro stations from 2 to 3 (the highest level), and then recompute the RE_{max}. The results show that the RE_{max} will decrease to 0.786 and 0.472 at the Haitiansan Road Station and Yishan Road Station, respectively, which confirms the fine effects achieved by improving the drainage capacity. In terms of risk aggravation, the two factors that matter are the maximum daily rainfall (H_1) and maximum daily runoff (H_3). Therefore, we need to be watchful of the possible increase in rainfall and runoff in the future induced by the changing climate.

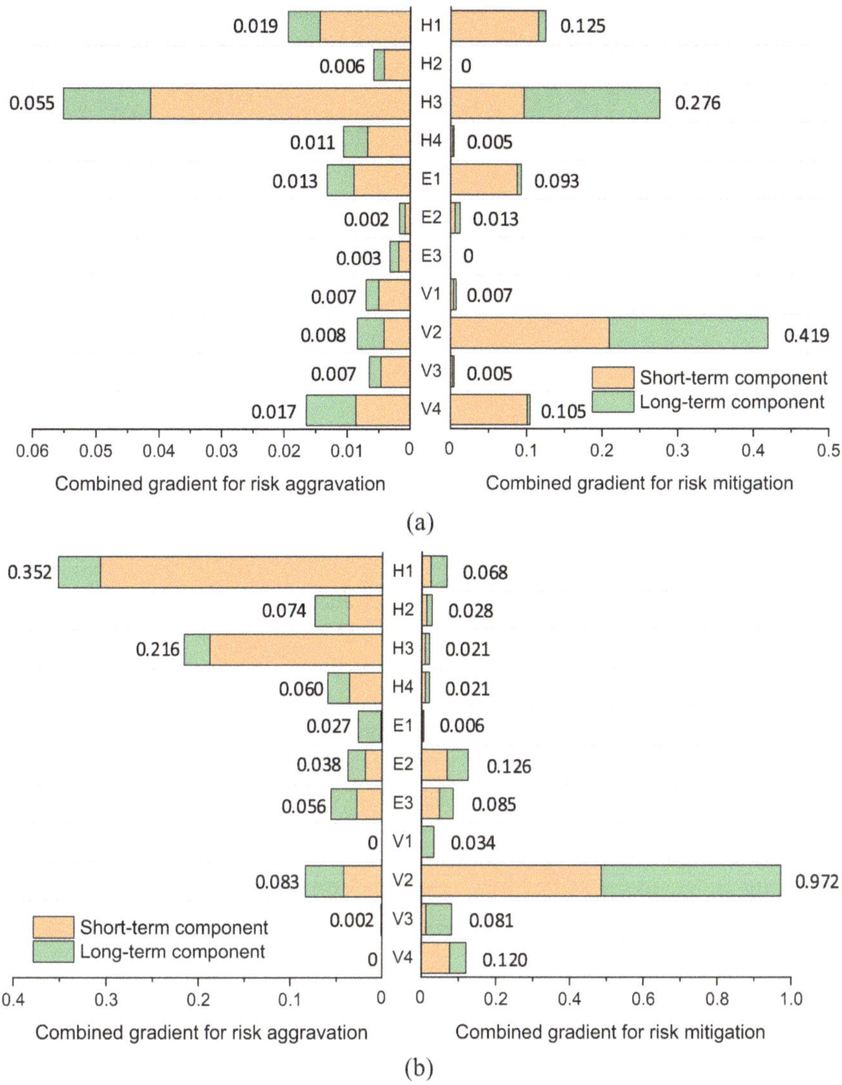

Fig. 5.9 Combined gradient for risk mitigation and aggravation: (**a**) the Haitiansan Road Station; (**b**) the Yishan Road Station

5.5 Merits Analysis

In this section, two baseline risk assessment methods that are slightly different from the proposed ER-based optimality approach are implemented to (1) demonstrate the necessity of considering the uncertainty of subjective parameters, and (2) verify the effectiveness of ER in evidence combination. The two baseline methods are: (1)

taking the interval mid-values as crisp values of the subjective parameters (i.e., without consideration of the uncertainty of subjective parameters); (2) using the weighted average method instead of ER to combine the evidence. As a result, Fig. 5.10 shows the RE of 378 metro stations in Shanghai without considering the uncertainty of subjective parameters; Figs. 5.11 and 5.12 give the RE_{max} and RUD of 378 metro stations in Shanghai obtained by the weighted average method. Table 5.2 summarizes the comparisons of risk calculation results between the proposed ER-based optimality approach and the two baseline methods at the Haitiansan Road Station and Yishan Road Station. The results are discussed as follows.

1. Without considering the uncertainty of subjective parameters, the risk assessment results tend to underestimate the flood risk, especially at those stations with large RUD values. As shown in Fig. 5.10, most metro stations in Shanghai have the RE values below 0.6. The highest RE is only 0.753, which occurs at the Heshahangcheng Station (marked out in Fig. 5.10a). At the Haitiansan Road Station and Yishan Road Station, the two typical metro stations mentioned in Sect. 5.4.3, the RE without considering the uncertainty of subjective parameters are 0.585 and 0.747 (see Table 5.2), respectively, far lower than the RE_{max} given by the proposed ER-based optimality approach. At the peripheral stations that have large RUD, such underestimations are much more substantial. Specifically, the RE values of the eastern metro stations suffering from high rainfall and runoff are lower than 0.6 in general, and nearly 30 western metro stations have RE values that even approach 0. Thus, it is likely to cause erroneous judgments on the flood risk and inadequate preparations for possible flood events at these metro stations if we do not consider the uncertainty of subjective parameters.
2. Compared with the weighted average method, ER could provide an evidence combination result with stronger discriminative power. As shown in Fig. 5.11, nearly all metro stations have the RE_{max} within the range of [0.4, 0.7]. That is to say, the flood risks at different metro stations do not show obvious distinctions. From Table 5.2, the RE_{max} obtained by the weighted average method at the

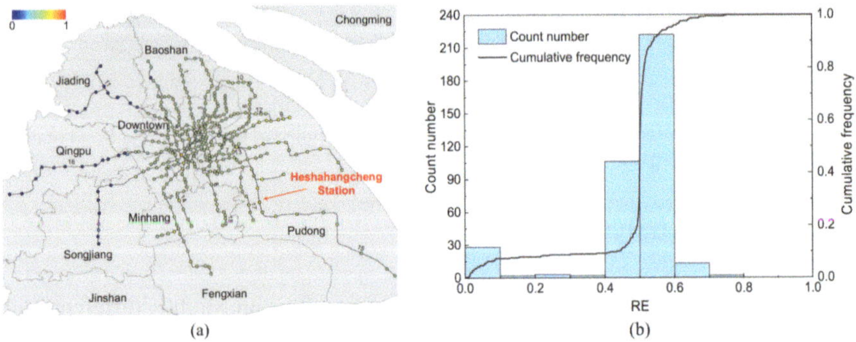

Fig. 5.10 RE of 378 metro stations in Shanghai without considering the uncertainty of subjective parameters: (**a**) spatial distribution map; (**b**) histogram with cumulative frequency

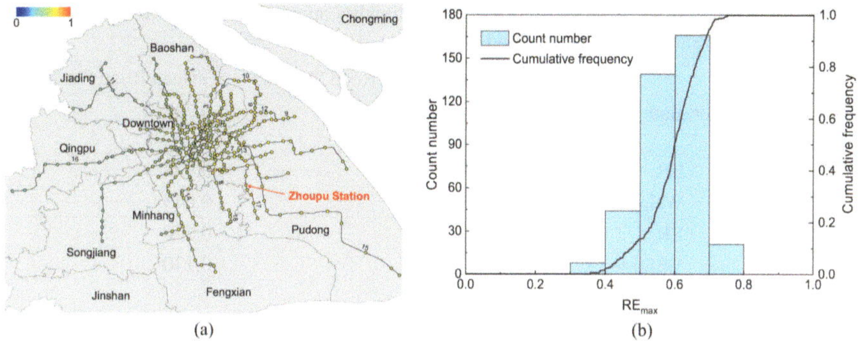

Fig. 5.11 RE_{max} of 378 metro stations in Shanghai obtained by the weighted average method: (**a**) spatial distribution map; (**b**) histogram with cumulative frequency

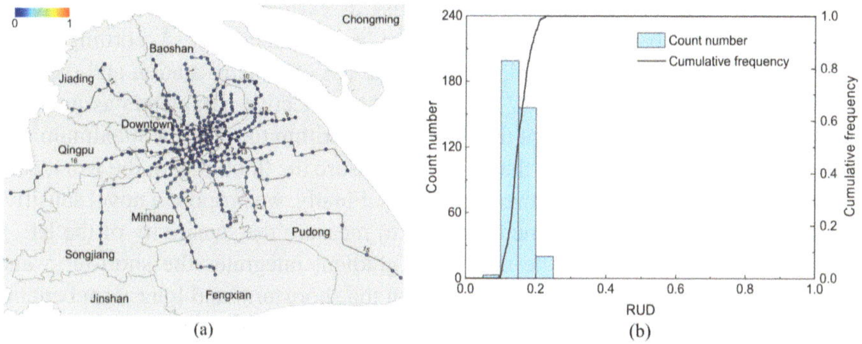

Fig. 5.12 RUD of 378 metro stations in Shanghai obtained by the weighted average method: (**a**) spatial distribution map; (**b**) histogram with cumulative frequency

Haitiansan Road Station and Yishan Road Station are 0.587 and 0.643, respectively, far below the results of ER. This demonstrates the advantage of ER in evidence combination for risk assessment, which can yield conservative results for high-risk stations and discriminate the risks of different stations effectively. In addition, it is seen from Fig. 5.12 that a large proportion of metro stations have the RUD values lower than 0.2. As listed in Table 5.2, the RUD values at the Haitiansan Road Station and Yishan Road Station are only 0.160 and 0.130, respectively. This indicates that the RE acquired through the weighted average method is not sensitive to the uncertainty of subjective parameters. Normally, we would prefer a method whose results show low RUD, which implies the robustness of the assessment model. However, given that the RE_{max} obtained by the weighted average method is not sufficiently informative, the lower RUD is correspondingly unable to provide valuable information.

Table 5.2 Comparisons between the ER-based optimality approach and the two baselines at the Haitiansan Road Station and Yishan Road Station

Item		Haitiansan Road Station	Yishan Road Station
RE without uncertainty of subjective parameters		0.585	0.747
RE_{max}	ER	0.996	0.958
	Weighted average	0.587	0.643
RUD	ER	0.746	0.790
	Weighted average	0.160	0.130

5.6 Chapter Summary

In this chapter, an optimality approach based on ER, which could consider the uncertainty of subjective parameters, is proposed to assess and mitigate the flood risk of metro stations. Firstly, we set up a risk assessment structure containing 11 influential factors based on hazard, exposure, and vulnerability. According to the ER rules, the function relationship between the RE of a metro station and the subjective parameters is established. Regarding the subjective parameters as intervals instead of crisp values, we adopt the SaNSDE algorithm to solve the maximum and minimum of RE, and then take the RE_{max} to measure the flood risk in the worst case under the uncertainty of subjective parameters. Finally, we propose a novel sensitivity indicator termed the combined gradient to measure the sensitivity of the RE_{max} value to the factor values. The combined gradient integrates the short-term and long-term gradients, which can consider both the short-term and long-term benefits of risk mitigation measures. The proposed ER-based optimality approach is implemented for a total of 378 metro stations in Shanghai to verify its feasibility and effectiveness.

In view of the case study, the conclusions can be summarized below: (1) The metro stations in eastern Shanghai have much higher flood risks than those in the central and western areas, which is due to the unbalanced rainfall and runoff distributions. Some specific metro stations in central Shanghai are also at high risk of being flooded. The risk assessment results are basically consistent with the government flood-prevention decisions and the past flood event, which verifies the validity of the proposed approach to some extent. (2) It is feasible and reasonable to use the combined gradient to identify key factors and give effective measures for flood risk mitigation at the metro stations correspondingly. At the Haitiansan Road Station and Yishan Road Station, improving the drainage capacity (V_2) is the most significant measure to reduce the flood risk, while the possible increase in the maximum daily rainfall (H_1) and maximum daily runoff (H_3) may be the two factors that cause further aggravation of flood risk. (3) The comparisons with the two baseline methods demonstrate the advantages of the proposed ER-based optimality approach. Compared with the baseline without the uncertainty of subjective parameters, the optimality approach can give reasonable and conservative assessment results for the flood risk. Compared with the weighted average method, ER is capable of combining evidence with stronger discriminative power.

The main novelty of this chapter can be summarized in the two points below. First, this chapter conducts a flood risk assessment that considers the uncertainty of all the involved subjective parameters. Nearly all MCDA-based risk assessments require the domain experts to give some subjective judgments beforehand, which are naturally uncertain due to the divergence among experts' opinions. The proposed ER-based optimality approach takes such uncertainty into account and produces an overall flood risk for a metro station in the worst case under the uncertainty of subjective parameters. Second, this chapter proposes a novel sensitivity indicator termed the combined gradient that considers both the short-term and long-term benefits of risk mitigation measures. Using the combined gradient, we can identify those factors important to flood risk mitigation so that the most effective flood-prevention strategies can be advised.

References

1. O'Donnell EC, Thorne CR (2020) Drivers of future urban flood risk. Philos Trans R Soc A Math Phys Eng Sci 378:20190216
2. Bucar RCB, Hayeri YM (2022) Quantitative flood risk evaluation to improve drivers' route choice decisions during disruptive precipitation. Reliab Eng Syst Saf 219:108202
3. Hou J, Mao H, Li J, Sun S (2019) Spatial simulation of the ecological processes of stormwater for sponge cities. J Environ Manag 232:574–583
4. Lyu H-M, Shen S-L, Zhou A-N, Zhou W-H (2019) Flood risk assessment of metro systems in a subsiding environment using the interval FAHP-FCA approach. Sustain Cities Soc 50:101682
5. Lyu H-M, Sun W-J, Shen S-L, Arulrajah A (2018) Flood risk assessment in metro systems of mega-cities using a GIS-based modeling approach. Sci Total Environ 626:1012–1025
6. Xu T, Xie Z, Zhao F, Li Y, Yang S, Zhang Y et al (2022) Permeability control and flood risk assessment of urban underlying surface: a case study of Runcheng south area, Kunming. Nat Hazards 111:661–686
7. Fahad MGR, Nazari R, Motamedi MH, Karimi M (2022) A decision-making framework integrating fluid and solid systems to assess resilience of coastal communities experiencing extreme storm events. Reliab Eng Syst Saf 221:108388
8. Liu J, Liu P, Wang X, Liu S, Wang G, Wu S et al (2021) Design and construction of sponge city facilities in Changzhen depot of Shenzhen metro. J Shenzhen Univ Sci Eng 38:1–110
9. He R, Zhang L, Tiong RLK (2023) Flood risk assessment and mitigation for metro stations: an evidential-reasoning-based optimality approach considering uncertainty of subjective parameters. Reliab Eng Syst Saf 238:109453
10. Lin Z, Hu S, Zhou T, Zhong Y, Zhu Y, Shi L et al (2022) Numerical simulation of flood intrusion process under malfunction of flood retaining facilities in complex Subway stations. Buildings 12:853
11. Kusano K, Ishigaki T, Ozaki T (2021) Characteristics of flooding damage through the subway network by pluvial flood. J Jpn Soc Civ Eng Ser B1 (Hydraul Eng) 77:I_511–I_516. (in Japanese)
12. Lyu H-M, Shen S-L, Yang J, Yin Z-Y (2019) Inundation analysis of metro systems with the storm water management model incorporated into a geographical information system: a case study in Shanghai. Hydrol Earth Syst Sci 23:4293–4307

13. Lyu H-M, Shen S-L, Zhou A, Yang J (2019) Perspectives for flood risk assessment and management for mega-city metro system. Tunn Undergr Space Technol 84:31–44
14. Lyu H-M, Zhou W-H, Shen S-L, Zhou A-N (2020) Inundation risk assessment of metro system using AHP and TFN-AHP in Shenzhen. Sustain Cities Soc 56:102103
15. Wang G, Liu Y, Hu Z, Zhang G, Liu J, Lyu Y et al (2021) Flood risk assessment of subway systems in metropolitan areas under land subsidence scenario: a case study of Beijing. Remote Sens 13:637
16. Wang G, Liu L, Shi P, Zhang G, Liu J (2021) Flood risk assessment of metro system using improved trapezoidal fuzzy AHP: a case study of Guangzhou. Remote Sens 13:5154
17. UR (2022) Understanding risk community: what is disaster risk? https://understandrisk.org/vizrisk/what-is-risk/. Accessed 25 Aug 2022
18. Yu C, Liu M, Xu X, Shi Y (2019) The urban rain-flood risk division based on the cloud model and the entropy evaluation method—taking Changzhou as an example. J Phys Conf Ser 1168:032087
19. Zhang W, Sun H, Fang W, Zhu C, Jia G (2023) Trust evaluation and decision based on DS evidence theory: early models and future perspectives. IEEE Access 11:16032–16041
20. Jiang W, Deng X. D-S evidence theory for information modelling and application. Beijing: Science Press; 2018. (in Chinese)
21. Du Y-W, Zhong J-J (2021) Generalized combination rule for evidential reasoning approach and Dempster–Shafer theory of evidence. Inf Sci 547:1201–1232
22. Xu S, Hou Y, Deng X, Chen P, Ouyang K, Zhang Y (2021) A novel divergence measure in Dempster–Shafer evidence theory based on pignistic probability transform and its application in multi-sensor data fusion. Int J Distrib Sens Netw 17:15501477211031473
23. Hsieh F-S (2022) Trust-based recommendation for shared mobility systems based on a discrete self-adaptive neighborhood search differential evolution algorithm. Electronics 11:776
24. Chang H (2021) Energy-saving design and implementation in metro weak current systems: a case study of the Kunming metro system. Urban Rail Transit 7:301–333
25. SM (2021) Shanghai Metro: important announcement of operation information: service suspension (in Chinese). https://m.weibo.cn/1742987497/4662834027562911. Accessed 4 Oct 2022
26. Lyu H-M (2019) Risk assessment methods and countermeasures for floods of metro system in subsiding environment. Shanghai

Chapter 6
Fire Hazard Analysis

6.1 Introduction

The metro system undertakes increasingly important tasks to transporting large passenger flows. The fire threat has become a major concern in metro stations with severe consequences because of the presence of intensive crowds. Typically, most of the metro stations are located underground with few exits and entrances. Due to the heavily crowded population and the situation of underground space, the magnitude of injuries and casualties can be overwhelmingly disastrous. Thus, a large risk comes into being where a number of evacuees may be stranded in fire accidents in metro stations. For instance, on January 12, 2015, a fire accident at the L'Enfant Plaza station platform in Washington, D.C. U.S., caused one death and 91 injuries [1]. The fire hazards and accidents have raised great public attention. The performance-based design has been commonly applied in prescriptive fire safety codes for safety consideration [2].

The basic idea in the life safety assessment under fire conditions is to determine when occupants can escape safely before a dangerous situation occurs. In performance-based fire safety engineering design, the concept of Available Safe Egress Time (ASET) for fire safety assessment has been widely used by fire safety engineering practitioners. ASET is defined as the time interval between the detection time and the occurrence of a hazard. Required Safe Egress Time (RSET) is the time needed to exit the fire when warned, which should be less than the determined ASET [3]. Thus, it proves that the safety exit becomes a task that satisfies the relationship of SI (Safety Index) = ASET/RSET > 1, where the space is threatened by the fire effect.

The possibility of a safe escape is the most crucial aspect of a building fire scenario, and SI can be used as a significant indicator to reflect and assess the risk magnitude associated with a fire scenario. To accurately assess the safety risk of a building, fire remains a challenging problem pending open studies, due to the fact:

L. Zhang et al., *Resilience in Urban Transit Systems*, Advances in 21st Century Human Settlements, https://doi.org/10.1007/978-981-95-4322-9_6

(1) A large number of factors are involved in calculating ASET and RSET; (2) The accuracy and availability required for fire data are lacking. To take into account critical factors representing fire, building, and human characteristics, a hybrid approach that incorporates the ASET//RSET concepts, numerical simulation, and multi-attribute decision analysis (MADA) is proposed to support the assessment of fire safety risk in complex buildings.

The disparity between RSET and ASET plays a decisive role in regard to safety risk assessment and prevention measures required to support the possibility of a safe escape in the case of fire. Many flaws have been raised with the original ASET/ RSET approach since the end of the last century. One significant criticism comes from the simulation of the more complex real world. The ASET / RSET concept was originally derived from a simple two-zone model of a single compartment, which is a one-dimensional approach because the conditions are assumed to be uniform throughout the room. As the advancement of complex three-dimensional (3D) simulation models yields very detailed results, the smoke layer may no longer be uniform and may only form a range of conditional areas that are difficult to maintain on a portion of the compartment area. Therefore, in these cases, continuing to apply the ASET/RSET method may no longer be appropriate. While these deficiencies still exist, the use of credible "worst case" in fire safety analysis can help disperse the need for probabilistic (stochastic) assessment of fire safety. In order to address the uncertain, stochastic, and dynamic nature of the fire behavioral performance, this chapter aims to propose an improved ASET/RSET approach that is capable of discovering worst fire scenarios in a 3D environment.

The main research questions are: (1) How to define a fire scenario in a 3D model setup? (2) How to calibrate the calculation of ASET and RSET in terms of different untenability criteria? and (3) How to assess the magnitude of the fire safety risk in different scenarios? In this chapter, an integrated framework that utilizes numerical simulation, ASET, RSET, and MADA is proposed to discover the worst scenario in case of a fire [4]. Specifically, numerical simulation models are used to simulate changes in fire propagation behavior over time in all locations in a building, while MADA is used to consider multiple untenability criteria for the ASET calibration and to fuse the evacuation performance at all the bottlenecks in the evacuation route, in order to make a more comprehensive evaluation of fire safety risk among different fire scenarios. A metro station in China is used as a case study to demonstrate the applicability and feasibility of the proposed approach.

6.2 ASET and RSET

In recent years, in order to ensure safe escapes from building fires, many fire codes have been released worldwide, including NFPA 101 [5] in the U.S., NCC (National Construction Code) [6] in Australia, and GCFPBC (General Code for Fire Protection of Buildings and Constructions) [7] in Mainland China. Although those codes are in various forms in different countries, the fundamental objective is to provide support for occupant safety. For instance, in the U.S.'s NFPA 101, the root objective of the performance requirement is expressed as "any occupant who is not intimate with ignition shall not be exposed to instantaneous or cumulative untenable conditions". As a consequence, there is a need to develop specific fire safety strategies based on the clearly defined safety goals in order to facilitate the design and construction of safe and cost-effective buildings. The means of achieving the targeted objective, however, are not clearly stipulated.

In the study of fire evacuation in a building, the ASET/RSET concepts provide a good estimate of the safety escape performance in case of a fire, and thus, have been widely applied amongst fire safety engineering practitioners. Indeed, the field test and experiment are regarded as one of the most reliable research methods, and the field measured values are the most reliable. However, considering the danger and the high loss, it is impossible to conduct real fire conduct tests, especially for a fire accident in crowded public areas, such as a metro station. Therefore, in recent years, full-scale experiments have been conducted to study smoke propagation and safe evacuation in the event of a fire in metro stations [8].

With the rapid development of computer modeling and simulation technologies, several software programs have emerged to simulate the time of ASET and RSET [9]. For instance, the computational fluid dynamics model, Fire Dynamics Simulator (FDS) [10], can be used to identify fire scenarios under different fire sizes to determine ASET. CFAST (Consolidated Model of Fire and Smoke Transport) [11] is a two-zone fire model used to calculate the evolving distribution of smoke, fire gases, and temperature in the case of a fire so as to calculate the time of ASET. Other tools for estimating ASET include ASCOS (Analysis of Smoke Control Systems), ASET-B (Available Safe Egress Time—BASIC), ASMET (Atria Smoke Management Engineering Tools), and others. Several evacuation tools, including SIMULEX [12], PathFinder [13], etc., are available to determine RSET under different passenger loadings.

Compared to regular buildings, most of the metro stations are located underground with few exits and entrances, and thus, the magnitude of injuries and casualties can be overwhelmingly disastrous in case of a fire in metro stations. In addition, the metro station is characterized by a long and narrow enclosure, where the total length of the occupied space may exceed 100 m. The unsustainable condition area caused by the metro station fire may be relatively large, but it does not necessarily reach the full range of the compartment. That is to say, using the original ASET model proposed by Cooper will probably result in a larger ASET. This leads to the problem that safe egress from the fire effects can still satisfactorily achieve occupant safety performance goals in the event of a real fire in which the fire effect does not fully extend to the entire area. In recent years, a number of studies have been conducted to investigate the fire modeling and analysis in metro stations. For instance, Long et al. [14] conducted a study on emergency ventilation strategies for different fire scenarios in a double-island metro station using FDS, and validated the results through model experiments. Their findings indicated that optimizing the ventilation system can improve fire conditions.

In general, the ASET and RSET concepts continue to be the commonly accepted criteria available to assess the safety risk of a fire scenario. Due to the importance and specialty of metro stations in public transit systems, especially in countries with large populations, like China, more and more efforts have been made to investigate the impacts of ventilation strategy [14], shaft height [15], etc., on the fire escape performance in metro stations in recent years. A big gap exists in how to model and analyze stochastic nature of fire spreading behaviors using the ASET/RSET approach in a 3D environment with more complex building layouts, as more kinds of influential factors (e.g., building, fire, and human features) are involved. The main objective is to propose a hybrid improved ASET/RSET approach that is capable of (1) modeling the stochastic nature of fire spreading behaviors in a 3D environment; (2) calibrating the calculation of ASET and RSET in terms of different untenability criteria; and (3) assessing the magnitude of the fire safety risk among different scenarios.

6.3 Fire Hazard Analysis Framework in Metro Stations

The numerical simulation has become the predominant approach that is capable of performing fire safety modeling and what-if scenario analysis, as it can analyze the effects of fire and extreme temperatures on materials and offer detailed and reliable representations of real-life scenarios. Aiming to identify the worst fire scenario to diffuse the need to undertake a probabilistic risk assessment of fire safety, a systematic approach is proposed to quantify the magnitude of the fire safety risk in metro stations. The overall workflow is shown in Fig. 6.1. In the proposed approach, the following three main steps are adopted, including (1) fire hazard identification; (2) fire hazard simulation; and (3) fire hazard assessment.

6.3.1 Fire Hazard Identification

A fire risk has two components balanced against each other, where one is the possibility of a fire occurring and the other is the magnitude of consequences of that fire. Identification of the fire risk in metro stations is a process of recognizing where fire hazards exist and defining their characteristics, aiming to determine fire scenarios with a high probability and serious consequences. From a controllable and practical perspective for fire risk mitigation in underground metro stations, the following three factors, refined from the fire, building, and human features, respectively, are mainly considered in identifying possible fire scenarios with a high-potential risk. In general, among the three factors, the fire location is given super consideration when identifying high-risk fire scenarios, while the heat release rate and the number of occupants are assigned their maximum values in the identified fire scenarios.

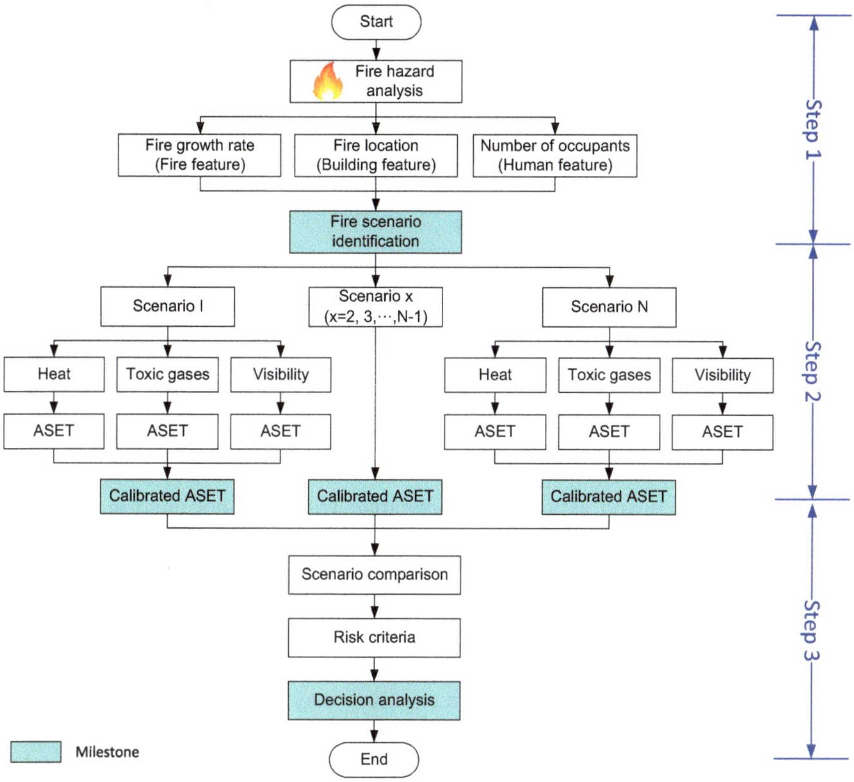

Fig. 6.1 Workflow of the proposed approach to fire safety assessment

1. Heat release rate (fire feature). This is an important factor in determining fire mortality because many fatal events are characterized by rapid development after initial discovery. It can be determined by a formula based on exponential growth that varies according to the flame growth coefficient of the burning material [16].
2. Fire location (building feature). The fire location indicates the place where the fire ignites in the building, which is closely related to the building layout. Different locations of the fire source are expected to exert different effects on the estimation of ASET. To some extent, building layout can illustrate the degree of complexity in building design. To save much time in a fire escape, more attention should be paid to the area (or location) where the fire occurs frequently, particularly for those complex buildings, such as a metro station.
3. Number of occupants (human feature). Human behaviors can be uncertain, dynamic, and even involve interactions among different humans in case of a fire escape. If the density of people is high, passengers cannot move forward and waste more time. In this chapter, we assume that there are N occupants who are to be evacuated during the fire scenarios, where N is equal to the maximum num-

ber of occupants specified in the codes or the number of occupants at rush hours. This is because the objective is to compare the magnitude of the fire safety risk across different scenarios under extreme conditions, in order to discover the worst fire scenario.

6.3.2 Fire Hazard Simulation

At this step, the simulation technique is used to model fire scenarios identified at the last step and monitor the diffusion of hazards yielded by the fire through the simulation process. In general, the main classes of hazards yielded by fire in underground metro stations include heat (temperature), toxic gases (carbon monoxide), and smoke obscuration (visual impairment), which are selected to represent the smoke diffusion process, as described below.

1. Heat. The prolonged exposure of more than 15 min to hot environments may cause heat stroke. Even for short durations, exposure to the hot smoke at high temperatures may cause skin pain or skin and respiratory tract burns. In addition, the increased humidity in the air may limit evaporation and thus lower the tolerance time.
2. Toxic gases. In building fires, exposure to toxic gases will cause incapacitation (loss of consciousness), and severe exposure may even cause death. The most common asphyxiant gas is carbon monoxide (CO), and the CO concentration is a major concern in fire safety risk assessment. The problem is that very low levels of CO can cause incapacitation or even death.
3. Smoke obscuration. The soot contained in the smoke can obscure light, and thus, reduce the visibility. The reduction in visibility is not directly life-threatening; however, it may reduce the walking speed of the occupants. In the smoke, combustion gases include acid gases (e.g., HF, HCl, HBr, SO_2, and NO_x) and organic irritant gases (e.g., acrolein, formaldehyde, crotonaldehyde) [17]. If the occupants are located at a long distance away from an exit and the visibility drops significantly, they may be unable to find their way out of the building. In both cases, it may lead to an increased exposure time to heat and toxic gases, which should be taken into account.

In performance-based design, ASET is commonly estimated by accounting for the time to reach the tenability limit in fire models. Actually, various tenability criteria have been proposed around the world; however, there is no common set that has been universally accepted. Consequently, building designers in different countries and districts should justify the criteria they use, where some follow high values and some follow low values. For instance, in Hong Kong, China, the common tenability limits on life safety for occupants and firemen follow lower values [5]. In order to consider all the tenability criteria, in terms of heat, toxic gases, and smoke obscuration, for the estimation of ASET, the earliest time to reach the tenability limit at a specific location in the building is considered as the calibrated ASET there. In other words, the minimum ASET regarding different tenability criteria should be selected as the calibrated ASET, which can be termed the "minimum principle". In this chapter, the "minimum principle" is used as one critical evaluation criterion for performing MADA, aiming to calibrate the calculation of ASET across different untenability criteria.

6.3.3 Fire Hazard Evaluation

In order to address the uncertainty underlying the data and assumptions in the analytical process, the allowance of design uncertainties can be expressed as a safety index. The concept of the safety index has been used in fire safety engineering in order to account for uncertainties and risk in performance-based design. It permits the designer to make a conservative assessment while allowing for a smaller margin of error. Safety Index, denoted by SI, is given as follows.

$$\mathrm{SI} = \frac{\mathrm{ASET}}{\mathrm{RSET}} \geq \xi \tag{6.1}$$

where ASET refers to the time between the ignition and the moment when the environment becomes untenable to occupants, while RSET refers to the time from ignition to the moment the affected occupants are able to reach a place of safety. The indicator ξ refers to the threshold of the value of SI. In current codes on performance-based fire engineering design, there is no clear recommendation on how to choose an appropriate safety index threshold [18]. Nonetheless, the value of the threshold should be more than one but not more than three, and typically between 1.5 and 2.

Generally, SI is used as a risk criterion to evaluate the safety of occupants in a building [19], which can provide a good measure for analyzing the magnitude of safety risk in different fire scenarios. A higher SI value indicates an increased level of safety in a specific fire scenario, while an increased level of risk is indicated by a lower SI value. At the last step, the calibrated ASET at one specific location in the building is obtained through the "minimum principle", with all the tenability criteria, including heat, toxic gases, and smoke obscuration, taken into account. Indeed, the allowable time at the key evacuation route should be much more important in the fire evacuation performance. Given that several evacuation routes are available, how to fully consider the fire escape performance in different evacuation routes is a challenging problem. To address this issue, this chapter proposes a more comprehensive indicator by using MADA to assess the safety risk across different fire scenarios.

In case of a fire, the occupants will rush for exits of the underground metro stations. In the evacuation process, the occupants may need to pass through doors, stairs, or corridors to reach the exit. When many occupants choose the same evacuation route, congestions may occur due to the high occupant density in certain areas.

Those areas, such as doors, stairs, and exits, can be formed as the "bottlenecks" in the fire escape process, which play a significantly important role in the efficiency of the fire evacuation performance. The bottlenecks consist of a parallel-connection system. Thus, the magnitude of the resilience (or the safety risk) of the overall system in case of fire can then be calculated based on the safety performance at individual bottlenecks. In order to fuse the evacuation performance at all the bottlenecks in the evacuation route for achieving comprehensive evaluation and decision analysis, one indicator, SI_{Total} representing the magnitude of the safety risk of the overall parallel-system, is proposed for MADA purposes in this chapter. Given the life safety escape performance at the ith bottleneck is represented by SI_i, ($i = 1, 2, \ldots, M$), the magnitude of the safety risk for the overall parallel-system consisting of M bottlenecks can be given as follows.

$$SI_{Total} = \left(\prod_{i=1}^{i=M} SI_i\right)^{1/M} \tag{6.2}$$

where, SI_i is the calculated safety index at the ith ($i = 1, 2, \ldots, M$) bottleneck that represents the escape performance in case of the fire, M is the total number of the identified bottlenecks in the building, and SI_{Total} is the magnitude of the safety risk of the overall parallel-system. The lower the value of SI_{Total}, the worse (i.e., the more dangerous) the fire scenario. Given that several fire scenarios are identified at the first step, SI_{Total} can be used as an indicator to make a comparison among different scenarios and detect the worst fire scenario, in order to help diffuse the need to undertake a probabilistic assessment of the fire safety. Accordingly, corresponding fire control measures and strategies can be developed for risk mitigation in fire escape.

6.4 Experimental Studies

6.4.1 Data and Resources

In recent years, the construction of metro tunnels has developed rapidly around the world, especially in developing countries such as China. In order to demonstrate the applicability and feasibility of the proposed approach in this chapter, one realistic metro station in the Wuhan metro system in China, denoted as the YJW station, is used as a case to model and analyze different fire scenarios.

The YJW station is one station on the Metro Line 2 in the Wuhan Metro System in the City of Wuhan, China. This metro station went into operation on December 28, 2012. Many schools, universities, residential communities, and business centers are located near the YJW station. As a result, the volume of the passenger flow at this station is very large in rush hours. This is an underground two-story island station, with a train platform on the second basement and a lobby on the ground floor. Figure 6.2 shows the lobby layout of the YJW station. The train station is 241.30 m

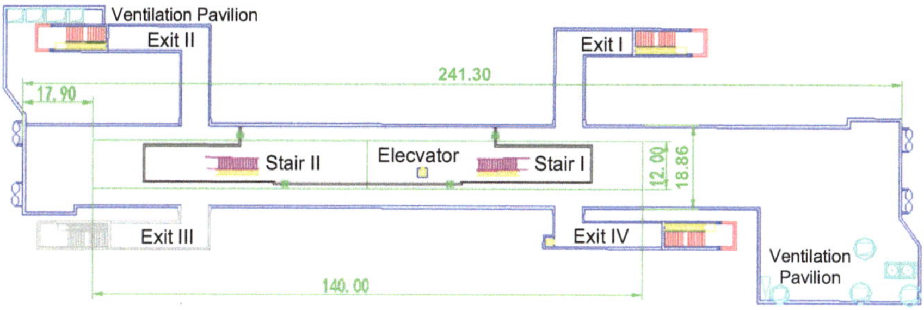

Fig. 6.2 Layout of the hall floor at the YJW station (unit: m)

long and 18.86 m wide. The trains are running on the platform floor. The passengers first get out of the trains on the platform floor, and then go up to the hall floor by using Stair I (in the east) or Stir II (in the west). A vertical elevator is installed to connect the platform floor to the hall floor. Considering the elevator shall be out of service in case of a fire, the elevator is treated as an obstacle in the fire simulation model in this chapter. There are four exits, namely Exits I, II, III, and IV (see Fig. 6.2), which connect the hall floor to the outside.

Fire hazard analysis aims to propose possible fire scenarios with a high probability and serious consequences by considering the factors of the heat release rate, fire location, and occupants. In case of a fire at the YJW station, the occupants at the platform floor will first rush to the stairs (i.e., Stair I or II) and then find the exits (i.e., Exit I, II, III, or IV) to reach a safe place, while the occupants at the hall floor will rush to the exits directly. That is to say, the stairs and exits should be considered as "bottlenecks" in the fire escape process. The ignition of fires at those locations will exert a direct impact on the efficiency of the occupant evacuation performance. Actually, in public areas at the metro station, most of the combustibles come from carry-on luggage of passengers and their clothes. Distributed combustibles will be gathered when passengers rush to certain areas. Thus, the probability of the fire ignition at congestion areas, such as stairs and exits, should be higher than that in loose areas. By referring to the fire design documents of the YJW station, the information regarding the heat release rate and the number of occupants during rush hours can be obtained. Considering the symmetry of the layout at the YJW station, as shown in Fig. 6.2, a total of four scenarios are identified in this chapter. Table 6.1 illustrates the description of those four fire scenarios.

Fire Dynamics Simulator (FDS) [10], as a Computational Fluid Dynamics (CFD) model of the fire-driven fluid flow, solves well all important fire mechanisms through a form of the Navier-Stokes equations appropriate for low-speed and thermally-driven flow with an emphasis on smoke and heat transfer from fires. Thus, this chapter employs FDS in investigating the smoke propagation in a metro station. The core algorithm is an explicit predictor-corrector scheme that is second order

Table 6.1 Description of the identified four fire scenarios at the YJW station

Scenario	Description		
	Fire location	Heat release rate[a]	Number of occupants[b]
Scenario I	Entrance of Stair I at the platform floor	2 MW/m^2	1565
Scenario II	Exit of Stair I at the hall floor	2 MW/m^2	1565
Scenario III	Exit I at the hall floor	2 MW/m^2	1565
Scenario IV	Exit of Stair I at the hall floor	3 MW/m^2	1565

[a]The heat release rate is defined to be 2 MW/m^2 in normal cases (i.e., in Scenarios I, II, and III) according to the fire design documents of the YJW station. Scenario IV is an extreme scenario, where the heat release rate is defined to be 50% higher than that in Scenario II, in order to investigate the impact of the heat release rate on the fire safety risk
[b]The number of occupants at rush hours is specified to be 1565 by referring to the fire design documents

accurate in space and time. Based on the case background information and the description of the identified four fire scenarios in Table 6.1, the setup of those four fire scenarios at the YJW station is complete in FDS.

6.4.2 Scenario Simulation

In fire simulation studies, the definition of untenable criteria plays a very important role in the estimation of ASET. Although various tenability criteria have been proposed worldwide, there is no common agreement. As a result, the tenability criteria are justified and specified in the different building fire codes in different countries. In China, several building codes, such as GCFPBC (GB 55037–2022) [7] and Code for Design of Metro (GB 50157–2013) [20], can be referred to for the consideration of the untenable criteria in the fire design of metro station. Accordingly, the main untenable criteria in the fire design of metro stations in China can be summarized as (1) the temperature should be no more than 140 °C; (2) the CO concentration in air should be no more than 1500 ppm; and (3) the visibility should be no >10 m.

In previous simulation studies [21], the values of temperature, CO concentration, and visibility are basically measured at a height of 2.0 m. It is often argued that the values of fire hazards can be more appropriate if measured within the breathing zone, which can be taken at any height between 1.5 m and 2.0 m [21]. Considering the fact that the average height of Chinese male and female are 1.72 m and 1.60 m, respectively, the height for the measurement of fire hazard condition is set up to be 1.6 m during the simulation of the identified four fire scenarios in this chapter. In addition, the period of the occupant evacuation under a fire should be no more than 6 min according to the standards GB 50016-2006 and GB 50157-2013, and thus, the length of the numerical simulation in FDS is set up to 6 min. If none of the fire hazards reaches the tenability limit within 6 min, the value of ASET is confirmed to be 6 min in that scenario.

6.4.2.1 Scenario I

In Scenario I, the fire is ignited at the entrance of Stair I at the platform floor with a heat release rate of 2 MW/m^2. The smoke and heat are generated at the platform floor and then spread throughout the whole underground station. During the numerical simulation of the smoke diffusion process within 6 min, the content of the fire hazards, including temperature, CO concentration, and visibility, can be monitored anytime and anywhere at the station. Figure 6.3 illustrates cloud pictures of the fire smoke simulation over time under Scenario I at a height of 1.6 m. As aforementioned, the two stairs (Stairs I and II) and four exits (Exits I, II, III, and IV) are considered as "bottlenecks" in the fire escape process, and thus, the fire hazard conditions at those bottlenecks should be given super consideration and are recorded throughout the whole simulation process. Figure 6.4 shows the hazard evacuation process at Stair II, Exit I, and Exit II in Scenario I at a height of 1.6 m above the floor. Considering the symmetry of the layout at the YJW station, the hazard evacuation process at Exits III and IV should be the same as that in Exits I and II, respectively. In regard to Stair I, taking in account the psychology of the occupants and the actual situation of fire spreading behaviors, the occupants nearby Stair I will choose Stair I to go to the hall floor for escapes during the first 30 s because of the shortest evacuation route. After 30 s, the occupants will not venture their lives to go across Stair I, but go to Stair II for escape, as the fire at Stair I is growing very big then.

In accordance with the tenability limit specified in the fire design codes in China (see the horizontal dash line in Fig. 6.4), the values of ASET can be estimated. For

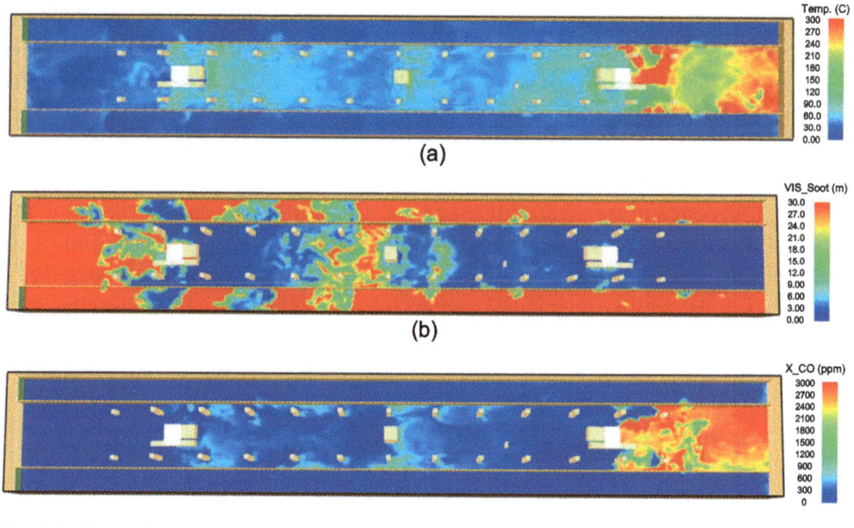

Fig. 6.3 Cloud pictures of fire smoke simulation in Scenario I at a height of 1.6 m: (**a**) temperature distribution at $T = 155$ s; (**b**) visibility soot distribution at $T = 167$ s; and (**c**) CO concentration at $T = 202$ s

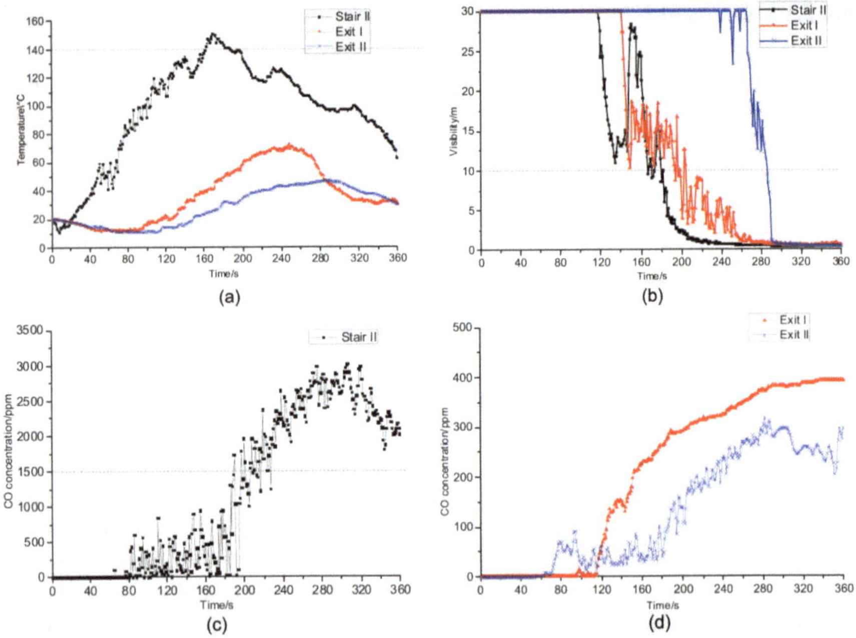

Fig. 6.4 Hazards evolution process during the smoke diffusion at the station within 6 min in Scenario I: (**a**) Temperature variation at Stair II, Exits I and II; (**b**) Visibility variation at Stair II, Exits I and II; (**c**) CO concentration at Stair II; and (**d**) CO concentration at Exits I and II

instance, in Stair II, the period when the temperature, visibility, and the CO concentration reach up to their tenability limits is 163 s, 167 s, and 202 s, respectively. In Exit I, both the temperature and the CO concentration do not reach their tenability limits within 6 min of simulation, and the values of ASET in terms of these fire hazards are both estimated to be 360 s. In regard to Stair I, as analyzed above, all occupants may not use Stair I for escape after 30 s of the ignition, and thus, the values of ASET in terms of those three fire hazards (including temperature, visibility, and CO concentration) are all estimated to be 30 s. Based on the minimum principle as proposed in the above methodology section, for a specific location, the earliest time to reach the tenability limit is considered as the final calibrated ASET there. The results of the final calibrated ASET at critical evacuation bottlenecks in Scenario I are presented in Table 6.2.

6.4.2.2 Scenario II

In Scenario II, the fire is ignited at the exit of Stair I at the hall floor with a heat release rate of 2 MW/m^2. In a similar numerical simulation process as mentioned in Scenario I, the distribution of the fire hazards at the whole station is monitored within 6 min. Figure 6.5 illustrates cloud pictures of the distribution of the CO

Table 6.2 Estimated ASET at critical evacuation bottlenecks in Scenario I

Hazards	Estimated ASET at critical evacuation bottlenecks (unit: s)					
	Stair I	Stair II	Exit I	Exit II	Exit III	Exit IV
Temperature	30	163	360	360	360	360
Visibility	30	167	194	286	286	194
CO concentration	30	202	360	360	360	360
Calibrated ASET	30	163	194	286	286	194

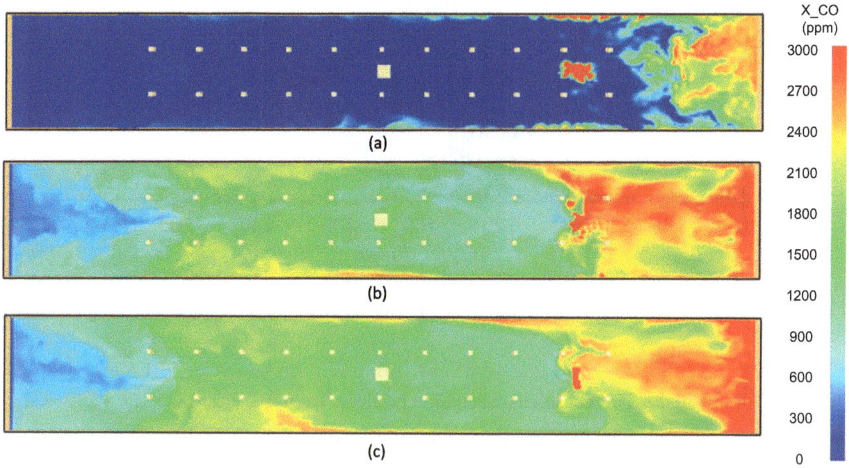

Fig. 6.5 Cloud pictures of the distribution of CO concentration at a height of 1.6 m above the ground in Scenario II at different: (**a**) $T = 165$ s; (**b**) $T = 258$ s; and (**c**) $T = 264$ s

concentration over time at a height of 1.6 m above the floor in Scenario II. The variation of the CO concentration in air at the evacuation bottlenecks, including Stair II, Exit I, and Exit II, is presented in Fig. 6.6. Considering the tenability limit of the CO concentration in air, the period between the ignition and the untenable moment, that is ASET, at Stair II, Exit I, and Exit II turns out to be 258 s, 165 s, and 264 s, respectively.

In the same way, through the fire simulation in Scenario II, taking the temperature as the tenability criteria, the values of ASET at Stair II, Exit I, and Exit II turn out to be 172 s, 115 s, and 360 s, respectively. Taking the visibility as the tenability criteria, the values of ASET at those three evacuation bottlenecks are 184 s, 84 s, and 247 s, respectively. Due to the symmetry of the layout at the YJW station, the values of ASET at Exits III and IV can be estimated accordingly. It should be noted in Stair I that, as similar to that in Scenario I, the values of ASET through those three tenability criteria should all be 30 s, as all occupants may not use Stair I for escape after 30 s from the fire ignition. Based on the minimum principle, the results of the final calibrated ASET at critical evacuation bottlenecks in Scenario II are presented

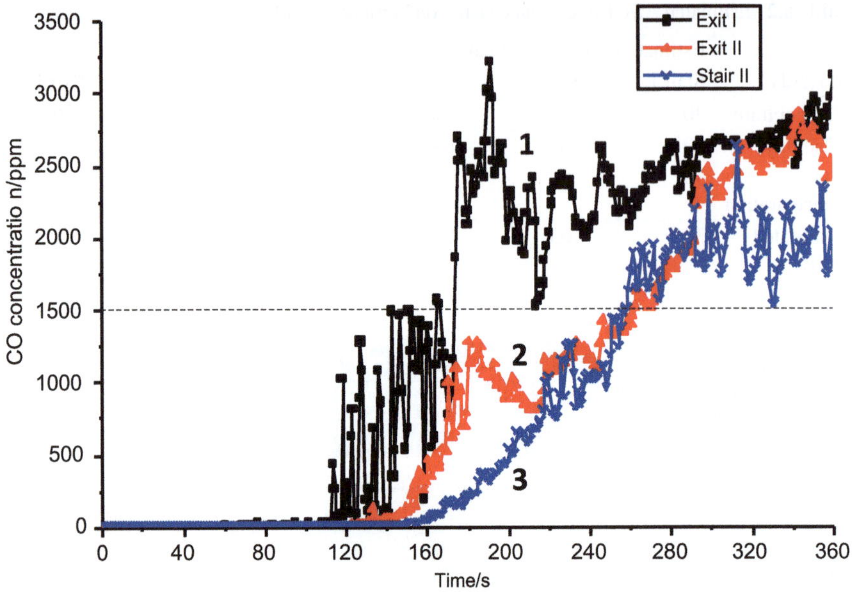

Fig. 6.6 Variation of CO concentration at critical evacuation bottlenecks within 6 min in Scenario II

Table 6.3 Estimated ASET at critical evacuation bottlenecks in Scenario II

Hazards	Estimated ASET at critical evacuation bottlenecks (unit: s)					
	Stair I	Stair II	Exit I	Exit II	Exit III	Exit IV
Temperature	30	172	115	360	360	115
Visibility	30	184	84	247	247	84
CO concentration	30	258	165	264	264	165
Calibrated ASET	30	172	84	247	247	84

in Table 6.3. It is clear that the periods to reach the tenability limit at Stairs I and II, Exits I, II, III, and IV are 30 s, 172 s, 84 s, 247 s, 247 s, and 84 s, respectively.

6.4.2.3 Scenario III

In Scenario III, the fire is ignited at Exit I at the hall floor with a heat release rate of 2 MW/m^2. Similar to that in Scenarios I and II, the Exit I is assumed to be open for evacuation only the first 30 s from the ignition of the fire, as the fire will grow to be too large to be safe for passing through after the first 30 s. During the fire smoke simulation within 6 min, the distribution of the fire hazards at the whole station can be monitored everywhere. Figure 6.7 shows the variation of the temperature at the evacuation bottlenecks, including Stair I, Stair II, Exit II, and Exit IV. Considering the tenability limit of the temperature (see the horizontal dash line in Fig. 6.7), the

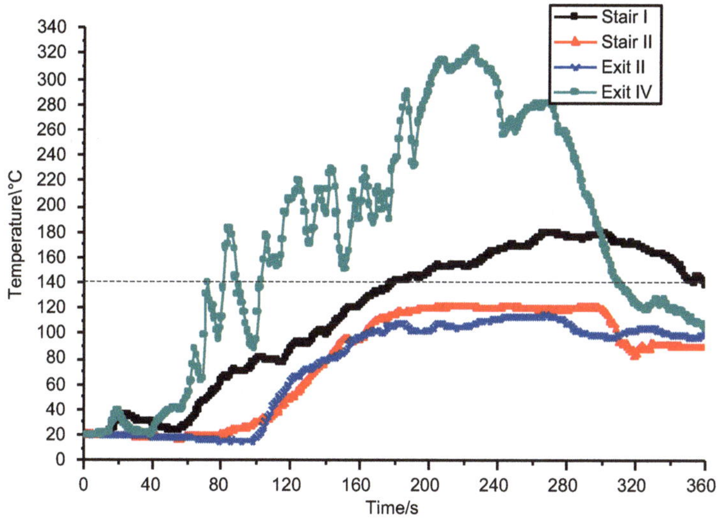

Fig. 6.7 Variation of the temperature at critical evacuation bottlenecks within 6 min in Scenario III

Table 6.4 Estimated ASET at critical evacuation bottlenecks in Scenario III

Hazards	Estimated ASET at critical evacuation bottlenecks (unit: s)					
	Stair I	Stair II	Exit I	Exit II	Exit III	Exit IV
Temperature	181	360	30	360	360	81
Visibility	96	182	30	223	223	113
CO concentration	185	282	30	243	243	137
Calibrated ASET	96	182	30	223	223	81

value of ASET at Stair I, Stair II, Exit II, and Exit IV turns out to be 181 s, 360 s, 360 s, and 81 s, respectively.

In the same way, through the fire simulation in Scenario III, taking visibility as the tenability criteria, the value of ASET at Stair I, Stair II, Exit II, and Exit IV is 96 s, 182 s, 223 s, and 113 s, respectively. Taking the CO concentration as the tenability criteria, the value of ASET at those four evacuation bottlenecks is 185 s, 282 s, 243 s, and 137 s, respectively. Considering the symmetry in the layout at the YJW station, the value of ASET at Exit III should be the same as that in Exit II. In general, the values of ASET at those six evacuation bottlenecks, including Stairs I and II, Exits I, II, III, and IV, can be estimated. In accordance with the minimum principle, the results of the final calibrated ASET at those six evacuation bottlenecks in Scenario III are presented in Table 6.4. It is clear that the periods to reach the tenability limit at Stairs I, II, Exits I, II, III, and IV are 96 s, 182 s, 30 s, 223 s, 223 s, and 81 s, respectively.

6.4.2.4 Scenario IV

In Scenario IV, the fire is ignited at the exit of Stair I at the hall floor with a heat release rate of 3 MW/m^2. The difference between Scenarios II and IV lies in the value of the heat release rate, which is increased by 50%. During the fire smoke simulation within 6 min in Scenario IV, the distribution of the fire hazards at the whole station can be monitored. Figure 6.8 illustrates the variation of the temperature at Stair II, Exit I, and Exit II in Scenario IV over time. Considering the tenability limit in terms of the temperature, the value of ASET at Stair II, Exit I, and Exit II is 111 s, 62 s, 117 s, respectively. In the same way as that in Scenario II, the values of ASET through different tenability criteria at the evacuation bottlenecks, including Stairs I and II, Exits I, II, III, and IV, can be estimated. According to the minimum principle, the results of the final calibrated ASET at those six evacuation bottlenecks in Scenario IV are presented in Table 6.5. It is obvious that the periods to reach the tenability limit at Stairs I, and II, Exits I, II, III, and IV in Scenario IV are 30 s, 111 s, 58 s, 117 s, 117 s, and 58 s, respectively.

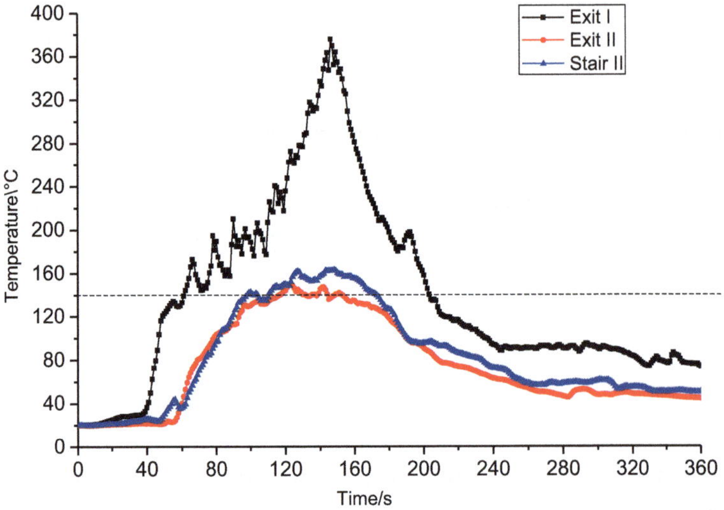

Fig. 6.8 Variation of the temperature at critical evacuation bottlenecks within 6 min in Scenario IV

Table 6.5 Estimated ASET at critical evacuation bottlenecks in Scenario IV

	Estimated ASET at critical evacuation bottlenecks (unit: s)					
Hazards	Stair I	Stair II	Exit I	Exit II	Exit III	Exit IV
Temperature	30	111	62	117	117	62
Visibility	30	116	58	134	134	58
CO concentration	30	147	92	156	156	92
Calibrated ASET	30	111	58	117	117	58

6.4.3 Analysis of Results

Comparison between ASET and RSET provides a basis for a life safety assessment. In order to analyze the life escape performance in case of the fire, the proposed indicator, SI_{Total}, provides a measure to evaluate the safety risk of different fire conditions. As shown in Tables 6.2, 6.3, 6.4, and 6.5, the final calibrated ASET at six evacuation bottlenecks is obtained in different fire scenarios at the YJW station, with all the tenability criteria taken into account. Equation (6.2) is then used to aggregate the magnitude of the safety risk in different scenarios. Taking Scenario I as an example, the calculation of SI_{Total} is given below.

$$\begin{aligned} SI_{\text{Total}} &= \left(\prod_{i=1}^{i=M} SI_i\right)^{1/M} = \begin{pmatrix} \dfrac{30}{RSET_1} \times \dfrac{163}{RSET_2} \times \dfrac{194}{RSET_3} \\ \times \dfrac{286}{RSET_4} \times \dfrac{286}{RSET_5} \times \dfrac{194}{RSET_6} \end{pmatrix}^{1/6} \\ &\approx 157.14 / \begin{pmatrix} RSET_1 \times RSET_2 \times RSET_3 \\ \times RSET_4 \times RSET_5 \times RSET_6 \end{pmatrix}^{1/6} \\ &= 157.14 / R \end{aligned} \tag{6.3}$$

where R refers to the geometrical mean of multiple RSETs, and $R = (RSET_1 \times RSET_2 \times RSET_3 \times RSET_4 \times RSET_5 \times RSET_6)^{1/6}$. The estimation of RSET is extracted from a number of sub-components relating to the procedural and evacuee performance. Based on the research work in [22], RSET can be formulated as follows.

$$RSET = t_d + t_n + t_p \tag{6.4}$$

where, t_d is the time from the fire ignition to detection, t_n is the time from the detection to notification, and t_p is the evacuation time that consists of pre-movement time (the time between the point when occupants become aware of the emergency and the point when they begin to move towards the exits) and travel time (the time required for occupants to travel to a place of safety).

Basically, in Eq. (6.4), t_d and t_n are determined according to the technological resources and do not include any reference to occupant activities [23]. It is apparent that the first two components of the RSET calculation ($t_d + t_n$) are procedural. The calculation of t_p, however, is very complex, as numerous factors are involved in its determination, such as building characteristics (particularly the building type, fire safety management systems, and the building layout) and occupant characteristics (particularly occupant numbers, alertness (waking or sleeping), and familiarity with the building) [24]. In order to simplify the calculation of the evacuation time t_p, Paul's formula [25] in Canada is proposed, which has been widely applied in emergency evacuation in public areas, as it considers both the time from the evacuation spot to the exit and the time that occupants go across the exit, and it is, thus, more consistent with the actual situation. Paul's formula is given below.

$$t_p = \frac{1}{fB}\left[N_a - \sum_{i=1}^{n}\int_0^{T_0} f(t)B(t)dt\right] + T_0 \tag{6.5}$$

where, f is the number of the out-flow people from the second door-ways (number/m/sec), B is the breadth of the second door-way (m), N_a is the total number of escaping occupants, n is the total number of the first door-ways, T_0 is the time when the crowd formed the normal current at the second door-ways or the staircases from the panic place, $f(t)$ is the number of out-flowing people from the first door-ways at time t (number/m/s), and $B(t)$ is the breadth of the first door-way at time t (m).

It is apparent that the estimation of RSET is mainly related to technological resources, building characteristics, and occupant characteristics. Given the same total number of escaping occupants in the same building, the value of RSET will be generally similar in different fire scenarios, as the technological resources and the setup of the building and occupant characteristics do not change significantly. As a result, in this case study, the values of $RSET_1$, $RSET_2$, $RSET_3$, $RSET_4$, $RSET_5$, $RSET_6$ will not change significantly in different fire scenarios. That is to say, the value of $R = (RSET_1 \times RSET_2 \times RSET_3 \times RSET_4 \times RSET_5 \times RSET_6)^{1/6}$ in Scenario I should be very similar to that in Scenarios II, III, and IV. To simplify the calculation process, in this case study, the values of SI_{Total} in Scenarios II, III, and IV can be obtained to be 114.23/R, 113.26/R, and 73.16/R, respectively. Table 6.6 illustrates the results of the evaluation of the fire safety risk in different fire scenarios.

It is clear that Scenario IV is the worst fire scenario among those four scenarios, indicating that the fire release rate plays a very significant role in the magnitude of the fire safety risk. Particularly, when comparing Scenario IV with Scenario II, a 50% increase of the fire release rate (from 2 MW/m^2 in Scenario II to 3 MW/m^2 in Scenario IV) can lead to a rough 36% decrease of SI_{Total} (from 114.23/R in Scenario II to 73.16/R in Scenario IV). Given the same fire release rate in Scenarios I, II, and III, Scenario III turns out to be the worst fire scenario among those three scenarios. It should be noted that the calculation of RSET may display slight differences due to the stochastic nature of the human escape behaviors and changes in the scenario setup. For instance, the number of exits is different across different scenarios at least 30 s after the ignition of the fire. Specifically, in Scenario III, there are only three exits after 30 s, while four exits remain in other scenarios (i.e., Scenarios I, II, and IV). Roughly, the true RSET value in Scenario III should be slightly higher than R, as calculated in Scenario I. That is to say, the true value of SI_{Total} in Scenario III should be lower than 113.26/R (as shown in Table 6.6). Anyhow, Scenario IV remains the worst fire scenario, given the same fire release rate (that is 2 MW/m^2) in Scenarios I, II, and III. That is to say, exits should be regarded as super bottlenecks with significant importance during the fire escape process.

Moving forward to the results in the case study, a better understanding of the fire hazard identification, simulation, and evaluation can be provided in this chapter. Much more attention should be paid to bottlenecks in the possible evacuation routes, such as the four exits on the hall floor in the case station. Spare exits could be very helpful in case of an emergency situation. Accordingly, sufficient materials,

Table 6.6 Evaluation of the fire safety risk in different fire scenarios

Scenario	Estimated ASET at critical evacuation bottlenecks (unit: s)						SI_{Total}	Rank
	Stair I	Stair II	Exit I	Exit II	Exit III	Exit IV		
Scenario I	30	163	194	286	286	194	157.14/*R*	4
Scenario II	30	172	84	247	247	84	114.23/*R*	3
Scenario III	96	182	30	223	223	81	113.26/*R*	2
Scenario IV	30	111	58	117	117	58	73.16/*R*	1

equipment, and staff for fire escape should be placed near those bottleneck areas ahead of time, in order to enhance the efficiency of the occupant evacuation performance in case of the fire. Furthermore, the fire release rate is a very important factor, as it largely contributes to the magnitude of the fire safety risk. This factor should be closely monitored and controlled for proactive safety management, particularly in the heavily congested areas.

6.5 Merits Analysis

The validation of a simulation model is a crucial aspect which should be considered with great concern. In order to further testify whether the model behaves sensibly and ensure the validation process is robust, the field test and experiment are employed for the model validation, since the results obtained from field experiment are considered more reliable than those from simulation models. Due to the fact that a real fire, particularly in a metro station, would be of high danger and great loss, only one scenario, Scenario I, is selected as the test background for the field experiment.

The experiment is conducted in the mid-night when the metro station is out of operation. In this field experiment, the temperature and the CO concentration sensors are first installed at Stair II, Exits I and II at the YJW station. Later, some small tests, such as the ignition of an A4 paper nearby the installed sensors, are conducted in order to make sure all the sensors work well and the data monitored by the sensors can be exported and are reliable. Next, a real fire with a heat release rate of 2 MW/m^2 is ignited at the entrance of Stair I (to be consistent with that in Scenario I) at the platform floor at the station. During the following 6 min, the data of the temperature and the CO concentration in the air are monitored and recorded by sensors located at Stair II, Exits I and II every second. Finally, a total of 360 records consisting of the temperature and CO concentration data at Stair II, Exits I and II are observed and then exported. Figure 6.9 illustrates the comparison between the simulated and observed results of fire hazards, including temperature and the CO concentration in air generated by the fire. The results of parameter values in the linear

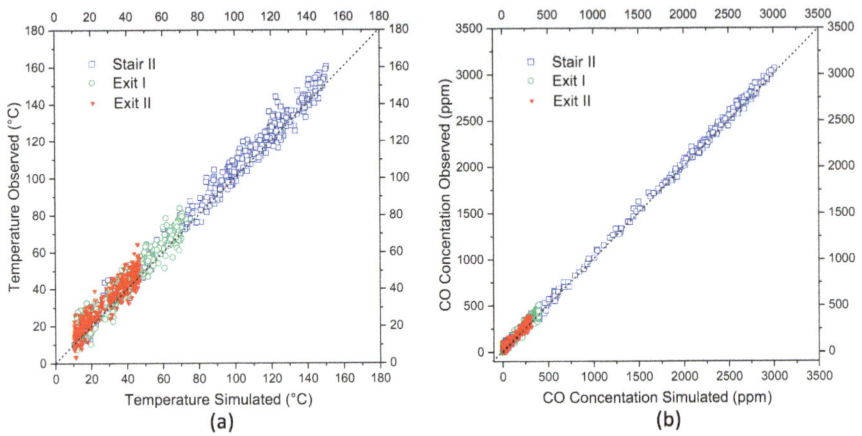

Fig. 6.9 Comparison between the simulated and observed fire hazards at the station: (**a**) temperature; and (**b**) CO concentration

Table 6.7 Results of parameter values in the linear regression lines between the simulated and observed fire hazards

Hazard	Location	Parameters in the linear regression line		
		Slope	Intercept	R^2
Temperature	Stair II	0.992	4.771	0.9831
	Exit I	0.993	4.021	0.9439
	Exit II	1.023	3.445	0.8863
CO concentration	Stair II	1.007	20.628	0.9992
	Exit I	1.056	14.445	0.9662
	Exit II	1.017	28.974	0.9250

regression lines between the simulated and observed fire hazards are presented in Table 6.7.

In general, as shown in Fig. 6.9, the observed results of the temperature and the CO concentration in the air are low when the simulated results are low, and similarly, the observed results are high when the simulated results are high. As seen in Table 6.7, regarding the temperature generated by fire, the fitted linear regression lines at Stair II, Exit I, and Exit II have a slope of 0.992, 0.993, and 1.023, respectively, which are very close to one and indicate that the established fire model in this chapter can simulate the fire smoke spreading process at the YJW station very well. Also, the values of R^2 regarding the fitted regression lines for Stair II, Exit I, and Exit II are 0.9831, 0.9439, and 0.8863, respectively, indicating a high fitting degree. Similarly, the regression lines related to the CO concentration at Stair II, Exit I, and Exit II also have a high fitting degree, with a value of slope ranging from 1.007 to 1.056. Particularly, the values of R^2 regarding the fitted regression lines for Stair II, Exit I, and Exit II are 0.9992, 0.9662, and 0.9250, respectively. In this way, the simulation model developed in this chapter is fairly validated, and can then be used to model the real fire development process in the real world, to a large extent.

6.6 Chapter Summary

With a growing number of metro stations running for operation, particularly in China, how to efficiently evacuate occupants in case of a fire has attracted wide public concern. The concept of safety index SI (Safety Index) = ASET/RSET has been commonly used in fire safety engineering designs to account for uncertainties and performance-based design. To assess the safety risk of a building fire remains a challenging problem, as numerous factors are involved in the estimation of ASET and RSET, and the required availability of data on fire is lacking a great deal. This chapter develops a systematic hybrid approach that integrates the ASET/RSET approach, numerical simulation technologies, and MADA, in order to quantify the magnitude of the fire safety risk and further identify the worst fire scenario. The contribution of this chapter lies in: (1) the body of knowledge by providing an improved ASET/RSET approach that is capable of taking multiple factors (i.e., fire, building, and human features) into account to assess the safety risk of fire conditions in a 3D environment with more complex building layouts; and (2) the state of the practice by providing a more accurate data-driven solution for the perception and discovery of the worst fire scenarios.

In light of the experimental study, some interesting findings are obtained to facilitate a better understanding of fire spreading behaviors and safety risk assessment. During the numerical simulation, it is appropriate to measure the hazards generated by the fire, such as the temperature, the CO concentration, and the visibility within the breathing zone. It is suggested that the height for the measurement of fire hazard conditions can be set up to be 1.6 m, considering the fact that the average height of Chinese male and female is 1.72 m and 1.60 m, respectively. Among the identified four most likely fire scenarios, Scenario IV becomes the worst fire scenario, associated with the lowest value of $SI_{Total} = 73.16/R$. The fire release rate plays a very significant role in the magnitude of the fire safety risk, as a 50% increase of the fire release rate (from 2 MW/m^2 in Scenario II to 3 MW/m^2 in Scenario IV) can lead to a rough 36% decrease of SI_{Total} (from $114.23/R$ in Scenario II to $73.16/R$ in Scenario IV). With the same fire release rate in Scenarios I, II, and III, Scenario III turns out to be more dangerous than the other two scenarios. Sufficient materials, equipment, and staff for fire escape should be placed near the four exits on the hall floor at the station. Field test and experiment results further indicate that the established fire model in this chapter can simulate the fire smoke spreading process at the station very well.

In this chapter, a comprehensive indicator, SI_{Total}, is proposed to quantify the magnitude of the overall safety risk of a building system in case of fire, where the estimation of ASET and RSET plays a very important role in ensuring the accuracy of the calculation of SI_{Total}. For comparison and simplification consideration, the estimation of a combination of multiple RSETs is replaced by a variable $R = (\text{RSET}_1 \times \text{RSET}_2 \times \text{RSET}_3 \times \text{RSET}_4 \times \text{RSET}_5 \times \text{RSET}_6)^{1/6}$ in the case study. This is a "special strategy" employed in this chapter in order to make a comprehensive evaluation in a very efficient manner. Indeed, the estimation of RSET mainly involves technological resources, building features, and occupant characteristics. Particularly for the occupant characteristics, the number (or density), flow speed, distribution, route selection strategies, and moving incentives of the occupants in buildings (e.g., metro stations) should have a large impact on the estimation of RSET. For instance, a higher occupant density is more likely to have a higher value of RSET, as the congestion may occur in certain areas and waste much time in the evacuation.

References

1. Külekçi G, Özsoy D (2024) Fire safety in metro systems: risks, solutions, and global standards. EJONS Int J 8:516–530
2. Gernay T (2024) Performance-based design for structures in fire: advances, challenges, and perspectives. Fire Saf J 142:104036
3. Schröder B, Arnold L, Seyfried A (2020) A map representation of the ASET-RSET concept. Fire Saf J 115:103154
4. Zhang L, Wu X, Liu M, Liu W, Ashuri B (2019) Discovering worst fire scenarios in subway stations: a simulation approach. Autom Constr 99:183–196
5. Association NFP (2024) NFPA 101 life safety code. National Fire Protection Association
6. Board ABC (2022) National construction code. Australian Building Codes Board
7. Development MoHaU-R (2022) General code for fire protection of buildings and constructions (GB 55037-2022). Beijing, China
8. Li X, Zhang L, Shang B, Fang X, Tao Y, Ma Y et al (2025) Study on full-scale experimental and numerical investigation of fire-induced smoke control in an underground Double-Island Subway Station. Int Commun Heat Mass Transf 163:108697
9. Sutrisno HH, Kusumohadi CS, Fadhila F (2025) Effort to Improve Aset and Rset Time Based on Software Simulation In Pt. X Industry. J Phys Conf Ser 2973:012002
10. Sajid Z, Yang Y, You P, Deng H, Cheng X, Danial SN (2022) An explorative methodology to assess the risk of fire and human fatalities in a Subway Station using fire dynamics simulator (FDS). Fire 5:69
11. Jang H-Y, Hwang C-H (2024) Evaluation of available safety egress time (ASET) in performance-based design (PBD) using CFAST. Fire 7:108
12. Bae Y-S, Choi M (2025) Optimizing key evacuation features for safer egress in complex buildings with underground connections: a simulation-based approach to resilient and sustainable design. Sustainability 17:3731
13. Chen H, Zhang C, Zhang J, Shu Y, Qi X, Jiang C (2024) Simulation of fire emergency evacuation in a large-passenger-flow subway station based on the on-site measured data of Shenzhen Metro. Transp Saf Environ 6:tdad006
14. Long Z, Zhong M, Chen J, Cheng H (2023) Study on emergency ventilation strategies for various fire scenarios in a double-Island subway station. J Wind Eng Ind Aerodyn 235:105364
15. Lan B, Li Y-R, Li P-C, Gong H-F (2022) Numerical simulation of the chimney effect on smoke spread behavior in subway station fires. Case Stud Therm Eng 39:102446
16. Tang D, Beattie K (1997) Integrated prediction of fire, smoke and occupants evacuation of buildings. Dublin Institute of Technology/IES Limited
17. Guillaume E (2022) Evaluation of gas phase: Smoke and toxicity analysis. In: Analysis of flame retardancy in polymer science. Elsevier, pp 161–190

18. Babrauskas V, Fleming JM, Don Russell B (2010) RSET/ASET, a flawed concept for fire safety assessment. Fire Mater 34:341–355
19. Qu L, Chow W (2013) Common practices in fire hazard assessment for underground transport stations. Tunnell Underground Space Technol 38:377–384
20. Commission BMP (2013) Code for design of metro (GB 50157-2013). China Building Industry Press, Beijing, China
21. Poh W (2010) Tenability in building fires: limits and design criteria. Fire Aust Spring:24–26
22. DiNenno PJ (2008) SFPE handbook of fire protection engineering. SFPE
23. Gwynne SMV, Purser D, Boswell DL (2011) Pre-warning staff delay: a forgotten component in ASET/RSET calculations. In: Pedestrian and evacuation dynamics. Springer, pp 243–253
24. Committee SPaS (2003) Application of fire safety engineering principles to the design of buildings. Code of practice (BS 7974:2001). BSI, London, UK
25. Pauls J (1987) Calculating evacuation times for tall buildings. Fire Saf J 12:213–236

Chapter 7
Passenger Evacuation Analysis

7.1 Introduction

The metro system has become a vital component of urban transportation due to its high efficiency, environmental benefits, and ability to bypass surface traffic congestion. However, rising ridership has led to growing safety and security challenges. In recent years, incidents such as system failures, fires, and malicious attacks have triggered large-scale evacuations under constrained conditions. These emergencies highlight the vulnerability of underground transit systems, where confined spaces and high passenger density significantly increase evacuation risks. Rapid and safe passenger evacuation is critical to minimizing casualties and avoiding secondary disasters, especially when infrastructure is partially compromised [1]. Consequently, assessing metro system vulnerability from the standpoint of evacuation has become a key aspect of safety management.

In response to these concerns, both researchers and practitioners have focused on improving emergency evacuation in metro stations. Simulation-based methods are commonly employed to model evacuation dynamics, evaluate the effects of behavioral factors such as walking speed and panic, and assess the role of guidance systems in improving passenger flow [2, 3]. These efforts have significantly advanced our understanding of evacuation processes. However, many studies focus on general evacuation planning rather than evaluating the system's vulnerability under complex or constrained emergency scenarios.

Recent studies increasingly apply simulation-based models to evaluate station capacity, assess evacuation time, analyze bottlenecks, and explore route selection under different hazard scenarios [4–6]. These studies commonly employ key performance indicators such as evacuation time, maximum passenger density, and exit utilization to measure evacuation effectiveness. In parallel, attention has been given to behavioral and environmental factors, including walking speed, information guidance, leadership positioning, and the use of vertical transport systems. While

L. Zhang et al., *Resilience in Urban Transit Systems*, Advances in 21st Century Human Settlements, https://doi.org/10.1007/978-981-95-4322-9_7

these efforts have advanced evacuation planning, most studies concentrate on optimizing specific strategies or improving metrics under fixed conditions. In practice, evacuation outcomes are shaped by diverse and often interacting factors, including evacuee behavior, architectural layout, guidance systems, obstacle configuration, and crowd dynamics [7]. These relationships are frequently non-linear and uncertain. This complexity poses a significant challenge for assessing system vulnerability and developing adaptive evacuation strategies that remain effective under uncertain or constrained conditions.

To address this issue, this chapter proposes a hybrid evaluation framework that integrates simulation-based modeling, a random forest meta-model, and a multi-indicator comparison process. This approach enables the identification of key influencing factors and the analysis of their impact on critical performance indicators such as evacuation time, passenger density, and renovation cost. This research contributes to the field by addressing the current gap in evaluating evacuation complexity under constrained conditions and offering a flexible methodology to support data-driven decision-making for metro station safety planning.

7.2 Passenger Evacuation Under Disaster

Passenger evacuation evaluation plays a critical role in understanding and mitigating system vulnerability during emergencies in metro stations. As metro environments are characterized by confined spaces, high passenger density, and diverse structural constraints, evacuation dynamics are inherently complex and scenario-specific. To evaluate evacuation performance and guide decision-making under such conditions, researchers have developed a wide range of methodologies [8]. These approaches can be broadly classified into three categories: macroscopic flow-based simulation methods, microscopic behavior-based agent models, and hybrid methods.

1. **Macroscopic methods.** Macroscopic flow-based simulation methods treat crowds as continuous flows governed by movement rules and parameters. These models assume homogeneous passenger behavior and apply flow dynamics such as density-speed relationships to estimate evacuation time, identify congestion zones, and evaluate route usage. Tools like Simulex and the SFPE model in Pathfinder simulate crowd dispersal based on parameters such as exit width, walking speed, and corridor capacity [9, 10]. While computationally efficient and useful for rapid assessments, these models lack the ability to capture emergent individual behaviors, complex interactions, or adaptive responses to evolving conditions.
2. **Microscopic methods.** Microscopic behavior-based agent models simulate each passenger as an autonomous agent with individual goals, perceptions, and decision-making capabilities. These models incorporate elements such as the social force model, herding behavior, and obstacle avoidance to represent

evacuation dynamics realistically [11]. Commonly used platforms include AnyLogic, FDS + Evac, and the steering mode in Pathfinder [12]. By modeling interactions at the individual level, these approaches can capture emergent phenomena such as queuing, congestion spillovers, and route-switching behavior under stress. However, they are computationally intensive and sensitive to behavioral parameter settings.

3. **Hybrid methods.** Hybrid methods have emerged as a promising direction for evaluating evacuation strategies in complex environments. These methods first employ simulations to generate evacuation performance data across a range of scenarios. Then, surrogate models are trained to approximate the relationships between key input variables and performance indicators [6]. This approach enables rapid evaluation and supports large-scale sensitivity analyses or optimization without repeatedly invoking the full simulation engine, significantly improving computational efficiency [13]. They enable multi-indicator or multi-objective analysis, allowing decision-makers to explore trade-offs between competing objectives, which is particularly valuable in metro station scenarios involving partial blockages, infrastructure constraints, or dynamic passenger behavior.

In summary, evacuation evaluation methods have evolved from simplified flow-based models to sophisticated hybrid frameworks that combine simulation and machine learning. While macroscopic and microscopic methods provide foundational insights, hybrid approaches enhance flexibility, scalability, and decision support capacity. Building on these developments, this study adopts a hybrid strategy evaluation approach that integrates agent-based simulation, random forest meta-modeling, and multi-scenario comparison. This framework supports a more systematic assessment of evacuation vulnerabilities and strategy effectiveness in metro stations.

7.3 Passenger Evacuation Evaluation Framework

To evaluate the vulnerability of metro stations from the perspective of passenger evacuation, this study adopts a hybrid approach, as illustrated in Fig. 7.1. The process begins with a comprehensive analysis of evacuation dynamics and the development of an evaluation framework based on key influencing factors. A metro station model is then constructed, and evacuation scenarios are simulated under varying conditions. Using the simulation results, a meta-model is developed to capture the relationship between influencing factors and evacuation performance indicators. Finally, multiple representative evacuation strategies are evaluated across scenarios to compare their effectiveness and support decision-making under different vulnerability conditions.

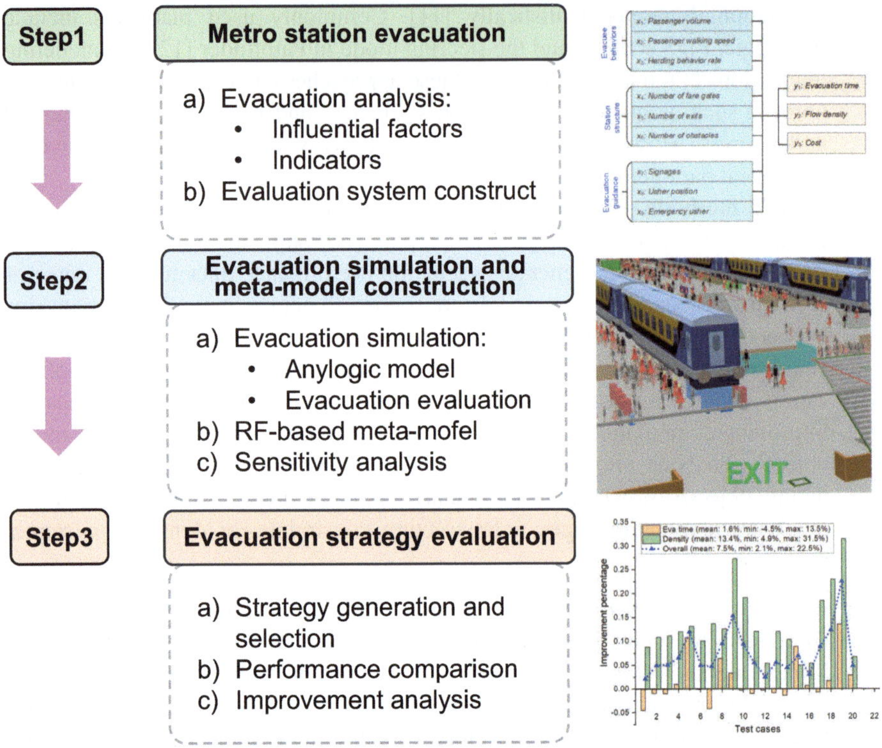

Fig. 7.1 Workflow of the proposed model for passenger evacuation evaluation

7.3.1 *Passenger Evacuation Influential Factors*

Evacuation performance in metro stations can vary significantly across scenarios due to distinct environmental and structural conditions. It is therefore essential to identify key factors influencing the evacuation process under different event types. To enable a comprehensive evaluation of evacuation performance, an index system comprising three core indicators (total evacuation time, crowd density, and renovation cost) was developed, as shown in Fig. 7.2. Evacuation time reflects the efficiency and safety of the response, as prolonged durations increase the likelihood of panic-induced crowd turbulence and potential harm. Crowd density is a key indicator of risk, particularly at bottlenecks such as exits, stairs, and turning points; the maximum density observed during simulation is used as a proxy for crowd pressure and hazard potential. Renovation cost is included to reflect the financial feasibility of evacuation improvement strategies, especially given the spatial and economic constraints of centrally located metro stations.

Structural constraints also differ among stations. To account for contextual variations among metro stations, expert consultations and on-site investigations were conducted to identify key factors affecting evacuation effectiveness. Nine

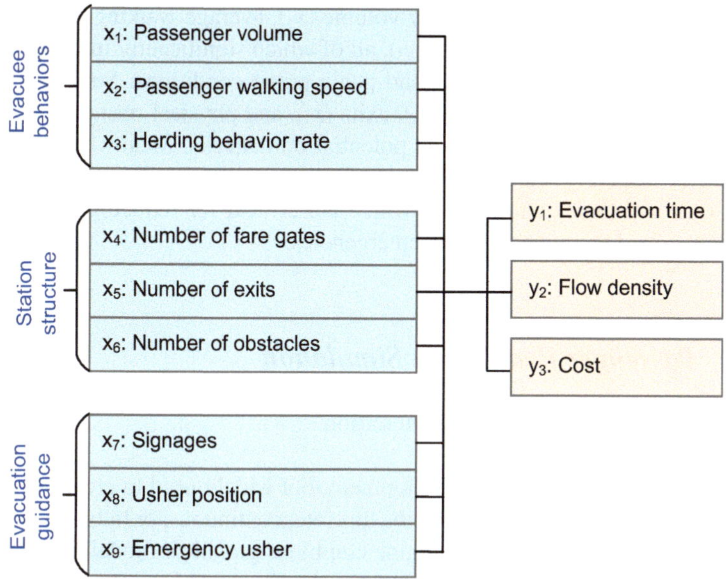

Fig. 7.2 Established input-output framework for the evacuation evaluation analysis

Table 7.1 Influential factors for the evaluation of emergency evacuation in metro stations

Aspect	Indicator	Description
Evacuee behaviors	x_1: Passenger volume	The number of passengers, deciding the flow speed, and density
	x_2: Passenger walking speed	Different walking speeds of evacuees according to the age distribution
	x_3: Herding/small group behavior rate	The social behavior of maintaining consistent speed with companies, choosing the same evacuation path, and exit
Station structure	x_4: Number of fare gates	Number, width, etc.
	x_5: Number of exits	Number, width, etc.
	x_6: Number of obstacles	Isolation fence, maintenance tables, etc.
Evacuation guidance	x_7: Signages	Evacuation signage (equipment), wall sign, hanging, ground sign (visibility, comprehensibility)
	x_8: Usher positions	Lighting conditions in the passages and exits
	x_9: Number of emergency ushers	Guidance for reducing the density in order to avoid possible danger

second-level influencing factors were identified across the three primary categories, serving as key inputs for evaluating the vulnerability of the metro station evacuation system. Table 7.1 presents the detailed definitions and classifications of all selected factors and evaluation indicators. These factors fall into three primary categories: evacuee behavior, station structure, and emergency guidance. Evacuee behavior includes variables such as passenger volume (x_1), average walking speed (x_2), and the proportion of herding behavior (x_3), all of which significantly influence evacuation dynamics under high-density and panic-prone conditions. Structural factors, such as the number of fare gates (x_4), exits (x_5), and physical obstacles (x_6), determine the internal flow capacity and potential bottlenecks. Meanwhile, emergency guidance is shaped by the effectiveness of signage (x_7), the strategic placement of ushers (x_8), and their quantity (x_9), which are critical for reducing confusion and managing crowd movement during emergencies.

7.3.2 Passenger Evacuation Simulation

7.3.2.1 Simulation Model for Evacuation

AnyLogic is a Windows-based development tool widely used to create virtual prototyping environments and simulate the discrete, continuous, or hybrid behaviors of complex systems. Its simulation engine combines graphical modeling with a code generator that translates models into Java code. The resulting simulation runs on a Java platform, with interfaces available to control execution and retrieve data via a text-based protocol over TCP/IP. AnyLogic enables rapid construction of system simulation models that incorporate external environments, physical equipment, and human operators. Owing to its broad adoption in both academia and industry, AnyLogic is employed in this study to develop the metro station evacuation simulation.

The Pedestrian Library in AnyLogic is designed to simulate the physical environment of densely populated spaces. It offers built-in functions for evaluating service point load efficiency and collecting data such as pedestrian flow density in specific areas. This library also provides an intuitive user interface for rapidly constructing flow-process diagrams. In addition, the Rail Library enables users to model and visualize various railway systems. It supports the integration of discrete events and agents to simulate activities such as passenger boarding and alighting, resource allocation, and other transport operations. Users can further customize models by building databases and flow-process diagrams based on specific requirements.

7.3.2.2 Meta-Model Construct

One major challenge in evaluating evacuation performance lies in the complex and often non-linear relationship between influencing factors and performance indicators. Advances in simulation and machine learning have made it feasible to approximate this relationship through meta-modeling. Common meta-modeling techniques include kriging, splines, support vector machines (SVMs), Bayesian networks (BNs), random forests (RF), and artificial neural networks (ANNs). Among these,

RF is well-suited for complex engineering problems due to its ability to model non-linear relationships without relying on statistical assumptions. Therefore, RF is adopted in this study to map input variables to evacuation performance metrics.

An RF model comprises an ensemble of decision trees, each generating a prediction based on input features. The purpose of a regression tree in RF is to find a series of splits that are deduced from one of d variables, x^k (for a $k \in \{1,\ldots,d\}$), and the threshold value, τ, that can divide the training data into two sets, i.e., nodes $\left\{i : x_i^k < \tau\right\}$ and $\left\{i : x_i^k \geq \tau\right\}$. Among all the possible splits, the split of a given node is decided by minimizing the sum of the homogeneity of its two corresponding child nodes, N_c^1 and N_c^2, which are given by Eq. (7.1):

$$\min \sum_{i \in N_c^i} \left(y_i - \overline{y}^{N_c^i}\right)^2 \tag{7.1}$$

where $\overline{y}^{N_c^i} = \frac{1}{\left|N_c^i\right|} \Sigma i \in N_c^i y_i$ represents the mean value of the output variables of the observations belonging to N_c^i.

The growth of the tree would stop if the child nodes are homogeneous enough, or the number of observations in the child nodes is smaller than a pre-determined number. At the final split node, the tree stops branching, and the values represented by the final nodes are regarded as the prediction results. The average value of all the prediction results constructs the RF prediction model.

7.3.2.3 Prediction Performance and Sensitivity Analysis

After constructing the RF model, three indicators, mean absolute error (MAE), root mean squared error (RMSE), and R^2, are used to examine the accuracy of the developed RF model in this research. Eqs. (7.2) and (7.3) can be used to get the values of MAE and RMSE, respectively. Eq. (7.4) is used to calculate the value of R^2.

$$\text{MAE} = \frac{1}{N} \sum_{i=1}^{N} \left(f_i(x) - y_i\right) \tag{7.2}$$

$$\text{RMSE} = \sqrt{\frac{1}{N} \sum_{i=1}^{N} \left(f_i(x) - y_i\right)^2} \tag{7.3}$$

$$R^2 = 1 - \frac{\sum_{i=1}^{N} \left(y_i - f_i(x)\right)^2}{\sum_{i=1}^{N} \left(y_i - \overline{y}\right)^2} \tag{7.4}$$

where N is the total number of samples for testing, $f_i(x)$ is the predicted value, y_i is the values from evacuation simulations, and $\overline{y}$ is the mean of y_i.

To further examine how input variables influence evacuation performance, a sensitivity analysis is conducted using Shapley Additive Explanations (SHAP). SHAP is integrated into the proposed framework to quantify the contribution of each input factor within the RF model. Widely adopted in the machine learning community, SHAP offers fast, model-specific implementations and provides insights into both positive and negative impacts of features. This enables a deeper understanding of how each variable drives variation in the output indicators. Additionally, SHAP is straightforward to implement. Let P denote the set of all input features and S the set of non-zero elements. The SHAP value is computed using Eq. (7.5).

$$\varphi_i = \sum_{S \in P \setminus \{i\}} \frac{|S|!\left(K-1-|S|\right)!}{K!}\left[f_x\left(S \cup \{i\}\right) - f_x\left(S\right)\right] \tag{7.5}$$

where S is a subset of P, $f_x(S) = E[f(x)|x_s]$, which stands for the expected output of the original model with the feature in set S.

7.3.3 Passenger Evacuation Strategy Assessment

7.3.3.1 Identification of Effective Evacuation Strategies

In this study, three key indicators are used to evaluate the effectiveness of evacuation strategies at metro stations: total evacuation time (T), passenger density during evacuation (D), and renovation cost (C). These indicators collectively reflect safety performance and implementation feasibility. Since lower values are preferable for all three indicators, the overall selection rule can be expressed as Eq. (7.6). In addition, two safety thresholds are introduced to ensure basic evacuation requirements. Specifically, evacuation is considered acceptable if the total time does not exceed $S_t = 360$ s, and the passenger density remains below $S_d = 3.6$ unit/m^2. These safety conditions can be expressed as constraints Eq. (7.7). The renovation cost C is calculated as Eq. (7.8):

$$\begin{cases} \min T \\ \min D \\ \min C \end{cases} \tag{7.6}$$

$$\begin{cases} T \le S_t \\ D \le S_d \end{cases} \tag{7.7}$$

$$C = v_4 \times c_4 + v_5 \times c_5 + v_6 \times c_6 + v_7 \times c_7 + v_9 \times c_9 \tag{7.8}$$

where v_4–v_9 represents the variation of the influential factors x_4 to x_9, and c_4 to c_9 is the cost required for the renovation. It is noted that the variation of x_1, x_2, x_3, and x_8 does not require extra cost.

To ensure comparability among indicators with different scales, normalization is applied using Eq. (7.9). A distance-based metric. D_u, is then used to evaluate how close a solution is to the ideal point (0, 0, 0) using Eq. (7.10), which represents minimum values for all indicators. Once effective strategies are selected based on the shortest distance, the original (unnormalized) values of indicators can be recovered by Eq. (7.11).

$$y_i = \frac{Z_i - Z_{i\min}}{Z_{i\max} - Z_{i\min}} \tag{7.9}$$

$$D_u = \sqrt{(y_{1u})^2 + (y_{2u})^2 + (y_{3u})^2} \tag{7.10}$$

$$Y_i = (Z_{i\max} - Z_{i\min}) \times y_i + Z_{i\min} \tag{7.11}$$

where y_i is the normalized value, Y_i is the original value of ith objective, $Z_{i\max}$ is the $Z_{i\min}$ are the maximum and the minimum values for the ith objective, respectively, D_u is the distance of the uth solution, and y_{1u} to y_{3u} are the values of the three objectives of the uth solution on the Pareto front.

In this study, the RF-based meta-model is developed to approximate the relationship between influencing factors and evacuation outcomes. The NSGA-III algorithm is then applied to generate a diverse set of strategy combinations under various scenarios. Based on the evaluation process, a subset of representative and effective evacuation strategies is identified. Figure 7.3 demonstrates the framework for integration of the RF-based meta-model and NSGA-III.

7.3.3.2 Comparative Analysis of Evacuation Performance

To quantify the effectiveness of selected evacuation strategies, a comparative analysis is conducted against a baseline scenario. The analysis focuses on two core performance indicators: total evacuation time and passenger density. Although renovation cost is considered in the overall evaluation, it is excluded here as it does not directly reflect evacuation efficiency or safety. The overall improvement degree (ID) is calculated by averaging the improvements in evacuation time (ID_T) and passenger density (ID_D) as shown in Eqs. (7.12)–(7.14).

$$\text{ID} = \text{ID}_T * 0.5 + \text{ID}_D * 0.5 \tag{7.12}$$

$$\text{ID}_T = \begin{cases} \dfrac{T - S_t}{T} * \alpha + \dfrac{S_t - T_O}{S_t} * \beta & \text{if } T \geq S_t \\ \dfrac{T - T_O}{T} * \beta & \text{if } T < S_t \end{cases} \tag{7.13}$$

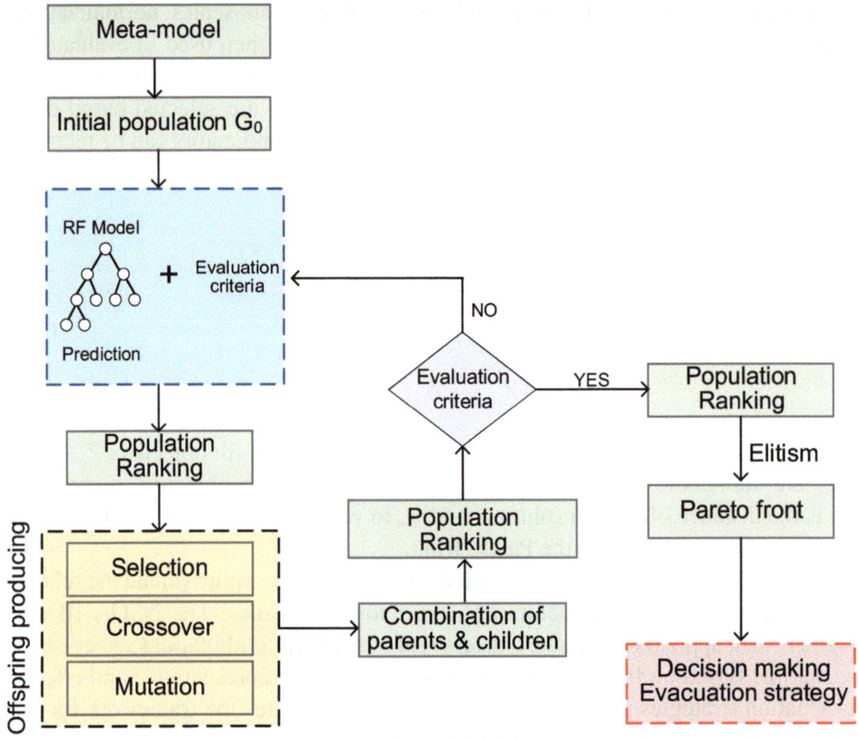

Fig. 7.3 Framework of the integration of RF and NSGA-III

$$\mathrm{ID}_D = \begin{cases} \dfrac{D - S_d}{D} * \alpha + \dfrac{S_d - D_O}{S_d} * \beta & \text{if } D \geq S_d \\ \dfrac{D - D_O}{D} * \beta & \text{if } D < S_d \end{cases} \tag{7.14}$$

where T_O and D_O are the obtained optimal solutions for the evacuation time and density, respectively. The coefficients α and β reflect the relative contribution of improvements above and below the safety thresholds S_t and S_d, respectively. In this study, $\alpha = 0.8$ and $\beta = 0.2$, emphasizing improvements above the threshold. For instance, reducing evacuation time from 380 s to 360 s is considered more critical than from 340 s to 320 s, even though both yield a 20 s reduction.

7.4 Experimental Studies

7.4.1 Data and Resources

Mass Rapid Transit (MRT) is a rail transit system in Singapore, which was first constructed in 1987 and now covers most of the city. Despite having one of the most developed railway networks, the rapidly increasing ridership is forcing the government to upgrade the metro systems continuously. Due to the growing and development demand in Singapore, the investigation for evacuation in the MRT system becomes an indispensable task, since it is closely related to safety issues. Thus, this research chooses one of the metro stations in Singapore to explore the evacuation evaluation.

Jurong East (JE) MRT station, an elevated interchange in western Singapore, connects the North-South and East-West lines. The station comprises two levels: Level 1 serves as the concourse, while Level 2 accommodates three platforms, facilitating transfers between four train lines. Due to rapid population growth, the station frequently experiences overcrowding, especially during peak hours, with ridership exceeding 20,000 passengers per hour. Originally not designed to handle such volumes, the station now operates under high, and sometimes extreme, passenger densities, underscoring the urgency of evaluating evacuation safety.

7.4.2 Passenger Evacuation Modeling

To simulate evacuation events at the JE MRT station, a 3D model is developed using AnyLogic, as shown in Fig. 7.4. The station has two levels: Level 1 is the concourse with fare gates, ticketing machines, and service centers; Level 2 includes three platforms facilitating train transfers on two lines. The model incorporates both internal and external exits, comprising four main external exits, three fare gates (internal exits), six pairs of escalators, six stairs, four emergency stairs, and two emergency internal exits.

To capture the dynamic nature of evacuation, four typical scenarios are defined based on varying passenger volumes and panic levels. Panic in large crowds, especially in metro stations, is often associated with increased walking speed and herding behavior. Under panic, passengers tend to accelerate and follow others blindly, particularly in high-density situations. Therefore, panic level is modeled using average walking speed and herding rate, with higher values indicating more severe panic. The four defined scenarios are summarized in Table 7.2. Scenario I is under a regular volume and a low panic level. Scenario II is under a high passenger volume and a low panic. Scenario III is under a regular passenger volume but with a high panic. Scenario IV is under a high passenger volume and a high level of panic. Figure 7.5 illustrates a simulated evacuation using a heatmap of passenger flows.

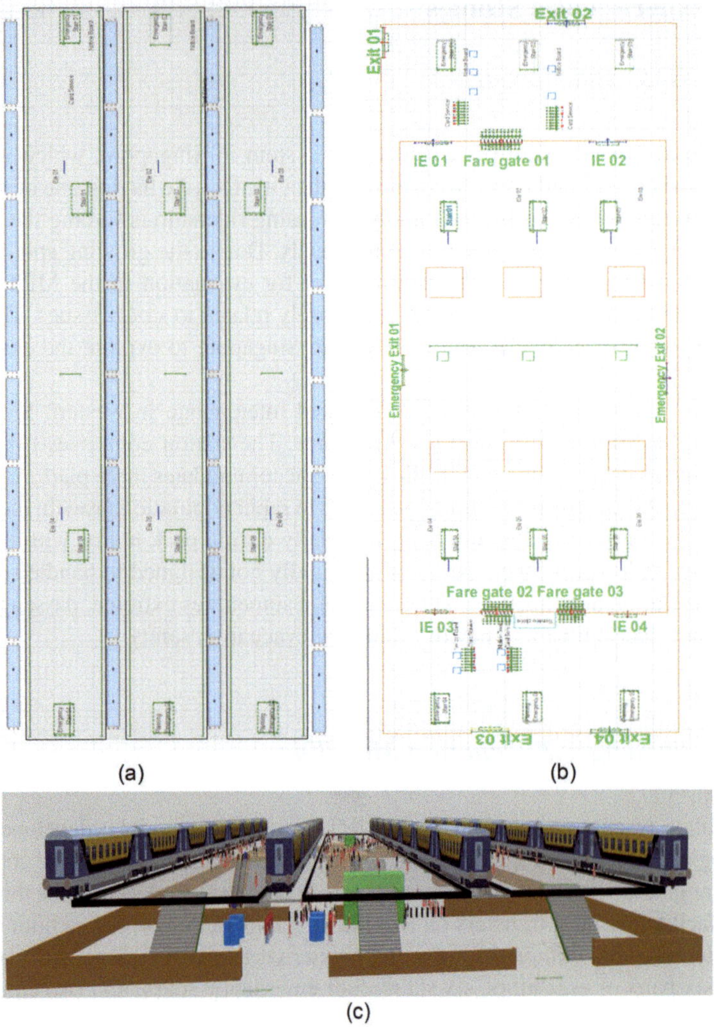

Fig. 7.4 Constructed model of JE MRT station: (**a**) Second floor; (**b**) First floor; (**c**) 3D view of the station

Simulation outcomes under each scenario are reported in Table 7.2. Detailed discussions are presented as follows.

1. Under Scenario I, which involves regular passenger volume and low panic, safe evacuation can be achieved. As shown in Table 7.2, the evacuation time is 247 s and the passenger density is 3.4 units/m^2, both within acceptable safety thresholds. However, increases in either volume or panic compromise evacuation performance. In Scenario III, where only panic is elevated, the evacuation time is reduced to 151 s due to faster movement, but density rises to 3.9 units/m^2,

Table 7.2 Settings and the evacuation evaluation results for the four scenarios

Factors	Scenarios			
	I	II	III	IV
x_1	1000	3000	1000	3000
x_2	Normal	Normal	Fast	Fast
x_3	Normal	Normal	High	High
x_4	20	20	20	20
x_{5_1}	3	3	3	3
x_{5_2}	4	4	4	4
x_6	All	All	All	All
x_7	Low	Low	Low	Low
x_8	External exits	External exits	External exits	External exits
x_9	4	4	4	4
Evaluation results				
Eva-time (y_1)	247	416	151	477
Max density (y_2)	3.4	5.4	3.9	6.2

Note: x_{5_1} represents the internal exits, including the three sets of fare gates indicated in Fig. 7.4b and other potential internal exits (IE01 to IE04) that could be used for the optimization, x_{5_2} represents the four external exits, exits 01, 02, 03 and 04 as indicated in Fig. 7.4b. "All" for the factor x_6 means that all the obstacles, railings, service desks, cleaning, and maintenance tools are present

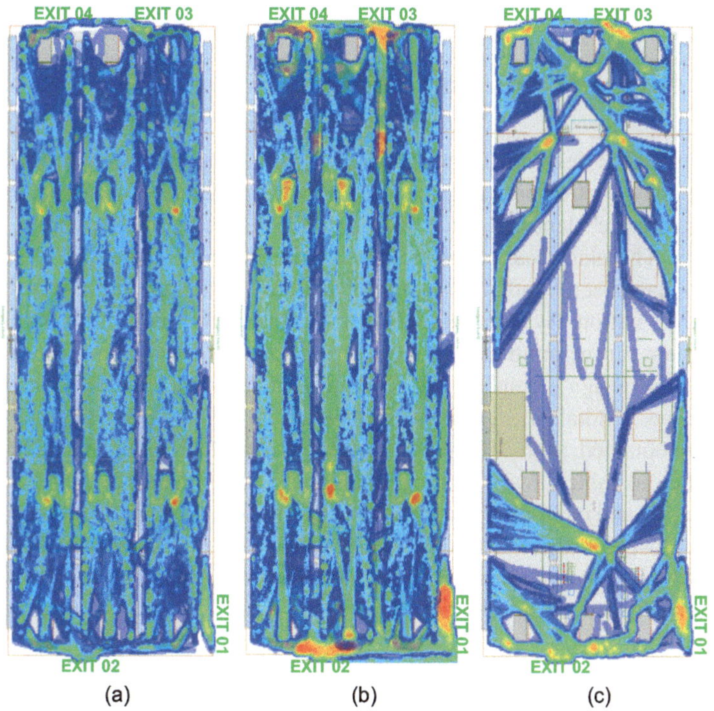

Fig. 7.5 Heatmap for the evacuation process of Scenario I: (**a**) 40 s since the evacuation began; (**b**) 140 s since the evacuation began; (**c**) 240 s since the evacuation began (dark red represents a higher density and light blue represents a lower density)

exceeding the safety limit. Although evacuation is more rapid, crowding becomes a concern. In Scenario II, with higher passenger volume but low panic, the evacuation time increases to 416 s and density reaches 5.4 units/m^2, both above safety thresholds. Scenario IV, with both high volume and high panic, results in the worst performance: evacuation takes 477 s and density rises to 6.2 units/m^2. The prolonged duration and severe congestion pose significant safety risks, highlighting the necessity of optimization to ensure safe evacuation under high-stress conditions.

2. The "faster-is-slower" effect is observed in high-volume scenarios. While panic does not necessarily hinder evacuation under normal crowd levels, it becomes detrimental when passenger volume is high. The key difference between Scenarios I and III lies in the presence of panic in Scenario III. Although density increases, evacuation time is significantly reduced due to accelerated movement. However, in Scenario IV, where both passenger volume and panic are high, panic leads to a longer evacuation time, rising from 416 s to 477 s, and a dangerously high density of 6.2 units/m^2. This outcome results from impatience and insufficient guidance, which trigger irregular congestion at fare gates and exits. Such clogging substantially increases density and delays evacuation. These findings suggest that managing panic levels during evacuation is crucial, and appropriate guidance should be provided to maintain orderly movement.

7.4.3 Passenger Evacuation Prediction

A meta-model is developed to establish the relationship between key influencing factors and evacuation performance indicators. As previously discussed, nine influential factors are identified, each with multiple levels defined based on practical feasibility. The classification of these levels is shown in Table 7.3. All proposed measures have been validated by experts and are considered implementable at the target metro station. By combining different levels of the nine factors, various retrofit strategies can be simulated to support evacuation optimization. An orthogonal experimental design is employed to generate sufficient data for accurate meta-model construction. A total of 260 evacuation simulation results from the metro station model are obtained, with 240 used for training the prediction model and 20 reserved for validation. Table 7.6 shown the prediction results of the meta-model. The detail of simulation and meta-model enabled predictions for two objectives is compared in Fig. 7.7. Based on the constructed meta-model, SHAP sensitivity analysis is performed to quantify the contribution of each factor to the evacuation performance indicators. The results are illustrated in Figs. 7.8 and 7.9 and discussed in the following section.

Table 7.3 Experiments under the rule of orthogonal experimental design

Factor	Levels				
x_1	1 (1000)	2 (1600)	3 (2400)	4 (3000)	5 (3600)
x_2	1 (Normal)	2 (Fast)	3 (Hurry)	–	–
x_3	1 (10%)	2 (15%)	3 (20%)	4 (25%)	5 (30%)
x_4	1 (20)	2 (25)	3 (30)	4 (35)	5 (40)
x_{5_1}	1	2	3	4	–
x_{5_2}	1~12 (see Table 7.4)				
x_6	1 (All)	2 (Railings)	3 (Boards and service desks)	4 (Maintenance tools)	5 (None)
x_7	1 (Low effective)	2 (Normal)	3 (High effective)	–	–
x_8	1~18 (see Table 7.5)				
x_9	1 (4)	2 (6)	3 (8)	4 (10)	5 (12)

Note: The code for different levels of x_{5_2} is presented in Table 7.4. The code for different levels of x_8 is presented in Table 7.5. "–"indicates that it is not applicable (for example, the factor x_2 only has three levels)

Table 7.4 Code description for different levels of the external exits x_{5_2}

Code	Combination
1	Exits 1, 2
2	Exits 1, 3
3	Exits 1, 4
4	Exits 1, 3, 4
5	Exits 1, 3, 6
6	Exits 1, 3, 5
7	Exits 1, 2, 3, 4
8	Exits 1, 2, 5, 6
9	Exits 1, 4, 5, 6
10	Exits 1, 2, 4, 5, 6
11	Exits 1, 2, 3, 4, 5
12	Exits 1, 2, 3, 4, 5, 6

Note: Exits 5 and 6 refer to the emergency exits, which means that the passenger can get through exits 5 and 6 to evacuate successfully by setting the emergency exits as external exits

1. The constructed meta-model demonstrates high predictive accuracy for evacuation outcomes. As shown in Fig. 7.7, the RF-based meta-model produces well-fitted regression lines for both evacuation time and density. According to Table 7.6, the RMSE, MAE, and R^2 for evacuation time on the training set are 19.3, 30.4, and 0.94, respectively. For evacuation density, the RMSE and MAE are 0.41 and 0.39, with an R^2 of 0.92. On the test set, the RMSE, MAE, and R^2 for evacuation time are 32.5, 37.6, and 0.91, respectively, while those for density are 0.85, 0.41, and 0.91. These low error values and consistently high R^2 scores

Table 7.5 Code description for different levels of the ushers and positions x_8

Code	Number of ushers	Positions of ushers
1	4	1, 3, 4, 6
2	4	7, 8, 9, 10
3	4	11, 12, 13, 14
4	4	15, 16, 17, 18
5	6	1, 2, 3, 4, 5, 6
6	6	1, 3, 15, 16, 17, 18
7	6	4, 6, 15, 16, 17, 18
8	8	7, 8, 9, 10, 15, 16, 17, 18
9	8	11, 12, 13, 14, 15, 16, 17, 18
10	10	1, 2, 3, 4, 5, 6, 7, 8, 9, 10
11	10	1, 2, 3, 4, 5, 6, 11, 12, 13, 14
12	10	1, 2, 3, 4, 5, 6, 15, 16, 17, 18
13	12	1, 2, 3, 4, 5, 6, 15, 16, 17, 18, 19, 20
14	10	7, 8, 9, 10, 15, 16, 17, 18, 19, 20
15	10	11, 12, 13, 14, 15, 16, 17, 18, 19, 20
16	12	7, 8, 9, 10, 11, 12, 13, 14, 15, 16, 17, 18
17	12	1, 3, 7, 8, 9, 10, 15, 16, 17, 18, 19, 20
18	12	4, 6, 11, 12, 13, 14, 15, 16, 17, 18, 19, 20

Note: Position numbers, 1 to 20, are indicated in Fig. 7.6 representing the combinations of covering all potential clogging areas. The emergency exits 01 and 02 in Fig. 7.6b and the emergency stairs 05 and 06 in Fig. 7.6a (indicated by the numbers 19 and 20 in parentheses) are not available in the actual station, and they are built in the simulation model to assess the influence on the evacuation optimization

confirm the reliability and accuracy of the constructed meta-model in predicting evacuation performance.

2. Passenger volume is the most influential factor affecting both evacuation time and density, while the rankings of other factors vary slightly between the two indicators. As shown in Fig. 7.8, the number of exits and herding behavior rank second and third in importance for evacuation time. In contrast, Fig. 7.9 indicates that the usher position and the number of internal exits are the second and third most important factors for evacuation density. The strong influence of exit quantity on evacuation time is expected, given its direct impact on flow capacity. Herding behavior, however, exhibits a non-linear relationship with evacuation time. As demonstrated in Fig. 7.8b, moderate herding can reduce evacuation time under low passenger volume, while excessive herding under high volume leads to delays. During emergencies, passengers may lack awareness of optimal routes, making proper guidance essential. Internal exits also play a key role in dispersing passenger flow and thus help reduce evacuation density.

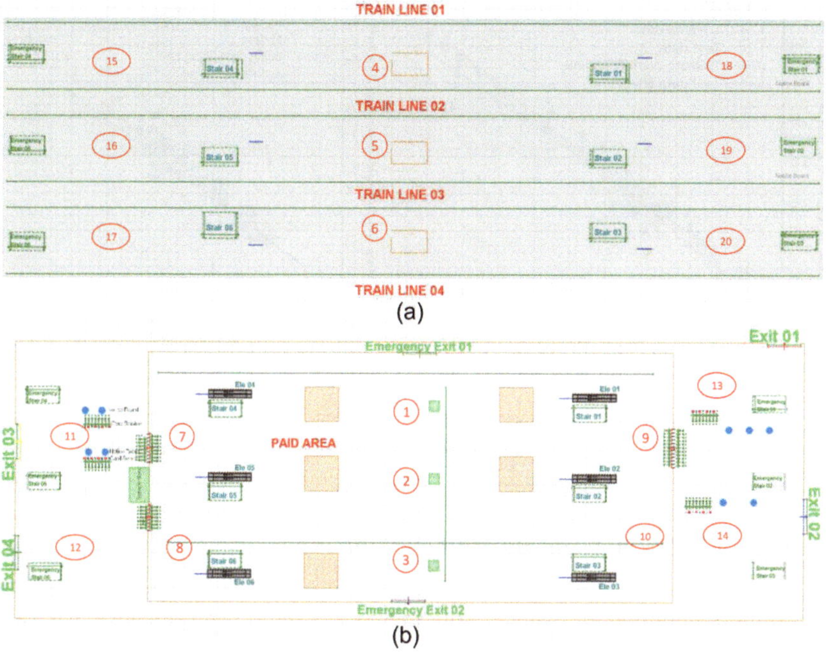

Fig. 7.6 Possible optimization plans: (**a**) Second floor; (**b**) First floor

Table 7.6 Prediction performance test results of the RF-based meta-model between training and test data

Objective	Training data			Test data		
	MAE	RMSE	R^2	MAE	RMSE	R^2
y_1	19.3	30.4	0.94	32.5	37.6	0.91
y_2	0.41	0.39	0.92	0.85	0.41	0.91

3. The numbers of fare gates, obstacles, and signage are the three least influential factors affecting evacuation time and density. As shown in Figs. 7.8 and 7.9, contrary to intuition, neither signage nor the number of fare gates significantly impacts evacuation performance. One possible explanation is that effective emergency guidance may offset the limited contribution of static signage. Regarding fare gates, their narrow width restricts throughput under high passenger volumes, making additional gates less effective and potentially not cost-efficient. Similarly, the obstacle factor shows no clear influence on evacuation outcomes. These findings suggest that limited staff resources should prioritize real-time evacuation guidance rather than obstacle removal.

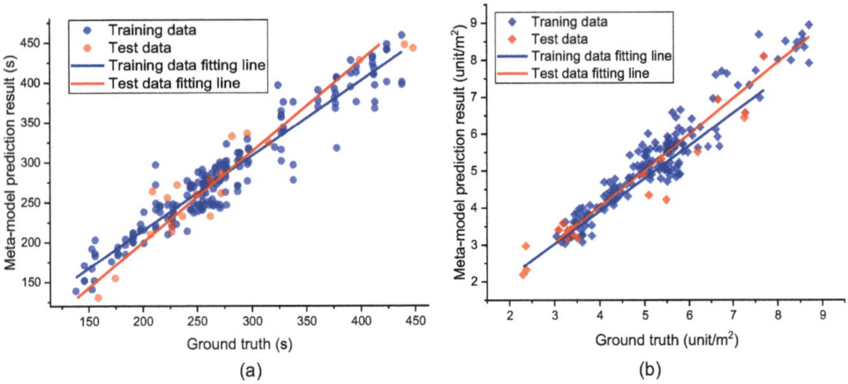

Fig. 7.7 Scatterplot of simulation and meta-model enabled predictions for the two objectives of evacuation time and density in both training data and test data: (**a**) Evacuation time (y_1); (**b**) Density (y_2)

7.4.4 Passenger Evacuation Decision

Based on the established evaluation framework, a set of representative evacuation strategies was applied to Scenarios II through IV for comparative analysis. These scenarios met basic safety thresholds and were selected for in-depth assessment. To reflect realistic evacuation conditions, 20 test cases were designed by varying key influencing factors such as passenger volume, average walking speed, and herding behavior rate. The evaluation process followed the strategy identification framework illustrated in Fig. 7.3.

For each test case, multiple candidate strategies were generated using the RF-based meta-model integrated with the NSGA-III algorithm. The analysis aimed to explore the performance of various strategy combinations under different conditions. Core evaluation indicators included evacuation time, passenger density, and renovation cost. Simulation parameters are summarized in Table 7.7, and the following assumptions were applied: (1) Stairwell movement speed is set at 0.6 m/s when crowd density exceeds 3.5 persons/m^2; (2) Counterflow is permitted on platforms but not in stairwells; (3) Extreme emergencies such as stampedes are excluded; (4) Passengers respond immediately to alarms; (5) Escalators stop during evacuation but remain usable as stairs; (6) Structural constraints prevent widening of escalators and staircases.

Comparative results for Scenarios II to IV are presented in Table 7.8. Figure 7.10 visualizes the distribution of strategy outcomes across the evaluation space, enabling comparative analysis across all scenarios. Figure 7.11 summarizes how selected strategies adjust key input factors and the associated renovation costs. Figure 7.12 presents an overall comparison of the 20 test cases, focusing on the variation in evacuation performance achieved by different strategies.

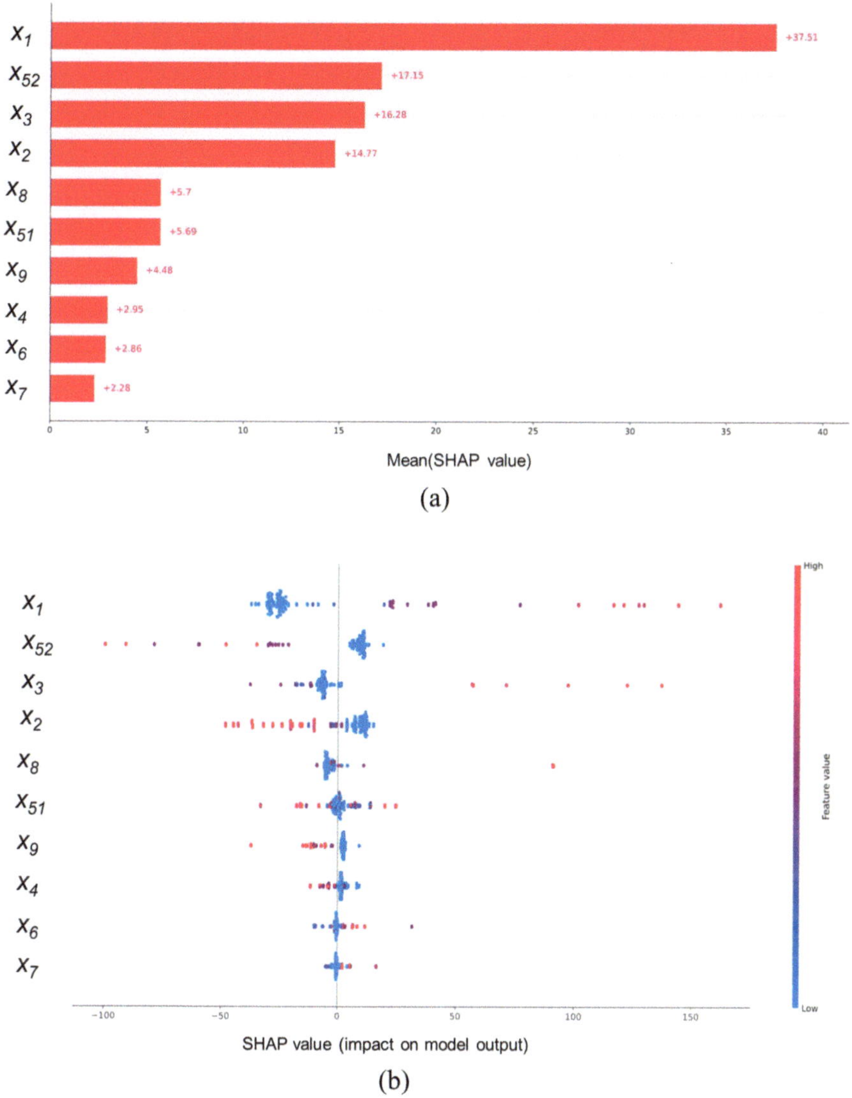

Fig. 7.8 Sensitivity analysis results for the nine factors: (**a**) Contributions of factors to the evacuation time; (**b**) Relationship between factors and the evacuation time

1. The comparative results indicate that high-volume scenarios benefit more significantly from strategy adjustments. As shown in Table 7.8, the selected strategy for Scenario III (regular volume) achieved a modest improvement of 2.8%. In contrast, Scenarios II and IV (high volume) saw improvements of 19.9% and 27.2%, respectively. However, these gains involve trade-offs. For instance, in Scenario III, passenger density decreased substantially (from 3.9 to 3.1 persons/m^2), but total evacuation time increased slightly (from 151 to 176 s), highlighting

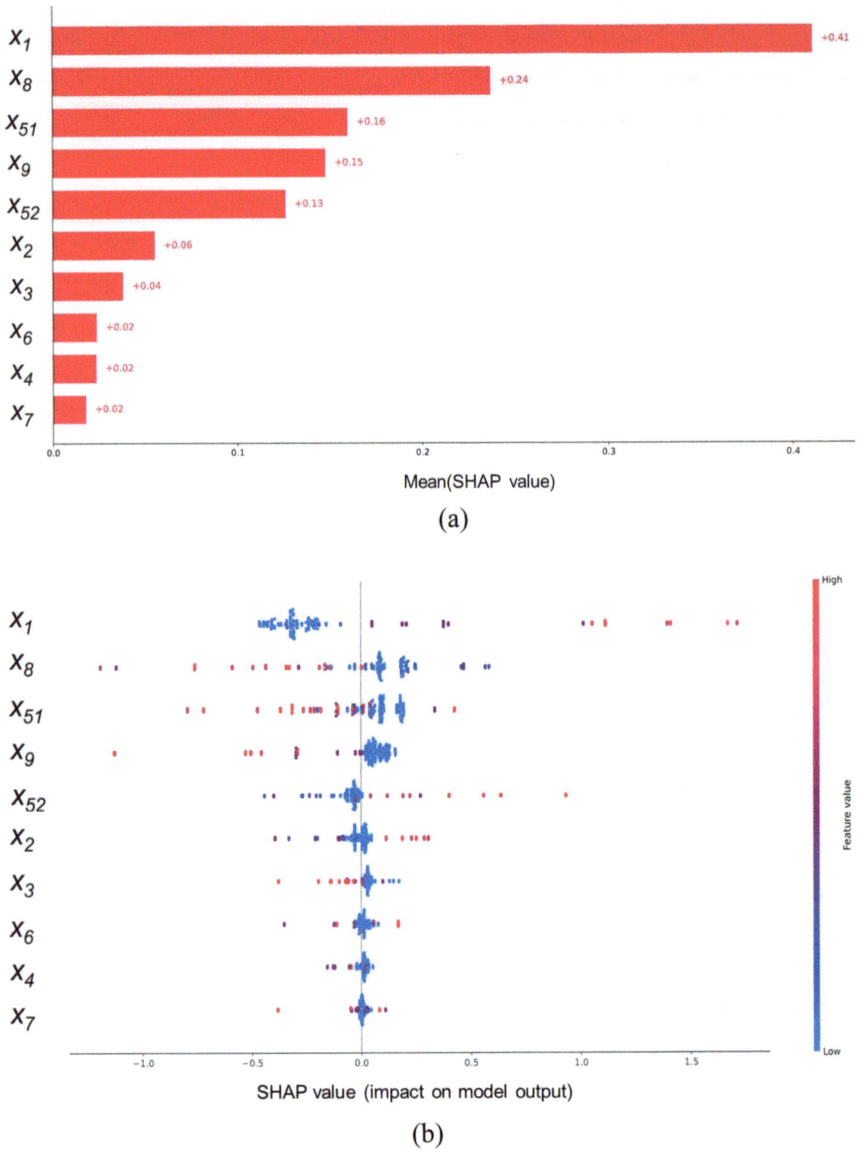

Fig. 7.9 Sensitivity analysis results for the nine factors: (**a**) Contributions of factors to the evacuation density; (**b**) Relationship between factors and the evacuation density

a common conflict between performance indicators: lower density may come at the cost of longer evacuation times. When renovation cost is considered alongside time and density, the results reveal a multidimensional trade-off. As shown in Fig. 7.10, some strategies that perform well in terms of time and density are excluded due to prohibitively high costs. For Scenarios II and IV, evacuation

Table 7.7 Parameters for conducting the NSGA-III algorithm

Code	Description	Value
n	Number of decision variables	9
M	Number of objective functions	3
Cn	Constraints	2
In	Initial population size	100
N	Number of generations	600
Cr	Cross-over rate	0.8
Mr	Mutation rate	0.2
c_4	Cost for the renovation of x_4	1000 $
c_{5_1}	Cost for the renovation of x_{5_1}	3000 $
c_{5_2}	Cost for the renovation of x_{5_2}	10,000 $
c_6	Cost for the renovation of x_6	3000 $
c_7	Cost for the renovation of x_7	2000 $
c_9	Cost for the renovation of x_9	100,000 $

Note: Costs for the influential factors are based on the local conditions

Table 7.8 Evacuation performance of representative strategies under Scenarios II to IV

	Before optimization		After optimization			
Scenario	Time (s)	Density (unit/m^2)	Time (s)	Density (unit/m^2)	Cost ($)	ID
I	247	3.4	–	–	–	–
II	416	5.4	337	3.4	310,000	19.9%
III	151	3.9	176	3.1	19,000	2.8%
IV	477	6.2	348	3.5	316,000	27.2%

time was reduced from 416 s to 337 s and from 477 s to 348 s, respectively, while density dropped from 5.4 persons/m^2 to 3.4 persons/m^2 and 6.2 persons/m^2 to 3.5 persons/m^2, respectively. These improvements, however, came with renovation costs of 310,000 and 316,000 USD, respectively. These findings underscore the importance of evaluating evacuation strategies across multiple performance dimensions.

2. Across the 20 test cases, the selected strategies achieved an average improvement of 7.5%. As shown in Fig. 7.12, all cases demonstrated measurable performance enhancements. On average, evacuation time decreased by 1.6%, while passenger density improved by 13.4%. Improvements were more pronounced in cases where baseline conditions exceeded safety thresholds, with maximum reductions in evacuation time and density reaching 13.5% and 31.5%, respectively. In several cases, evacuation time slightly increased to allow more effective reductions in density, as illustrated in Fig. 7.12b, reaffirming the trade-off between safety and efficiency metrics. Moreover, higher investment does not necessarily yield greater performance gains. For example, in Case 4, the highest renovation cost (467,000 USD) resulted in only a 6.5% improvement. These findings emphasize the importance of holistic evaluation, considering both safety

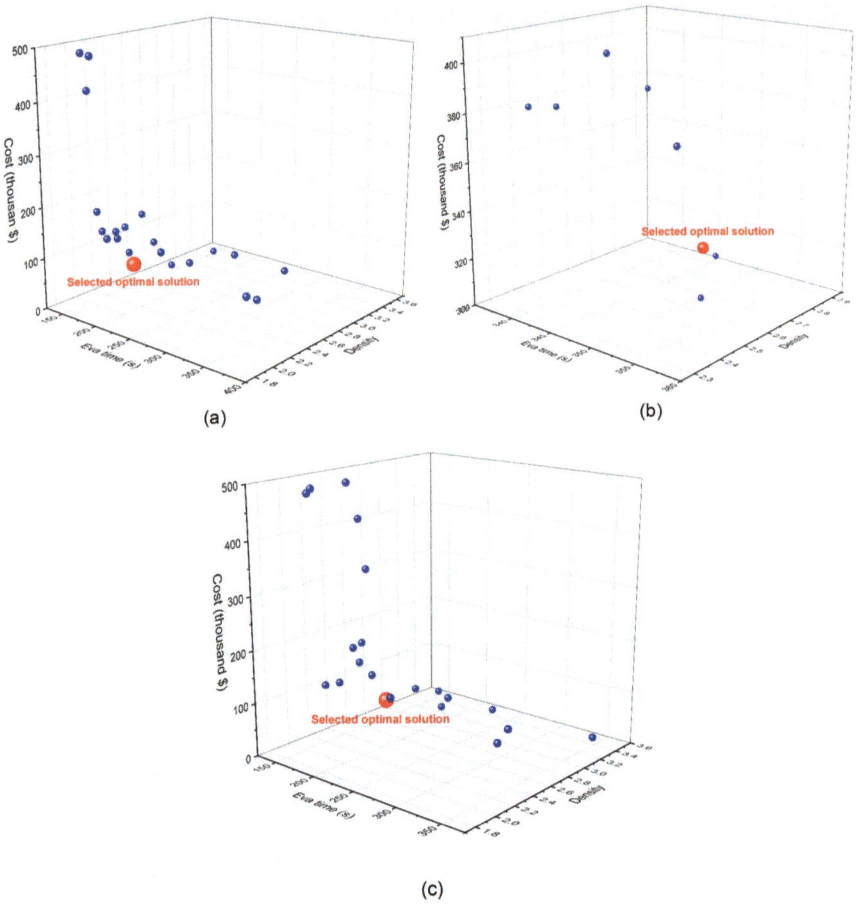

Fig. 7.10 Performance distribution of evacuation strategies under different scenarios: (**a**) Scenario II; (**b**) Scenario III; (**c**) Scenario IV

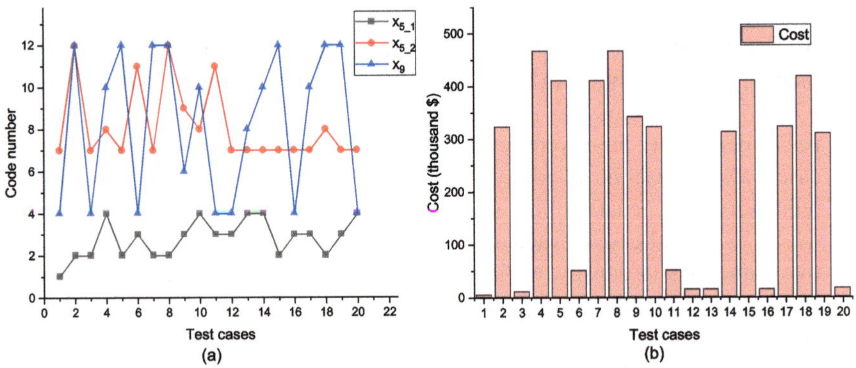

Fig. 7.11 Adjustments of influential factors and renovation costs in representative strategies: (**a**) Updated values of x_{5_1}, x_{5_2}, and x_9; (**b**) Values of the cost

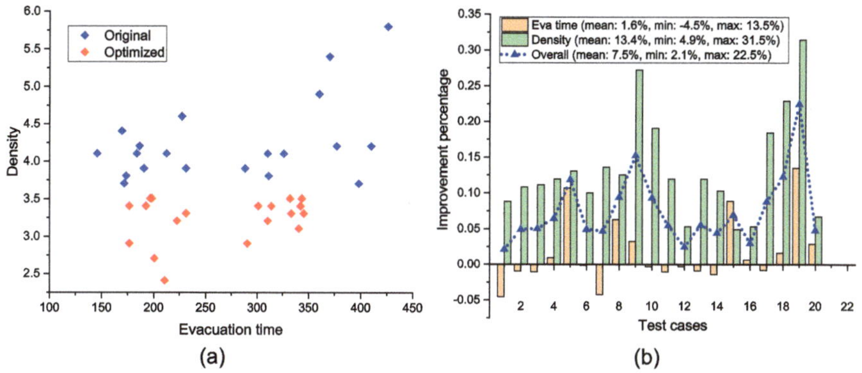

Fig. 7.12 Comparison of evacuation performance between baseline and selected strategies: (**a**) Evacuation time and density comparison; (**b**) Improvement degree

performance and resource constraints. Despite the inherent trade-offs, all selected strategies led to positive improvements, indicating that safe evacuation is achievable through context-specific strategy adjustments.

7.5 Merits Analysis

As demonstrated in previous sections, safe evacuation at MRT stations is achievable under regular passenger volumes. However, under high-volume conditions, evacuation becomes substantially more challenging. In such cases, tailored strategies have been shown to improve safety outcomes. Notably, earlier scenarios did not account for additional emergency constraints such as blocked exits or route disruptions. To assess the adaptability of evacuation strategies under more severe conditions, this section introduces a constrained scenario in which one major external exit is assumed to be closed. All four previously defined scenarios were re-evaluated using the same strategy identification framework.

The comparative results of evacuation strategies under constrained conditions are summarized in Table 7.9. For regular-volume Scenarios I and III, baseline evacuation times were 281 and 167 s, with corresponding densities of 4.4 and 5.1 persons/m^2. For high-volume Scenarios II and IV, times increased to 532 and 572 s, and densities reached 6.3 and 7.3 persons/m^2, respectively. After applying adjusted strategies to accommodate the blocked exit, evacuation time was reduced to 277 and 193 s in Scenarios I and III, while density dropped to 3.5 and 3.4 persons/m^2. For Scenarios II and IV, evacuation time improved to 427 and 472 s, with densities reduced to 4.1 and 4.3 persons/m^2. These findings suggest that even under constrained conditions, appropriately selected strategies can substantially enhance evacuation performance, particularly in high-risk, high-density scenarios.

Table 7.9 Evaluation results before and after optimization when one side of the station cannot be used

	Before optimization		After optimization			
Scenarios	Time (s)	Density (unit/m^2)	Time (s)	Density (unit/m^2)	Cost ($)	ID
I	281	4.4	277	3.5	14,000	7.7%
II	532	6.3	427	4.1	313,000	21.2%
III	167	5.1	193	3.4	16,000	10.8%
IV	572	7.3	472	4.3	419,000	23.4%

Table 7.10 Comparison of the improvement of the evacuation efficiency without and with constraints on the 20 test samples

		Improvement degree		
Condition	Objective	Mean	Min	Max
Without constraints (Case study)	Overall	7.5%	2.1%	22.5%
	Eva time	1.6%	−4.5%	13.5%
	Density	13.4%	4.9%	31.5%
With constraints (One exit closed)	Overall	10.8%	2.1%	23.7%
	Eva time	4.8%	−7.5%	17.7%
	Density	16.7%	4.4%	40.5%

1. The simulation results indicate that closing a major external exit substantially reduces the station's evacuation capacity, particularly under peak passenger volumes. As shown in Table 7.9, the baseline evacuation time and density under the constrained scenario reached 281 s and 4.4 persons/m^2 for Scenario I, and increased to 532 s and 6.3 persons/m^2 for Scenario II. Similar trends were observed in Scenario III (167 s, 5.1 persons/m^2) and Scenario IV (572 s, 7.3 persons/m^2). These results suggest that even moderate passenger volumes can result in excessive crowding under constrained conditions, while high-volume scenarios exceed safety thresholds for both time and density. After applying revised strategies, Scenarios I and III saw reduced densities (3.5 and 3.4 persons/m^2) and acceptable evacuation times. However, in Scenarios II and IV, despite improvements (evacuation times of 427 and 472 s and densities of 4.1 and 4.3 persons/m^2), safety thresholds were not met. This highlights that under extreme congestion and space limitations, internal adjustments alone may be insufficient. These findings align with expert opinions suggesting structural constraints limit evacuation efficiency during peak periods. Broader interventions, such as constructing a nearby relief station or adding a bus interchange to redistribute passenger flow, may be necessary to enhance the MRT system's resilience.
2. The comparative analysis further indicates that average performance improvements across the 20 test cases are more pronounced under constrained conditions. Although full compliance with safety thresholds is not achieved in all cases, the effectiveness of the proposed strategy evaluation framework remains evident. As shown in Table 7.10 and Fig. 7.13, the average improvement degree increased from 7.5% to 10.8% when one major exit was closed. Specifically, average improvements in evacuation time and passenger density rose from 1.6%

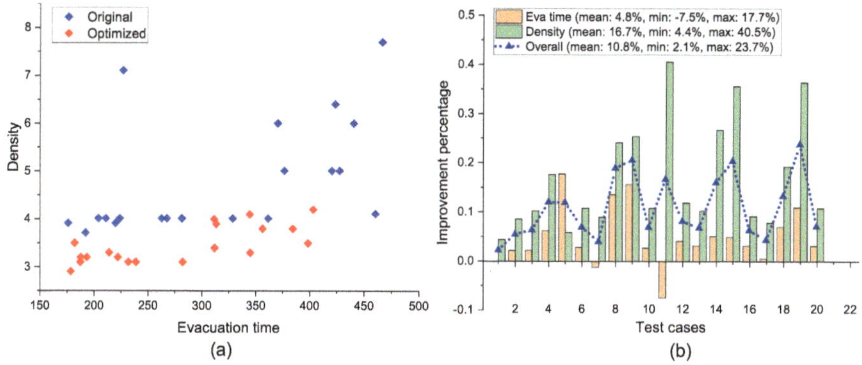

Fig. 7.13 Comparison of evacuation time and density between the original and optimized results and the improvement degree on the 20 test samples: (**a**) Evacuation time and density comparison; (**b**) Improvement degree

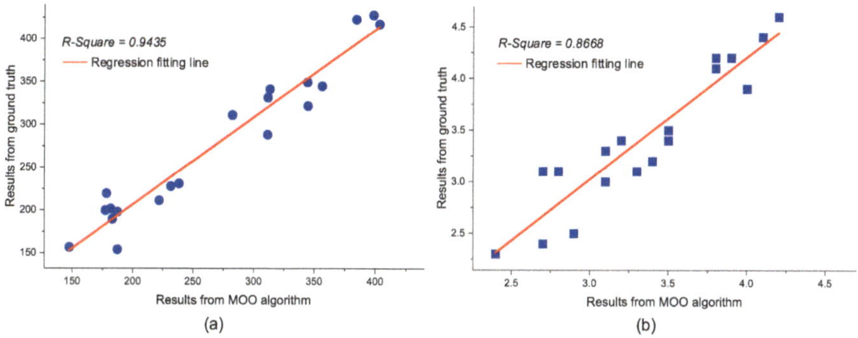

Fig. 7.14 Scatterplot of optimized results and simulated results for objectives of the evacuation time and density on the 20 test samples: (**a**) Evacuation time; (**b**) Evacuation density

to 4.8% and from 13.4% to 16.7%, respectively. The maximum improvement in a single case also increased from 22.5% (unconstrained) to 23.7% (constrained). These results demonstrate that, even under more severe emergency conditions, appropriately selected evacuation strategies can lead to meaningful performance gains. While full compliance may remain elusive, notable reductions in evacuation time and density help mitigate risk and support safer evacuation outcomes.

3. The reliability of the strategy evaluation framework was further validated through simulation-based testing. The selected evacuation strategies for all 20 test cases were reapplied to the original AnyLogic simulation models, and the resulting evacuation time and density were compared with predictions from the RF-based meta-model. As shown in Fig. 7.14, a high degree of consistency was observed between simulated and predicted values. The coefficients of determination (R^2) for evacuation time and density reached 0.9435 and 0.8668, respectively, indicating strong predictive accuracy. These findings confirm that the surrogate model reliably estimates evacuation performance across diverse

scenarios. Overall, the close agreement between simulated and predicted outcomes reinforces the robustness of the proposed strategy evaluation approach.

7.6 Chapter Summary

This chapter proposed a hybrid strategy evaluation framework. A 3D model of a metro station was developed to simulate evacuation processes, and a detailed analysis was conducted to identify the key influential factors and evaluation indicators related to evacuation efficiency and safety. The RF-based meta-model was used to approximate the relationships between input factors and evacuation outcomes, while NSGA-III helped generate a diverse set of candidate strategies that focus on comparing alternative evacuation strategies across multiple scenarios, with the aim of identifying representative and effective strategies for different levels of system vulnerability. This approach supports more informed and adaptive evacuation planning at metro stations.

A case study based on a real MRT station in Singapore was conducted to validate the applicability of the proposed framework. Key findings include: (1) Safe evacuation can generally be achieved under regular passenger volumes; however, under high passenger load and elevated panic levels, the density may rise to 6.2 persons/m^2, indicating a potentially hazardous situation. (2) Passenger volume has the greatest impact on both evacuation time and density, whereas fare gates, obstacles, and signage exert the least influence across both indicators. (3) The proposed evaluation framework effectively identifies strategies that enhance evacuation performance, achieving up to a 27.2% improvement under high passenger volume scenarios.

Overall, by combining simulation, meta-modeling, and scenario-based strategy generation, the proposed approach provides a robust and flexible tool for evaluating and enhancing metro station passenger evacuation planning. It contributes to improved safety outcomes by facilitating data-driven decision-making tailored to specific operational constraints and passenger conditions.

References

1. He R, Tiong RLK, Yuan Y, Zhang L (2024) Enhancing resilience of urban underground space under floods: current status and future directions. Tunn Undergr Space Technol 147:105674. https://doi.org/10.1016/j.tust.2024.105674
2. Zhang L, Liu M, Wu X, AbouRizk SM (2016) Simulation-based route planning for pedestrian evacuation in metro stations: a case study. Autom Constr 71:430–442. https://doi.org/10.1016/j.autcon.2016.08.031
3. Lu Y, Deng Y, Sun S (2023) Design for safety in metro station evacuation: a study of fire emergency evacuation behavior and simulation. Eng Constr Archit Manag 31:3479–3507. https://doi.org/10.1108/ECAM-12-2022-1169

4. Zhang Z, Yao X, Xing Z, Zhou X (2024) Simulation on passenger evacuation of metro train fire in the tunnel. Chaos, Solitons Fractals 187:115429. https://doi.org/10.1016/j.chaos.2024.115429
5. Yang X, Zhang R, Pan F, Yang Y, Li Y, Yang X (2022) Stochastic user equilibrium path planning for crowd evacuation at subway station based on social force model. Phys A Stat Mech Appl 594:127033. https://doi.org/10.1016/j.physa.2022.127033
6. Yang X, Yang Y, Qu D, Chen X, Li Y (2023) Multi-objective optimization of evacuation route for heterogeneous passengers in the Metro Station considering node efficiency. IEEE Trans Intell Transp Syst 24:12448–12461. https://doi.org/10.1109/TITS.2023.3292912
7. Zhang J, Huang D, You Q, Kang J, Shi M, Lang X (2023) Evaluation of emergency evacuation capacity of urban metro stations based on combined weights and TOPSIS-GRA method in intuitive fuzzy environment. Int J Disaster Risk Reduct 95:103864. https://doi.org/10.1016/j.ijdrr.2023.103864
8. Mandal T, Ramachandra Rao K, Tiwari G (2023) Evacuation of metro stations: a review. Tunn Undergr Space Technol 140:105304. https://doi.org/10.1016/j.tust.2023.105304
9. Xu H, Tian C, Li Y (2020) Emergency evacuation simulation and optimization for a complex rail transit station: a perspective of promoting transportation safety. J Adv Transp 2020:8791503. https://doi.org/10.1155/2020/8791503
10. Qin J, Liu C, Huang Q (2020) Simulation on fire emergency evacuation in special subway station based on pathfinder. Case Stud Therm Eng 21:100677. https://doi.org/10.1016/j.csite.2020.100677
11. Siyam N, Alqaryouti O, Abdallah S (2020) Research issues in agent-based simulation for pedestrians evacuation. IEEE Access 8:134435–134455. https://doi.org/10.1109/ACCESS.2019.2956880
12. Yang X, Zhou M, Dong H (2024) Passenger emergency evacuation in Subway Station systems: a bibliometric analysis and systematic review. IEEE Trans Intell Transp Syst 25:12912–12929. https://doi.org/10.1109/TITS.2024.3393016
13. Guo K, Zhang L, Wu M (2023) Simulation-based multi-objective optimization towards proactive evacuation planning at metro stations. Eng Appl Artif Intell 120:105858. https://doi.org/10.1016/j.engappai.2023.105858

Part III
System Resilience Enhancement

Chapter 8
Facility Operation Planning

8.1 Introduction

Flood hazard poses a significant threat to the urban transit systems (UTS). Since metro stations and tunnels are generally located underground in low-lying areas, floodwater is prone to intrude but difficult to drain. Additionally, it is difficult to evacuate passengers efficiently from a metro station under floods due to its relatively enclosed spaces, which poses a significant threat to human life safety [1]. Therefore, it has become particularly crucial to implement effective flood management for UTS.

Typically, flood management for UTS can be conducted in two aspects: pre-disaster preparations and post-disaster responses. Pre-disaster preparations include flood risk prediction, distribution of flood prevention materials, station closure decision-making, etc., while post-disaster responses usually involve passenger evacuation, restoration of flooded stations and section tunnels, etc. Currently, most research in this area focuses on post-disaster responses. Specifically, for passenger evacuation, a number of numerical simulations and physical experiments have been carried out [2–4]; regarding the post-flood restoration of metro systems, scholars have proposed different strategies to optimize the system resilience [5–7]. In terms of pre-disaster preparation, however, there is still limited research at present.

In fact, compared with the post-disaster responses, pre-disaster preparations are, if not more, equally important, as proper preparations can fundamentally prevent flooding from occurring or minimize its impacts. Targeting the metro flood issue, preparatory measures involve closing high-risk stations that may be flooded in advance, or deploying flood prevention materials to protect those high-risk stations so that they can still be operational during the flood. In practice, although closing all high-risk stations is a straightforward solution to mitigate the flood impacts, it can negatively affect people's travel. Conversely, protecting all those stations to ensure smooth public transportation comes at a significant cost to flood control resources.

L. Zhang et al., *Resilience in Urban Transit Systems*, Advances in 21st Century Human Settlements, https://doi.org/10.1007/978-981-95-4322-9_8

This brings about a critical decision-making question: after we identify the high-risk stations before a flood, given limited total resources, which stations should be allocated resources for protection, while the others should be closed? Up to now, no scholars have conducted profound studies on this question. Hence, we will try to fill the research gap in this chapter.

In this chapter, we propose a novel optimization framework based on network topology to optimize the closure-protection scheme for high-risk metro stations before a flood. First, we develop a time-weighted metro network model that considers line transfer. Secondly, we study the metro network performance under station closure due to floods. To evaluate the metro network performance under station closure, we construct the Integrated Network Performance (INP) based on both network efficiency and passenger utility. Lastly, a cost-performance-based greedy algorithm (CPGA) is adopted to solve the optimal closure-protection scheme, which can maximize the INP of the metro network under station closure. A case study on the Shanghai metro system is carried out to verify the effectiveness of the proposed framework.

8.2 Flood Management in UTS

With metro flood events becoming increasingly frequent and severe, a growing body of literature has investigated the flood management in UTS. Prior research in this domain mainly includes flood risk assessment, passenger evacuation, and post-flood system restoration. The literature on the above three aspects is briefly reviewed below.

1. **Flood risk assessment.** Flood risk assessment for UTS aims to identify areas that are vulnerable to floods within the system. Currently, research in this field is mainly based on numerical simulations or MCDA. Numerical-simulation-based assessment conducts flood process modelling in a metro system, based on which high-risk areas in the system can be identified. For instance, Lin, Hu, Zhou, Zhong, Zhu, Shi and Lin [8] employed the Volume of Fluid (VOF) method to simulate the flood intrusion process in an interchange subway station, finding that the property development floor was the highest-risk area. MCDA-based assessment establishes an index system for flood risk assessment of the metro system, and then applies MCDA methods to integrate all factors and generate an overall flood risk level. For example, Lyu, Shen, Zhou and Zhou [9] adopted the Interval Fuzzy Analytical Hierarchical Process incorporated with fuzzy clustering analysis to assess the flood risk of the Shanghai metro system, showing that more than 30% of metro lines in Shanghai were at a high risk of flooding.
2. **Passenger evacuation.** Timely passenger evacuation is a crucial task in the post-flood response to metro floods. As the foundation of flood evacuation research, Japanese scholars, represented by Taisuke Ishigaki, have conducted a series of full-scale physical model experiments, yielding a series of critical criteria for

safe evacuation that are widely adopted by the academic community [10]. Based on these critical criteria, researchers have further simulated the evacuation dynamics during metro floods and optimized evacuation plans. For instance, Yang, Dai, Li and Yang [11] used Pathfinder software for multi-agent modeling and proposed a two-stage method for passenger evacuation path optimization in a metro station under floods, which could simultaneously minimize total evacuation time, risk, and congestion. These studies provide constructive guidance for evacuation strategies during metro floods.

3. **Post-flood system restoration.** In this area, current research mainly investigates how restoration strategies can enhance the metro network resilience under floods. For example, Yadav, Chatterjee and Ganguly [5] compared three centrality-based recovery sequences and one optimization-based recovery sequence for disrupted metro stations. Bi, Schooling and MacAskill [12] suggested a post-flood restoration sequence for metro stations based on node importance, which was quantified by the number of satisfied passenger journeys and global efficiency.

In the aforementioned research areas, passenger evacuation and system restoration both fall under the category of post-flood responses. However, as mentioned earlier, pre-flood preparations are arguably more crucial than post-flood responses. Yet, research on pre-flood preparations remains limited. For station closure and protection, which are critical preparatory measures, no scholars have conducted modeling and optimization studies. Hence, current practices are mainly based on the subjective experience of metro managers, leading to decision biases. Therefore, this chapter aims to establish a model to quantitatively describe the closure-protection problem of metro stations and optimize closure-protection schemes under different flood scenarios. By doing so, we expect to address the research gap in the pre-flood preparations of UTS and thus contribute to improved metro flood management.

8.3 Facility Operation Planning Framework Under Flood

Figure 8.1 shows the framework proposed in this chapter for optimizing the closure-protection scheme. First, we develop a time-weighted metro network model that considers line transfer. Secondly, we study the metro network performance under station closure due to floods. Lastly, we formulate the optimization problem and solve it by using a cost-performance-based greedy algorithm (CPGA). The acquired optimal closure-protection scheme maximizes the INP of the metro network under station closure. The detailed method of each step in the framework is elaborated in the following sections.

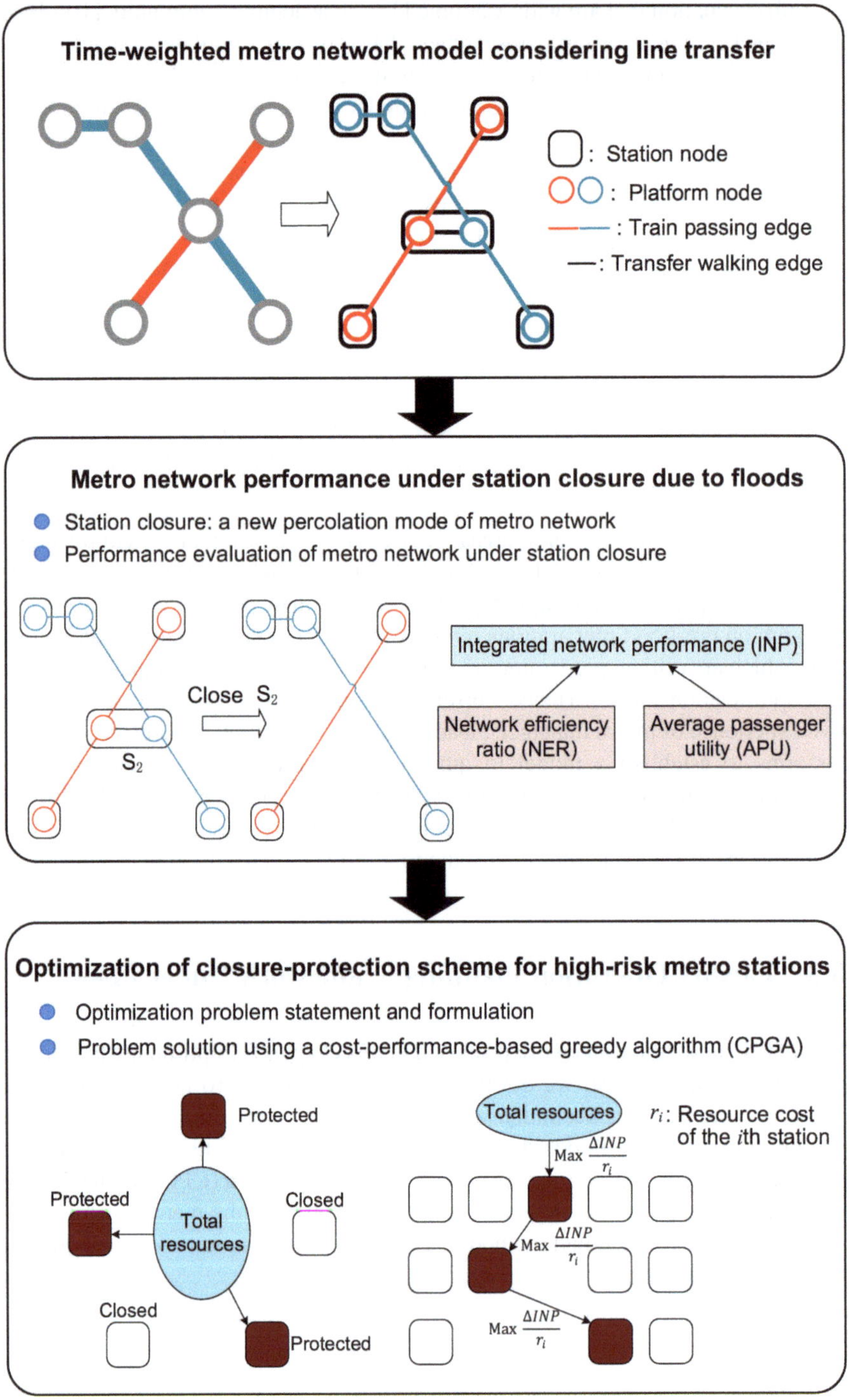

Fig. 8.1 General framework for optimizing the closure-protection scheme

8.3.1 *Time-Weighted Metro Network Model Considering Line Transfer*

Generally speaking, a transportation network can be modeled as a network containing nodes, edges, and weights. As shown in Fig. 8.2, this chapter establishes a time-weighted metro network $N = (S, P, E, T)$ that considers passengers' transfer walking time. In the proposed metro network model, each metro station is modeled as a station node S_i ($i = 1, 2, \ldots, N_S$, N_S is the number of stations in the metro network), which is represented by the rounded rectangle in Fig. 8.2. In general, metro stations can be classified into:

1. Transfer stations where multiple metro lines intersect and passengers can transfer from one line to another. Hence, we consider each transfer station node to contain multiple platform nodes, each of which represents a platform of a specific line through the station.
2. Non-transfer stations through which only one metro line passes. Each non-transfer station node only contains one platform node.

From the above descriptions, we know that defining a platform node necessitates both the metro station and the metro line to which it belongs. Denoting i and j as the number of stations and lines of the platform node, respectively, we can define the platform node as a subscript form $P_{i,j}$.

Between every two adjacent platform nodes, we establish a weighted undirected edge representing the path connecting the two platforms, and the weight equals the time that passengers spend on the path. In the proposed model, we have the following two types of edges:

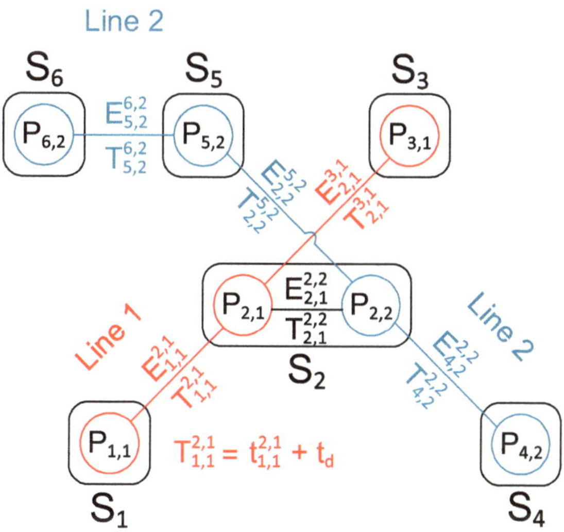

Fig. 8.2 Schematic diagram of the time-weighted metro network model

1. Train passing edge. This type of edge is the edge through which the train passes. It connects two platforms of two adjacent stations on the same line. For the edge connecting $P_{i_1,j}$ and $P_{i_2,j}$ on Line j, we herein denote it as $E_{i_1,j}^{i_2,j}$. The weight of the edge $E_{i_1,j}^{i_2,j}$, which is denoted as $T_{i_1,j}^{i_2,j}$, is calculated as Eq. (8.1).

$$T_{i_1,j}^{i_2,j} = t_{i_1,j}^{i_2,j} + t_d \tag{8.1}$$

   in which $t_{i_1,j}^{i_2,j}$ is the train's travel time between Stations S_{i_1} and S_{i_2} on Line j; t_d is the train's dwell time at one station, which is a constant herein.
2. Transfer walking edge. A transfer walking edge connects two platforms of different lines within one transfer station. Similarly, this type of edge can be denoted by E_{i,j_1}^{i,j_2}, with its weight expressed as T_{i,j_1}^{i,j_2} that represents the transfer walking time between Lines j_1 and j_2 within the transfer station S_i.

8.3.2 *Metro Network Performance Under Station Closure*

After establishing the time-weighted metro network model, in this section, we investigate the performance of metro network under station closure due to floods. First of all, to assess the resistance of a metro station to floods, we propose FRI that fuses the information from seven influential factors, including elevation, slope, etc., using an ER-based optimality approach. Correspondingly, we define the flood intensity index to measure the severity of a flood event, which can be estimated by hydro-meteorological experts before the flood occurs. It is assumed that all the stations with FRIs lower than the flood intensity index will be at high inundation risk; thus, they should either be closed or protected with certain flood control resources. The closure of some or all the high-risk stations will change the topology of the metro network; hence, we secondly propose a new percolation mode to simulate such network transformation caused by station closure. Lastly, to evaluate the metro network performance under station closure, we propose INP, which combines both network efficiency and passenger utility. Detailed methodologies involved are illustrated below.

8.3.2.1 Assessment of Metro Station's Resistance to Floods

In this chapter, the flood resistance assessment is conducted for each station in the metro network. It is considered that a metro station's resistance to floods is influenced by its seven factors [9, 13, 14]: elevation, slope, number of exits, population density, drainage capacity, road density within a radius of 500 m, and green area ratio within a radius of 500 m. Subsequently, an ER-based optimality approach proposed in Chap. 5 is utilized to fuse information from the seven factors and generate an FRI value to represent the station's resistance to floods.

Corresponding to FRI, we also define the flood intensity index to describe the severity of a rainfall-induced flood in terms of its destructive impact on metro stations. Hydrometeorological experts can forecast an upcoming flood event's flood intensity index. It is assumed that the spatial distribution of rainstorms is uniform; thus, the intensity index of one flood event is identical for all metro stations. When the FRI of a metro station is lower than the flood intensity index, we consider this station as a high-risk station; otherwise, it is classified as low-risk. This is an important premise of the following methodologies.

8.3.2.2 Station Closure: A New Percolation Mode for Metro Network

As mentioned, those high-risk stations will be flooded if we do not carry out any flood prevention measures. To avoid such a situation, in the practical operation of metro systems, metro managers generally deploy flood control resources to protect the high-risk stations (allowing them to continue operating even during floods) or simply close high-risk stations. When a high-risk station is closed, the trains will head directly from the previous station to the next station without stopping at the closed station. Correspondingly, passengers are unable to get on/off or transfer at this station. At the network level, station closure can be schematically shown in Fig. 8.3. As illustrated, when Station S_2 is closed, it is removed from the network; however, its adjacent stations on the same line are still connected in pairs. Because S_2 serves as a transfer point between Lines 1 and 2, with its closure, the two lines are no longer connected. This spawns a new percolation mode that is only applicable to metro networks. As more stations are closed, the metro network's topological performances, such as efficiency, giant cluster size, and average path length, will gradually degrade. Nevertheless, since station closure is less likely to cause the metro network to break into separate clusters, such performance degradation will occur at a slower rate compared to that caused by station disruption.

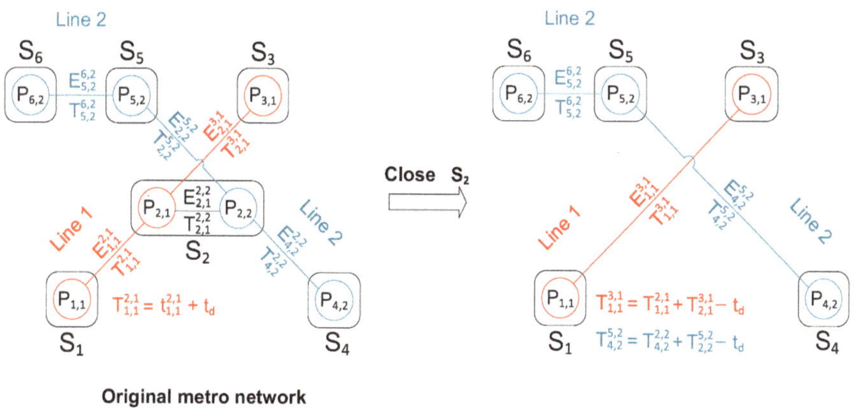

Fig. 8.3 Station closure in the time-weighted metro network

8.3.2.3 Performance Evaluation of Metro Network Under Station Closure

To comprehensively evaluate the performance of a metro network under station closure due to floods, in this section, we construct an integrated metric considering both network efficiency and passenger utility. Firstly, as verified in many previous studies [15, 16], network efficiency can be adopted to measure the global transportation efficiency of a metro network based on the shortest path between every two stations, written as Eq. (8.2):

$$\text{Eff} = \frac{1}{N_S\left(N_S - 1\right)} \sum_{i_1=1}^{N_S} \sum_{i_2=1, i_2 \neq i_1}^{N_S} \left(\frac{1}{\text{SP}_{i_1}^{i_2}} \right) \tag{8.2}$$

where N_S is the number of station nodes in the metro network under normal conditions; $\text{SP}_{i_1}^{i_2}$ is the weighted shortest path length from Stations S_{i_1} to S_{i_2}. As mentioned in Sect. 8.3.1, the edge weight in the metro network model consists of the train's travel time, dwell time, and the passenger's transfer walking time. Therefore, here $\text{SP}_{i_1}^{i_2}$ actually represents the shortest time from Stations S_{i_1} to S_{i_2}. Because one station may contain multiple platforms; hence, $\text{SP}_{i_1}^{i_2}$ is the minimum of the shortest time from one platform in S_{i_1} to one platform in S_{i_2}, written as Eq. (8.3):

$$\text{SP}_{i_1}^{i_2} = \min\left\{\text{SP}_{i_1,j_1}^{i_2,j_2}, j_1 \in \mathcal{L}_{i_1}, j_2 \in \mathcal{L}_{i_2}\right\} \tag{8.3}$$

where $\text{SP}_{i_1,j_1}^{i_2,j_2}$ is the shortest time from Platforms P_{i_1,j_1} to P_{i_2,j_2}; $\mathcal{L}_{i_1}$ and $\mathcal{L}_{i_2}$ are the sets of metro lines where S_{i_1} and S_{i_2} are located, respectively. Applying the Dijkstra algorithm [17], we can easily acquire the shortest path from one platform node to another.

Using Eqs. (8.2) and (8.3), we can calculate the efficiency of the metro network under normal conditions, i.e., Eff in Eq. (8.2). Under station closure, however, the network topology will change, which can further alter the shortest time between some stations and thus influence the network efficiency. To compare the network efficiency under station closure with that under normal conditions, we develop the Network Efficiency Ratio (NER), which is expressed by Eq. (8.4). NER ranges from 0 to 1, with higher values closer to 1 indicating better efficiency of the metro network.

$$\text{NER} = \frac{\text{Eff}'}{\text{Eff}} \tag{8.4}$$

where Eff' is the network efficiency under station closure.

As a useful evaluation metric, NER assesses the metro network performance from the perspective of network topology. However, it does not consider the passenger flow information and passengers' degree of satisfaction under station closure. From the perspective of passengers, the negative impacts of station closure are reflected in two aspects. On the one hand, those passengers who were supposed to enter or exit the closed station are now unable to complete their established travel route. On the other hand, the closure of a critical transfer station may result in a lengthier shortest path between some stations, thereby necessitating some passengers to endure extended travel time. The two impacts can be uniformly considered using the time delay, i.e., the difference in the shortest time before and after station closure:

$$\Delta SP_{i_1}^{i_2} = \mathrm{SP}_{i_1}^{\prime i_2} - \mathrm{SP}_{i_1}^{i_2} \tag{8.5}$$

where $\varnothing SP_{i_1}^{i_2}$ is the time delay from S_{i_1} to S_{i_2} under station closure; $\mathrm{SP}_{i_1}^{i_2}$ and $\mathrm{SP}_{i_1}^{\prime i_2}$ are the shortest time from S_{i_1} to S_{i_2} before and after the station closure, respectively.

Then, we consider that the passenger utility, i.e., the passengers' degree of satisfaction, is linearly negatively correlated with the time delay. When the time delay exceeds a certain limit, the passenger utility will be 0. We can thus construct the following relationship between the passenger utility and the time delay:

$$U_{i_1}^{i_2} = \begin{cases} -\dfrac{1}{t_e}\Delta SP_{i_1}^{i_2} + 1, \ 0 \le \Delta SP_{i_1}^{i_2} \le t_e \\ 0, \ \Delta SP_{i_1}^{i_2} > t_e \end{cases} \tag{8.6}$$

where $U_{i_1}^{i_2}$ is the utility of passengers traveling from S_{i_1} to S_{i_2}; t_e is a parameter. According to Eq. (8.6), when $\Delta SP_{i_1}^{i_2} = 0, U_{i_1}^{i_2} = 1$; when $\Delta SP_{i_1}^{i_2} \ge t_e, U_{i_1}^{i_2} = 0$. Here, we set t_e to 60 min in our following calculation.

Subsequently, we can utilize the automatic fare collection (AFC) records of the metro system to count the daily average passenger flow from one station to another, i.e., the daily average passenger flow of each origin-destination (O-D) pair. Then, we propose the Average Passenger Utility (APU) to calculate the average of all passengers' utility values, written as Eq. (8.7):

$$\mathrm{APU} = \frac{\sum_{i_1=1}^{N_S}\sum_{i_2=1, i_2 \ne i_1}^{N_S}\left(V_{i_1}^{i_2} \cdot U_{i_1}^{i_2}\right)}{\sum_{i_1=1}^{N_S}\sum_{i_2=1, i_2 \ne i_1}^{N_S} V_{i_1}^{i_2}} \tag{8.7}$$

where $V_{i_1}^{i_2}$ is the daily average passenger volume from S_{i_1} to S_{i_2}. As the range of $U_{i_1}^{i_2}$ is [0,1], APU also ranges from 0 to 1. A higher APU closer to 1 represents a higher level of passenger satisfaction.

Combining NER and APU, we construct the Integrated Network Performance (INP), written as Eq. (8.8):

$$\mathrm{INP} = \mu \cdot \mathrm{NER} + (1-\mu) \cdot \mathrm{APU} \tag{8.8}$$

where μ $(0 \le \mu \le 1)$ is the weight of NER. Different values of μ indicate different preferences for NER. In this chapter, we suggest $\mu = 0.5$ to treat both NER and APU equally.

Based on the above deduction of INP, it is easy to know that the value range of INP is also between 0 and 1, and a higher INP closer to 1 indicates better overall performance of the metro network. Besides, INP can be interpreted as a percentage representing the performance of a network relative to a normal network. To specify, if the INP of a metro network is 0.8, we can consider that this metro network's performance is 80% of the normal network. From this perspective, the change in INP can be intuitively understood as the variation of network performance expressed in percent points relative to the normal network. Thus, practitioners can use INP to assess the performance of a metro network in practice.

8.3.3 *Optimization of Closure-Protection Scheme for High-Risk Metro Stations*

8.3.3.1 Optimization Problem Formulation

Before a storm flood comes, we consider that the flood intensity index can be estimated by hydrometeorological experts. With each station's FRI determined using the ER-based optimality approach, we can identify a set of high-risk stations whose FRI values are lower than the flood intensity index. These high-risk stations are very likely to be flooded if no precaution is implemented; hence, each of the high-risk stations should be either closed or protected with certain flood control resources (e.g., sandbags, water barriers, and water pumps) so that they can still be operational during the flood. The ideal scenario would be to protect all high-risk stations to ensure smooth transportation for the public. However, due to limited resources, it may not be feasible to provide protection for all high-risk stations. In Sect. 8.3.2.3, we propose the INP as an evaluation metric for metro network performance. Therefore, we need to reasonably select the stations to protect so that the network's INP can be maximized.

Here, we assume that the resource quantities required to protect one station are proportional to (1) the flood intensity index and (2) the degree centrality of the station. Firstly, a higher flood intensity index indicates a more severe flood event and thus greater difficulty in station protection. Hence, one metro station requires more protective resources under a higher-intensity flood. Secondly, with the same flood intensity index, a large metro station requires more resources than a small one. In this chapter, the size of a metro station is measured by the station's degree centrality, which is defined as the total number of edges owned by all platform nodes under this station (transfer walking edges are not incorporated). Therefore, we consider that the required resources of one station are proportional to its degree centrality. By consulting experienced experts in metro emergency management, we learn that when the flood intensity index is 0.05, protecting a 1-degree-centrality station requires about 1 high-power water pump, 50 water-swelling sandbags (25 kg), 10 water barriers, 5 additional emergency personnel, and some other small tools, the sum of which is defined as one unit of resources in this chapter. Hence, based on the above premises, we have Eq. (8.9) below to calculate the number of resource units required by the station S_i:

$$r_i = \frac{F}{0.05} \times d_i \quad (i = 1,2,\ldots,N_S) \tag{8.9}$$

where r_i is the number of resource units required by Station S_i; F represents the flood intensity index; d_i is the degree centrality of S_i; N_S is the number of stations in the metro network.

To sum up, the optimization problem can be defined as: for a metro network $N = (S, P, E, T)$, given a high-risk station set $\mathcal{S}^h$, resource quantities r_i required by each high-risk station $S_i \in \mathcal{S}^h$, and the total resources r_T, determine a protected high-risk station set $\mathcal{S}^p$ and a closed high-risk station set $\mathcal{S}^c$, so that the INP of the metro network under station closure is maximized.

Based on the derivation process of INP in Sect. 8.3.2.3, we can also mathematically formulate this optimization problem as follows:

$$\max \ \mathrm{INP}\left(\mathcal{S}^p,\mathcal{S}^c\right) = \mu \bullet \mathrm{NER}\left(\mathcal{S}^p,\mathcal{S}^c\right) + (1-\mu) \bullet \mathrm{APU}\left(\mathcal{S}^p,\mathcal{S}^c\right)$$

$$= \mu \cdot \frac{\sum_{i_1=1}^{N_S}\sum_{i_2=1, i_2 \neq i_1}^{N_S} \frac{1}{\mathrm{SP}'^{i_2}_{i_1}\left(\mathcal{S}^p,\mathcal{S}^c\right)}}{\sum_{i_1=1}^{N_S}\sum_{i_2=1, i_2 \neq i_1}^{N_S} \frac{1}{\mathrm{SP}^{i_2}_{i_1}}} + (1-\mu)\frac{\sum_{i_1=1}^{N_S}\sum_{i_2=1, i_2 \neq i_1}^{N_S} \left[V^{i_2}_{i_1} \cdot U^{i_2}_{i_1}\left(\mathcal{S}^p,\mathcal{S}^c\right)\right]}{\sum_{i_1=1}^{N_S}\sum_{i_2=1, i_2 \neq i_1}^{N_S} V^{i_2}_{i_1}}$$

$$\text{s.t.} \ \mathcal{S}^p \cup \mathcal{S}^c = \mathcal{S}^h$$

$$\sum_{S_i \in \mathcal{S}^p} r_i \leq r_T \tag{8.10}$$

where $\mathrm{SP}'^{i_2}_{i_1}\left(\mathcal{S}^p,\mathcal{S}^c\right)$ and $U^{i_2}_{i_1}\left(\mathcal{S}^p,\mathcal{S}^c\right)$ are the shortest time and the passenger utility from S_{i_1} to S_{i_2} after we protect the stations in $\mathcal{S}^p$ while closing the stations in $\mathcal{S}^c$; $SP^{i_2}_{i_1}$ is the shortest time from S_{i_1} to S_{i_2} under normal conditions; $V^{i_2}_{i_1}$ is the daily average passenger volume from S_{i_1} to S_{i_2}.

8.3.3.2 Problem Solution via Cost-Performance-Based Greedy Algorithm (CPGA)

Essentially, the optimization problem presented in Sect. 8.3.3.1 is a nonlinear variant of the 0-1 knapsack problem. The 0-1 knapsack problem is a combinatorial optimization problem where a set of items with associated weights and values must be selected to fill a knapsack with a maximum weight capacity, while maximizing the total value of the selected items. In our research, these items correspond to the high-risk stations; an item's weight and value represent the required resources for protecting a station and its contribution to the metro network's INP, respectively, while the maximum weight capacity corresponds to the total available resources.

In this chapter, we adopt a cost-performance-based greedy algorithm (CPGA) as the solution algorithm, with the pseudocode presented in Algorithm 8.1. Its basic idea is to allocate resources to the stations that can generate the maximum INP gain using each unit of resource. Initially, we assume that all high-risk stations are closed, and the INP of the metro network is at its minimum. Then, we try protecting each closed station, calculating the ratio of the INP gain from protecting one station to the required resources of that station. This ratio represents the cost performance of protecting that station. After iterating over all closed stations, we select the station with the highest cost performance to protect and update the remaining available resources. Repeat the above steps until the remaining resources cannot fulfill the requirement of any closed station. Using the CPGA, we can conveniently obtain an approximate optimal closure-protection scheme for an optimization case.

Algorithm 8.1 CPGA

Input: High-risk station set $\mathcal{S}^h$; Total resources r_T
Required resources r_i of each high-risk station $S_i \in \mathcal{S}^h$;
Output: Protected station set $\mathcal{S}^p$; Closed station set $\mathcal{S}^c$

$\mathcal{S}^p \leftarrow \varnothing$
$\mathcal{S}^c \leftarrow \mathcal{S}^h$
for $k \leftarrow 1$ **to** $\left|\mathcal{S}^h\right|$ **do**
  $\mathrm{INP}_0 \leftarrow \mathrm{INP}\left(\mathcal{S}^p, \mathcal{S}^c\right)$
  $\Delta \leftarrow 0$
  for $S_i \in \mathcal{S}^c$ **do**
    if $\sum_{S_j \in \mathcal{S}^p \cup \{S_i\}} r_j > r_T$ **then**
      continue
    else
      if $\dfrac{\mathrm{INP}\left(\mathcal{S}^p \cup \{S_i\}, \mathcal{S}^c \setminus \{S_i\}\right) - \mathrm{INP}_0}{r_i} > \Delta$ **then**
        $\Delta \leftarrow \dfrac{\mathrm{INP}\left(\mathcal{S}^p \cup \{S_i\}, \mathcal{S}^c \setminus \{S_i\}\right) - \mathrm{INP}_0}{r_i}$
        $\mathcal{S}^{pc} \leftarrow \mathcal{S}^p \cup \{S_i\}$
  if $\Delta = 0$ **then**
    break
  else
    $\mathcal{S}^p \leftarrow \mathcal{S}^{pc}$
    $\mathcal{S}^c \leftarrow \mathcal{S}^h \setminus \mathcal{S}^p$

8.4 Experimental Studies

To verify the applicability and validity of the proposed framework, we take the Shanghai metro system in China (as of Apr 2015) as the studied case for demonstration purposes. Detailed elaborations on the data preparation and preprocessing and results analysis are presented in Sects. 8.4.1 and 8.4.2, respectively.

8.4.1 Data and Resources

As mentioned in Sect. 8.3.2.3, we need to estimate the daily average passenger flow of each O-D pair, i.e., $V_{i_1}^{i_2}$ in Eq. (8.7), to calculate APU and then acquire the INP of the metro network. In this chapter, the passenger flow of each O-D pair is retrieved from AFC records in Apr 2015, which were issued by the 2015 Shanghai Open Data Application (http://shanghai.sodachallenges.com/). Based on the AFC records, entry/exit behavior can be paired as a complete ride record according to the user ID. During this process, we exclude some data that only had one entry or one exit record. After data preparation and cleaning, we calculate the passenger flow of each O-D pair for each day in April 2015, and then average the passenger flow of each O-D pair over the month, thus getting the values of $V_{i_1}^{i_2}$ $\left(i_1, i_2 = 1,2,\ldots,N_S \text{ and } i_1 \neq i_2\right)$. Furthermore, we can easily calculate the daily average volume of passengers entering or exiting each metro station, as illustrated in Fig. 8.4.

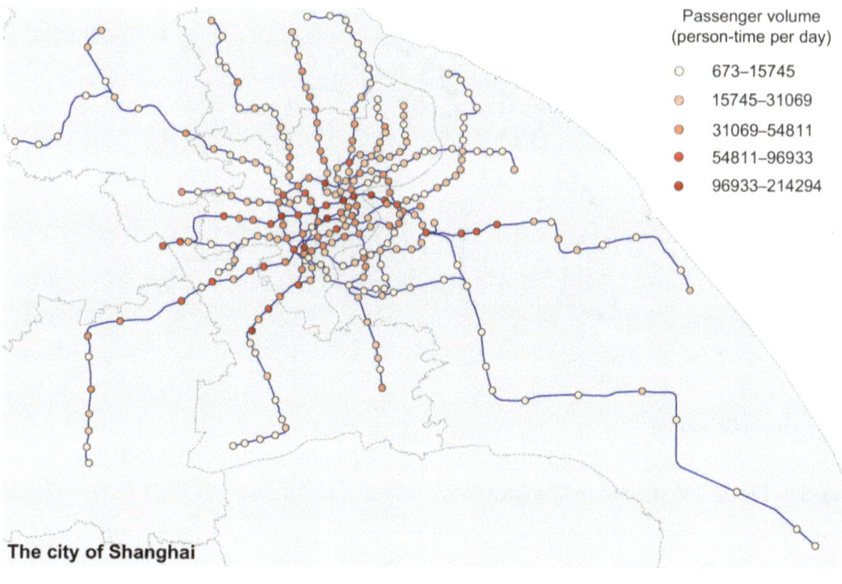

Fig. 8.4 Daily average passenger volume of metro stations in Shanghai in Apr 2015

According to the AFC records, a total of 286 metro stations were operational in Shanghai as of April 2015; hence, a metro network model with 286 station nodes can be established. Specifically, to determine the metro lines of each station in April 2015, the Shanghai Metro official website (http://service.shmetro.com/) can be accessed, allowing for the creation of platform nodes for each station. A series of edges is then added between platform nodes based on the metro network diagram. The finally established network model contains a total of 343 platform nodes and 401 edges. Besides, any relevant travel time or walking time information between stations or platforms can be obtained from the Gaode Map App, and the dwell time of a train at a station has been set to 0.5 min, on which basis we can calculate each edge's weight using the methodologies elucidated in Sect. 8.3.1.

To implement the optimization framework, we must also compute the FRI values for all 286 stations involved. This process requires us to obtain the values of the seven factors proposed in Sect. 8.3.2.1. To achieve this, we gather a series of raw data including (1) the digital elevation model including the elevation and slope data from Geospatial Data Cloud (http://www.gscloud.cn/) hosted by Chinese Academy of Sciences; (2) population density of Shanghai in 2015 from the WorldPop research program (https://www.worldpop.org/); (3) road network and green areas vectorized from the electronic map. Utilizing the ER-based optimality approach, we computed FRI for all the 286 stations, displaying the results in Fig. 8.5.

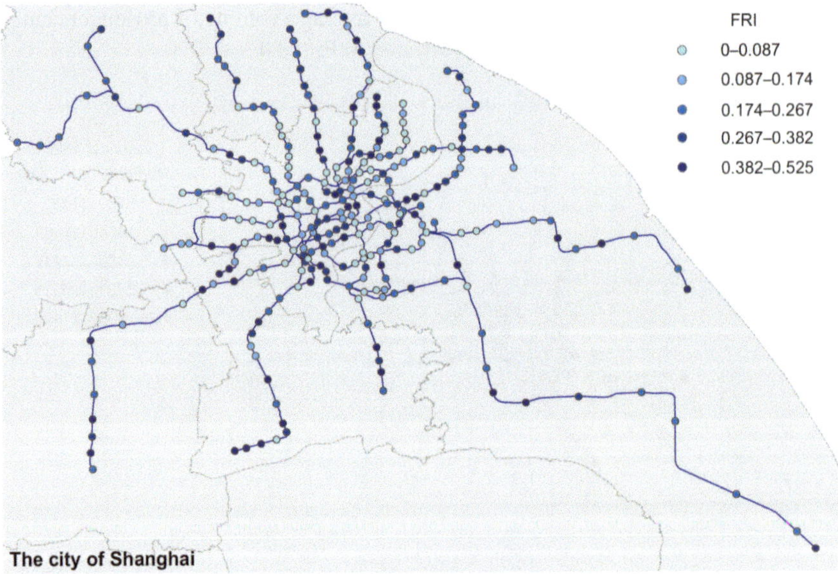

Fig. 8.5 FRI of 286 stations in Shanghai

8.4.2 *Optimization Results Under Different Flood Scenarios*

Based on the proposed methodology, we can calculate the number of high-risk stations as well as the required resources for protecting all high-risk stations under different flood intensity indexes, as shown in Fig. 8.6a, b. For different flood intensity indexes, we compute the INP of the metro network with all high-risk stations closed (see Fig. 8.6c), which serves as a baseline for optimization result analysis. According to Fig. 8.6a, when the flood intensity index is 0.1, there are 76 stations at high risk, accounting for 26.6% of the total stations; when the flood intensity index grows to 0.3, the number of high-risk stations becomes 198, constituting 69.2% of the total stations. Therefore, in this section, we define "flood intensity index = 0.1" and "flood intensity index = 0.3" as the representative scenarios of minor floods and major floods, respectively, and then optimize the closure-protection scheme for both scenarios with different total resources, with the results shown in Figs. 8.7 and 8.8.

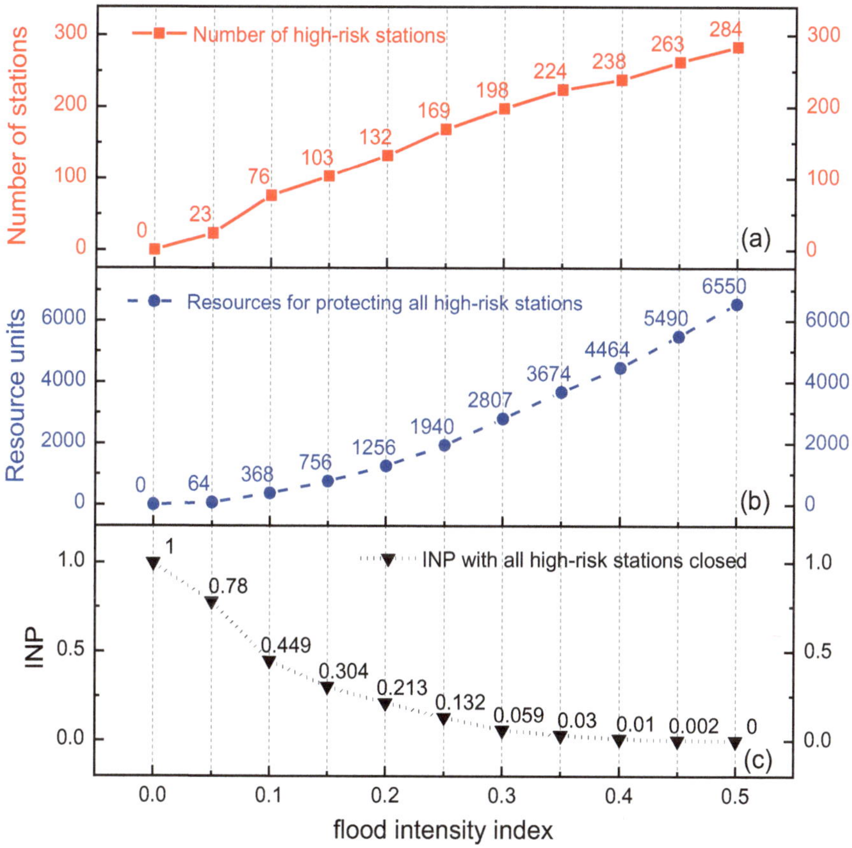

Fig. 8.6 Key variables under different flood intensity indexes: (**a**) number of high-risk stations; (**b**) resources for protecting all high-risk stations; (**c**) INP with all high-risk stations closed

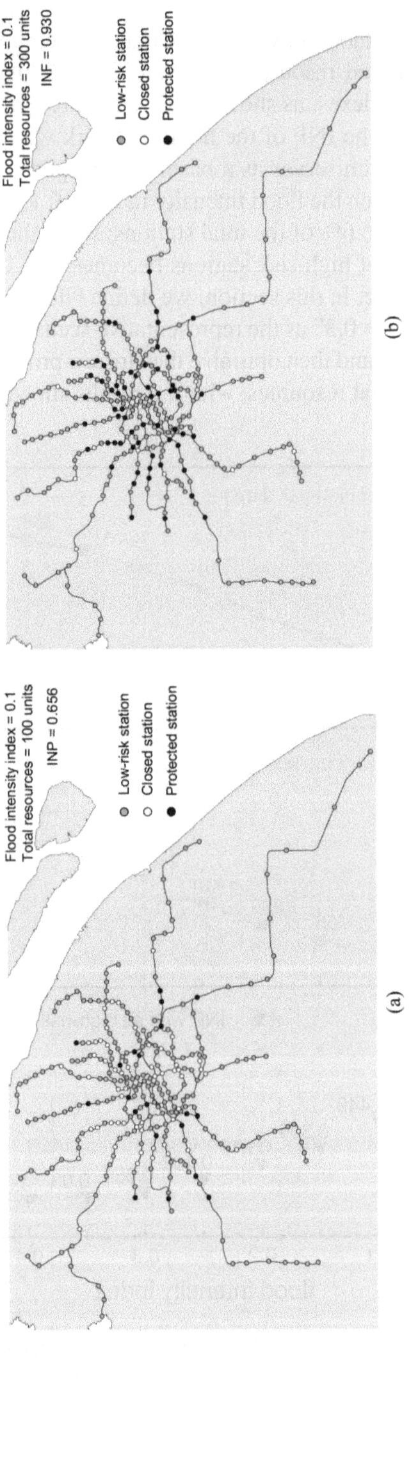

Fig. 8.7 Closure-protection schemes under minor floods (flood intensity index = 0.1): (**a**) total resources = 100 units; (**b**) total resources = 300 units

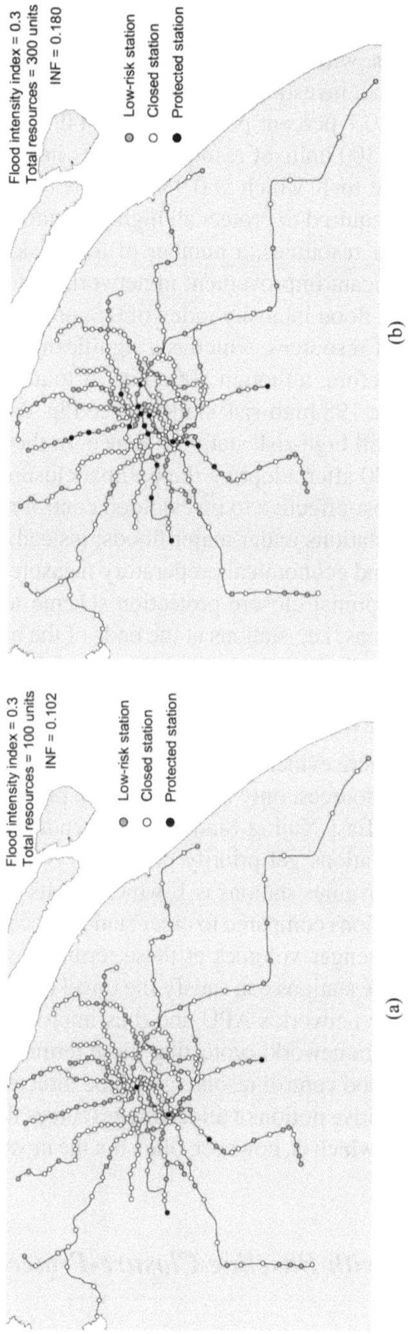

Fig. 8.8 Closure-protection schemes under major floods (flood intensity index = 0.3): **(a)** total resources = 100 units; **(b)** total resources = 300 units

According to Figs. 8.7 and 8.8, we can conclude that under minor floods, a small quantity of resources can significantly improve the performance of the metro network; during major floods, however, there is a limited absolute improvement in network performance even with a large investment of resources. As shown in Fig. 8.7, under a minor flood, investing 100 units of resources can increase the INP of the metro network by 20.7 percent points (from 0.449 in Fig. 8.6c to 0.656 in Fig. 8.7a), while investing 300 units of resources results in the INP of the network reaching a high value close to 1, which is 0.930. This is because when floods are minor, the total resources required to protect all high-risk stations are relatively low. Therefore, even with fewer resources, a number of high-risk stations can be protected, resulting in a significant improvement in network performance. In contrast, facing a major flood with a flood intensity index of 0.3, protecting all high-risk stations requires 2807 units of resources, which is a significant demand for both manpower and materials. Therefore, although 300 units of total resources are already considerable, only 18 of the 198 high-risk stations (see Fig. 8.8b) can be protected, which is less than 10% of all high-risk stations. Hence, in this case, the INP of the metro network is only 0.180 after adopting the optimal closure-protection scheme. Therefore, it may not be cost-effective to invest flood control resources to maintain the operation of high-risk stations under major floods; instead, closing all high-risk stations can be a feasible and economical preparatory measure.

Besides, the obtained optimal closure-protection scheme tends to prioritize the protection of terminus stations, i.e., stations at the ends of the metro lines, especially when the total resources are limited. As indicated in Fig. 8.7a, under a minor flood, the only two terminus stations among all high-risk stations, Jinyun Road Station and Shiguang Road Station, are both protected when the total resources are 100 units. This phenomenon is even more evident in Fig. 8.8a. Specifically, under a major flood with limited 100 units of resources, only five stations are protected, and two of them are terminus stations, i.e., East Xujing Station and Shendu Highway Station. The reason why the terminus stations get priority protection is mainly twofold: (1) the degree centrality of these terminus stations is 1, which results in less resource investment in protecting these stations compared to other stations according to Eq. (8.9); (2) based on Fig. 8.4, the passenger volumes at these terminus stations are relatively high; hence, protecting these stations can satisfy the travel demands of many passengers, thereby increasing the network's APU and thus improving its INP. Therefore, under the CPGA solution framework, protecting these terminus stations has higher cost performance, so the flood control resources will be directed towards them. This seems to contradict our intuitive notion of allocating resources first to those important multi-line transfer stations, which is, however, often not the most economical strategy.

8.4.3 Comparisons with Baseline Closure-Protection Strategies

To verify the superiority of the optimization framework, this section compares the INP results obtained from the proposed framework with those from three baseline strategies for station closure-protection decisions, including the random strategy, the degree priority strategy, and the betweenness priority strategy. The three baseline strategies are elaborated below:

1. Random strategy. This strategy randomly sorts all high-risk stations and then allocates resources according to the sorting order. If the remaining resources cannot meet the resource requirement of the current ranked station, they will be allocated to the next ranked station to achieve maximum utilization of resources.
2. Degree priority strategy. This strategy sorts all high-risk stations in descending order of their degree centrality, and subsequently allocates resources to these stations following the sorting order. The remaining resources will be allocated to the next-ranked station if it cannot meet the current station's requirement.
3. Betweenness priority strategy. A station's betweenness centrality refers to the number of O-D pairs whose shortest path passes through the station. The betweenness priority strategy first sorts all high-risk stations in descending order of betweenness centrality, and then allocates resources based on the sorting order. Like the previous two baselines, maximum utilization of all available resources is also ensured in this strategy.

Figure 8.9 gives the comparison of INP between the optimization framework and three baseline strategies, from which we can conclude that the optimization framework outperforms all three baselines. Among the four strategies, the random strategy performs the worst in most cases. The degree priority strategy is slightly better than the random strategy. The betweenness priority strategy performs better than the other two baselines in most cases, but still falls short compared to the optimization framework. This indicates that the optimization framework is the most effective in improving the metro network performance under floods. Furthermore, to quantitatively study the superiority of the optimization framework, we extract partial cases

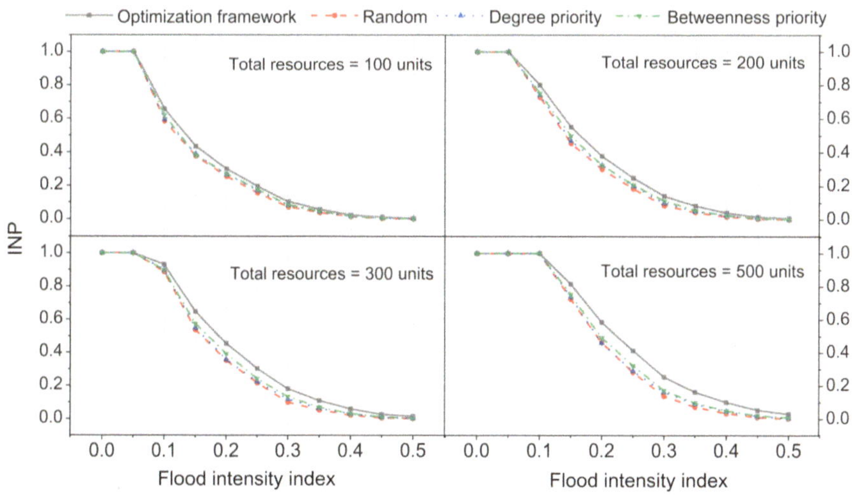

Fig. 8.9 Comparison of INP between the optimization framework and three baseline strategies

from Fig. 8.9 and list specific INP values in Table 8.1, then calculate the improvement of the optimization framework compared to the best baseline. It is seen that under minor floods (flood intensity index = 0.1), the optimization model can improve the INP of the metro network by 3.6 percent points using 100 units of resources. Under moderate floods (flood intensity index = 0.2), such an improvement becomes 6.2 percentage points with 300 units of resources available.

We also find that when the flood intensity is too small or too large, the advantage of the optimization framework over the baseline strategies is not very significant. However, when the flood is moderate, the optimization framework can yield a scheme with significantly higher INP than the baseline strategies. As listed in Table 8.1, the flood intensity indexes corresponding to the maximum differences in different total resource quantities are all around 0.2. This is because under these circumstances, the proportion of high-risk stations that can be protected by the available resources is moderate, and there are many feasible closure-protection schemes with large differences in INP, which provides ample room for optimization.

Table 8.1 Partial comparison of INP between the optimization framework and three baseline strategies

Total resources	Flood intensity index	INP					Improvement (percent point)
		Random	Degree Priority	Betweenness Priority	Best baseline	Optimization framework	
100	0	1	1	1	1	1	0
	0.1	0.583	0.595	0.62	**0.62**[a]	**0.656**	**3.6**
	0.2	0.254	0.265	0.269	0.269	0.299	3.0
	0.3	0.07	0.081	0.083	0.083	0.102	1.9
	0.4	0.013	0.014	0.015	0.015	0.022	0.7
	0.5	0	0	0	0	0.003	0.3
300	0	1	1	1	1	1	0
	0.1	0.885	0.898	0.897	0.898	0.930	3.2
	0.2	0.351	0.356	0.391	**0.391**	**0.453**	**6.2**
	0.3	0.099	0.117	0.133	0.133	0.180	4.7
	0.4	0.021	0.03	0.031	0.031	0.059	2.8
	0.5	0.001	0.002	0.002	0.002	0.014	1.2
500	0	1	1	1	1	1	0
	0.1	1	1	1	1	1	0
	0.2	0.461	0.458	0.493	**0.493**	**0.586**	**9.3**
	0.3	0.138	0.164	0.172	0.172	0.254	8.2
	0.4	0.034	0.046	0.05	0.05	0.101	5.1
	0.5	0.003	0.007	0.008	0.008	0.031	2.3

[a]The boldface entries denote the flood scenario where the optimization framework yields the greatest improvement over the best baseline for each total resource level

8.5 Merits Analysis

In this section, another two solution algorithms, i.e., the performance-based greedy algorithm and the genetic algorithm, are implemented for the same optimization problem, and the optimization results are compared with those by CPGA. Detailed workflows of the two baselines are given below:

1. Performance-based greedy algorithm. This baseline is also a greedy algorithm, but it allocates resources to the stations that generate the maximum INP gain without considering the station's resource requirement. Therefore, this algorithm may protect the stations, which greatly enhance the metro network's INP at the expense of considerable resource expenditure.
2. Genetic algorithm. Using the genetic algorithm to solve the presented problem, we need to first encode the closure-protection scheme as an array of length equal to the number of high-risk stations. Each element of the array is a binary integer, where 0 represents closure and 1 represents protection of a particular high-risk station. On this basis, we can adopt the typical genetic algorithm workflow. After a few generations of iteration, the population will gradually converge to an approximate optimal solution, with the best individual in the population being the final closure-protection scheme. In this process, to ensure that the individuals in the initial population and generated offspring make the most use of resources while satisfying the total resource constraint, we need to perform modification operations on these individuals.

For fair comparisons, all three algorithms are coded in the environment of Python 3.9.12 and implemented on a Windows 11 system with an AMD 6800H CPU (8 cores and 16 processing threads) and 16 GB RAM. As a result, Fig. 8.10a, b illustrate the optimal INP and run time of the three solution algorithms, respectively. The results are discussed as follows.

1. In all optimization cases, CPGA achieves the best optimal INP compared to the baseline solution algorithms. According to the relevant data in Fig. 8.10a, CPGA's optimal INPs are the highest in all 51 cases, which demonstrates its satisfactory optimization capability. In contrast, the genetic algorithm generates the worst optimization results in more than half of the cases. The performance-based greedy algorithm performs slightly better than the genetic algorithm but is still inferior to CPGA. It can also be seen that when the total resources are limited, CPGA's optimization results are only marginally superior to the baselines. As the total resources increase, however, CPGA's advantage becomes more and more significant.
2. When the total resources are <500 units, CPGA solves the problem quickly; however, if the total resources exceed 1000 units, the efficiency of CPGA will gradually decrease with the increase in total resources. As illustrated in Fig. 8.10b, when the total resources are within 500 units (including 500 units), CPGA completes every optimization case in less than 500 s. The longest run time is 377 s for the case of "total resources = 500 units, flood intensity

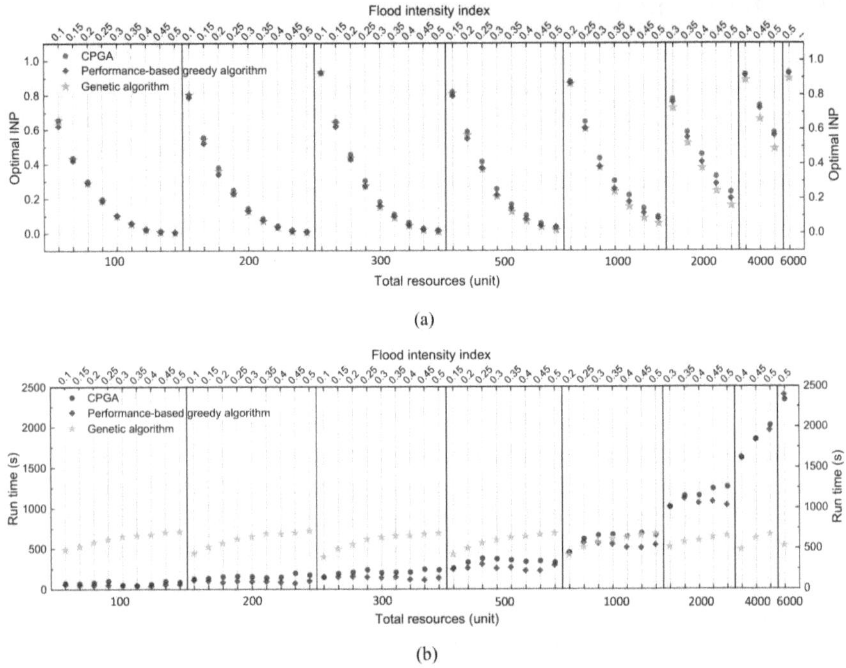

Fig. 8.10 Comparison between CPGA and two baseline solution algorithms: (**a**) optimal INP; (**b**) run time

index = 0.25". This efficiency is quite impressive, as we can formulate an optimal closure-protection scheme within a few minutes of determining the flood intensity index and the total resources available. As the total resources increase, more rounds are needed to allocate all the resources, which results in longer run time for the CPGA. However, given that the optimal INPs obtained by the performance-based greedy algorithm are lower than those of CPGA, we still consider CPGA to be the best solution algorithm for this optimization problem.

8.6 Chapter Summary

The flood problem has become a major threat to urban metro systems. From the perspective of pre-disaster preparations, this chapter proposes a new framework to optimize the closure-protection scheme for high-risk metro stations before the flood comes. To begin with, we develop a time-weighted metro network model that could consider line transfer, based on which the station closure is proposed as a new percolation mode of metro networks, and the INP is constructed to evaluate the network performance from both network efficiency and passenger utility. With every station's FRI determined by the ER-based optimality approach, we could identify a set of high-risk stations given the forecasted flood intensity index. For these high-risk stations, we employ CPGA to optimize the closure-protection scheme so that the INP of the metro network can be maximized. The proposed framework is implemented to verify the feasibility and effectiveness of the Shanghai metro system.

In view of the case study, the key findings can be summarized below: (1) The proposed optimization method can improve the performance of the metro network with a small quantity of resources under minor floods. During major floods, however, closing all high-risk stations is a more cost-effective strategy. (2) The obtained optimal closure-protection scheme prioritizes the protection of terminus stations with higher passenger volumes, particularly if the total resources are limited. The reason lies in the fact that protecting these stations can satisfy the travel demand of many passengers with a small quantity of resources, thereby increasing the network's APU and thus improving its INP at a lower cost. (3) Optimizing the closure-protection scheme for high-risk stations significantly improves the metro network's robustness during floods, with the degree of improvement increasing as the total resources rise. (4) In enhancing the integrated performance of the metro network, the proposed optimization model surpasses all three baseline strategies. Compared to the best baseline strategy, the betweenness priority strategy, the optimization method can improve the metro network's INP by 3.6 percent points with 100 units of resources under minor floods and 6.2 percent points with 300 units of resources under moderate floods.

The theoretical contributions of this chapter are twofold. First and foremost, our research stands among the initial studies to theoretically model pre-flood preparations for metro systems. In current practice, the decision of whether to close or protect a metro station in response to an upcoming flood relies primarily on the subjective experience of metro operators rather than quantitative analysis and optimization. Our model not only aids in pre-flood decision-making of emergency management departments, but also provides novel inspiration and insights for academic research in this area. Secondly, our research has, to some extent, enriched and refined relevant theories in network science. A new percolation algorithm has been proposed to simulate the station closure in a metro network. The constructed INP is used as a novel metric to evaluate the metro network performance from both the network efficiency and the passenger utility. These serve as a valuable reference for future research on disaster modelling of metro systems.

References

1. He R, Zhang L, Tiong RLK (2025) Responding of metro stations to upcoming floods: to close or to protect? Reliab Eng Syst Saf 256:110683
2. Higo E, Okada N, Hipel KW, Fang L (2017) Cooperative survival principles for underground flooding: vitae system based multi-agent simulation. Expert Syst Appl 83:379–395
3. Zheng Y, Li X-G, Jia B, Jiang R (2019) Simulation of pedestrians' evacuation dynamics with underground flood spreading based on cellular automaton. Simul Model Pract Theory 94:149–161
4. Tang Y, Zhou T, Zhong Y, Hu S, Lin J, Lin Z, Liu H, Liu B, Zhao Y, Wang Y, Lin H (2022) Risk assessment for critical flood height of pedestrian escape in Subway Station. Water 14:3409

5. Yadav N, Chatterjee S, Ganguly AR (2020) Resilience of urban transport network-of-networks under intense flood hazards exacerbated by targeted attacks. Sci Rep 10:10350
6. Zhang D-m, Du F, Huang H, Zhang F, Ayyub BM, Beer M (2018) Resiliency assessment of urban rail transit networks: Shanghai metro as an example. Saf Sci 106:230–243
7. Chen J, Liu J, Peng Q, Yin Y (2022) Strategies to enhance the resilience of an urban rail transit network. Transp Res Rec 2676:342–354
8. Lin Z, Hu S, Zhou T, Zhong Y, Zhu Y, Shi L, Lin H (2022) Numerical simulation of flood intrusion process under malfunction of flood retaining facilities in complex Subway stations. Buildings 12:853
9. Lyu H-M, Shen S-L, Zhou A-N, Zhou W-H (2019) Flood risk assessment of metro systems in a subsiding environment using the interval FAHP-FCA approach. Sustain Cities Soc 50:101682
10. Ishigaki T, Ozaki T, Inoue T, Shimada H, Toda K (2011) Drainage system, rainwater flooding and underground inundation in urban area. In: 12th International conference on urban drainage, Porto Alegre/Brazil. IWA
11. Yang X, Dai W, Li Y, Yang X (2024) An efficient evacuation path optimization for passengers in subway stations under floods. Tunn Undergr Space Technol 143:105473
12. Bi W, Schooling J, MacAskill K (2024) Assessing flood resilience of urban rail transit systems: complex network modelling and stress testing in a case study of London. Transp Res Part D: Transp Environ 134:104263
13. Lyu H-M, Sun W-J, Shen S-L, Arulrajah A (2018) Flood risk assessment in metro systems of mega-cities using a GIS-based modeling approach. Sci Total Environ 626:1012–1025
14. He R, Zhang L, Tiong RLK (2023) Flood risk assessment and mitigation for metro stations: an evidential-reasoning-based optimality approach considering uncertainty of subjective parameters. Reliab Eng Syst Saf 238:109453
15. Martello MV, Whittle AJ, Keenan JM, Salvucci FP (2021) Evaluation of climate change resilience for Boston's rail rapid transit network. Transp Res Part D: Transp Environ 97:102908
16. Huang H, Zhang D, Huang Z (2022) Resilience of city underground infrastructure under multi-hazards impact: from structural level to network level. Resil Cities Struct 1:76–86
17. Dijkstra EW (2022) A note on two problems in connexion with graphs. In: Edsger Wybe Dijkstra: his life, work, and legacy. ACM, pp 287–290

Chapter 9
Emergency Rescue Planning

9.1 Introduction

With the rapid expansion of urban transportation networks, metro systems have become a key indicator of a city's development and prosperity [1]. However, metro operations are occasionally disrupted by various incidents, among which fire disasters pose one of the most significant threats to passenger safety, infrastructure, and property. In the event of a metro fire, a swift and effective emergency response is crucial not only for minimizing casualties and economic losses but also for enhancing the resilience of metro operations. Therefore, the rapid identification and analysis of real-time fire emergency data are essential for formulating timely and effective countermeasures.

Nowadays, most of the metro fire emergency responses are based on experience, which tends to cause many deficiencies in the system, such as a lack of resource planning or a contradiction between supply and demand [2]. To mitigate biases arising from experience-based decision-making, computer simulation has emerged as a valuable tool for planning and modeling metro emergency responses. By adjusting parameters and executing multiple iterations, simulations can generate "virtual" datasets, enabling researchers to explore various emergency scenarios [3]. Several methods have been employed to simulate fire emergency responses, including the Social Force Model, Fire Dynamics Simulator, and Zero-Fidelity Simulation. However, these approaches struggle to effectively model the interactions among emergency response departments, passengers, and evolving fire conditions using real-time data. Consequently, a more advanced modeling approach is required to construct complex process models and integrate a broader range of real-time information.

Petri nets are a powerful mathematical tool for modeling dynamic discrete systems and hold significant potential for analyzing fire emergency response operations. They facilitate the definition of conditions, activation of activities and actions,

L. Zhang et al., *Resilience in Urban Transit Systems*, Advances in 21st Century Human Settlements, https://doi.org/10.1007/978-981-95-4322-9_9

and coordination among various departments. With continued advancements, several enhanced Petri net models have been developed, including Colored Petri Nets (CPN) and Timed Colored Petri Nets (TCPN) [4]. These advanced models have been widely applied in fire disaster management system analysis and design. However, as system complexity increases, Petri net structures can become excessively intricate. In such cases, Hierarchical Timed Colored Petri Nets (HTCPN) [5] offer a more structured and logical modeling approach.

In metro fire emergency response, a critical factor is response time, which comprises alarm response time, alarm processing time, dispatch time, transit time, start action time, and action time [6]. Another key aspect is the efficient allocation of emergency resources, which can minimize casualties and property damage while enhancing the effectiveness and cost-efficiency of rescue operations [7]. Both reducing response time and optimizing firefighter deployment are essential for a successful emergency response. However, these objectives often conflict. Overemphasizing response time reduction may lead to firefighter overuse and increased material consumption, whereas prioritizing minimal firefighter deployment may extend response time, exacerbating economic losses and casualties. Balancing these conflicting objectives is a critical challenge, necessitating multi-objective optimization (MOO). MOO enables the simultaneous optimization of multiple conflicting goals. The Skyline operator facilitates simulation-based MOO and decision-making by efficiently identifying optimal trade-offs in the dataset [8].

To effectively simulate the metro station fire emergency response process and optimize emergency planning by minimizing both response time and firefighter deployment, this chapter develops an HTCPN model for metro fire emergency response and simulates various scenarios using CPN Tools. The simulation dataset is then analyzed using the Skyline operator to determine the Pareto front of the multi-objective optimization (MOO) problem, while a fuzzy-based mechanism is employed to identify optimal compromise solutions. This approach addresses a research gap, as few existing studies simultaneously optimize time control and resource allocation in metro station fire emergency response. Consequently, the proposed method offers scientific guidance for real-time decision-making in metro fire emergency management.

9.2 Disaster Response Modelling Under Emergency

Disaster response modeling is essential for optimizing disaster management. These models provide quantitative insights into evacuation efficiency, resource allocation, and system resilience, enabling improved evacuation planning and risk mitigation. Given the complexity of metro systems and the time-sensitive nature of disaster response, various modeling techniques have been developed to address different aspects of disaster response. Notably, individual-based models, systemic models, and discrete event models have gained significant attention for their ability to capture varying levels of system complexity and decision-making processes.

1. **Individual-based models.** Individual-based models focus on simulating passenger behavior and emergency personnel actions at the micro level. Agent-Based Models (ABM) represent each passenger as an autonomous agent, allowing for realistic evacuation behaviors and crowd dynamics modeling [9]. ABM can effectively capture panic-driven movement, congestion effects, and human decision-making, but its computational cost increases significantly in large-scale scenarios. Alternatively, Cellular Automata (CA) models represent space as a discrete grid and use local movement rules to simulate pedestrian flow [10]. CA is computationally efficient and well-suited for high-density crowd scenarios, but it struggles with continuous movement representation and requires enhancements to incorporate dynamic environmental factors such as fire and smoke.
2. **Systemic models.** Systemic models analyze macro-level system behaviors and interactions, providing a broader perspective on disaster response. System Dynamics (SD) employs feedback loops and stock-flow relationships to model evacuation efficiency, fire growth, and emergency response coordination over time [11]. This approach is particularly useful for policy evaluation and long-term planning, though it lacks spatial granularity and cannot capture individual decision-making. Another systemic approach is Queuing Theory (QT). QT can be integrated into systemic approaches to model certain aspects of system operations, particularly in analyzing service node efficiency [12]. QT is valuable for estimating average waiting times and identifying potential bottlenecks under steady-state conditions. However, these models assume steady-state flows and do not account for time-dependent parameters and panic dynamics in emergency scenarios.
3. **Discrete event models.** Discrete event models focus on the sequential progression of critical events during an emergency, making them particularly useful for time-sensitive disaster response planning. Petri net models offer a graphical and mathematical framework to represent concurrent events, processes, and resource constraints. A Petri net consists of places, transitions, and tokens that flow through the net, enabling transitions when preconditions are met. This structure is well-suited to model a metro emergency, where numerous events (fire ignition, alarm activation, train stoppage, evacuation start, firefighting, etc.) occur in sequence or in parallel [13]. As research on Petri nets has progressed, various advanced models have been introduced, such as CPN and TCPN, which enhance scalability and real-time decision-making in emergency response modeling [14].

To sum up, metro station fires are highly dynamic, requiring real-time resource allocation optimization based on evolving fire conditions and the progress of emergency response tasks. However, existing research primarily focuses on evacuation optimization and operational strategy adjustments, with limited attention to the integrated optimization of time and resources. To address this gap, we propose an approach that integrates HTCPN simulation with MOO to balance response time and firefighter deployment. The HTCPN, developed based on multi-layer event modeling, can support dynamic optimization of emergency resources by integrating real-time sensor data, thereby improving decision-making under uncertain conditions.

9.3 Simulation-Based Optimization for Rescue Planning

To optimize both response time and firefighter resource allocation in metro fire emergency response, this chapter proposes a multi-objective optimization (MOO) model based on simulation analysis. Figure 9.1 illustrates the workflow of the proposed approach. First, a Hierarchical Timed Colored Petri Net (HTCPN) model of metro station fire emergency response is developed and validated. Next, response time and firefighter dispatch are simulated under various scenarios using CPN Tools. Finally, the Skyline operator is applied to extract the Pareto front of the MOO problem, and a fuzzy-based mechanism is employed to determine the optimal compromise solutions.

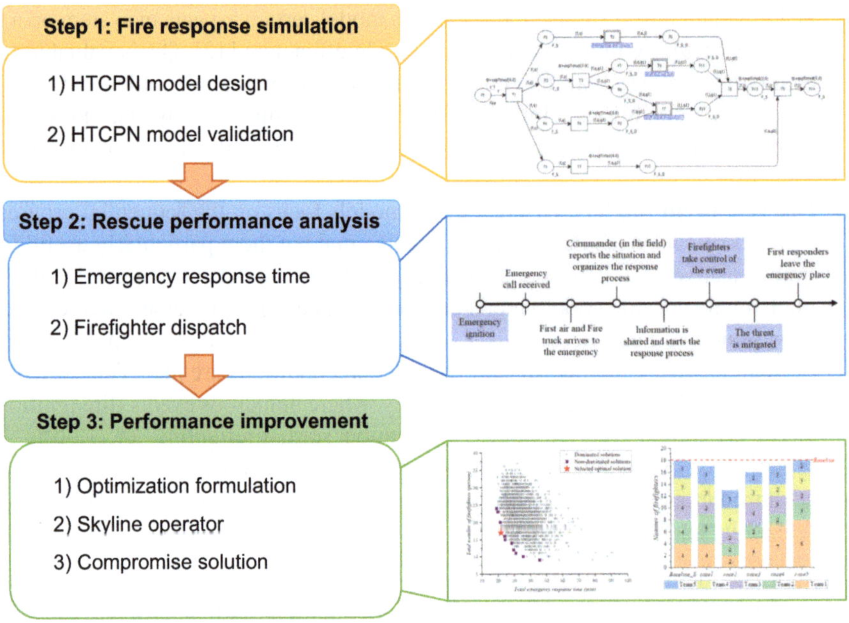

Fig. 9.1 Workflow of the proposed model for fire emergency response optimization

9.3.1 *Fire Response Simulation*

Due to different environmental characteristics, the emergency response procedures in different scenarios may be different. Metro station emergency response has its own characteristics. This section introduces the basic principles of HTCPN and the emergency response process of the metro system considering the IoT. How to check the correctness and rationality of the constructed Petri net model through structural validation is also expounded.

9.3.1.1 HTCPN Model Design

Petri nets, introduced by Carl Adam Petri in the 1960s, serve as mathematical representations for modeling discrete parallel systems. Places and transitions are the two basic components of Petri nets, representing two different types of elements. In Petri nets, tokens are distributed within a directed graph, where their allocation corresponds to model states and trigger events. According to predefined rules, transitions are enabled when their input places contain the required number of tokens. Once enabled, transitions trigger events by retrieving tokens from input places and transferring them to output places.

HTCPN is an extension of CPN and TCPN, where the structure of the model can be organized hierarchically. A hierarchical representation of the specification is a modular representation of the model. In this modular representation, a set of submodels is composed together to form one big model. Therefore, the specification is represented as an abstract model that can be refined towards an elaborated model at any period. HTCPN uses hyper-transitions in the abstract model to hold in the submodels. The refinement of the abstract model passes through the unfolding of these hyper-transitions. The general structure of HTCPN is shown in Fig. 9.2.

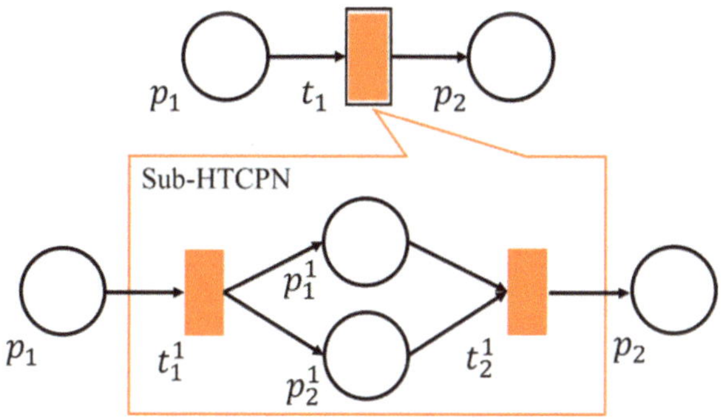

Fig. 9.2 The general structure of HTCPN

9.3.1.2 HTCPN Model Development

Metro stations are critical components of urban infrastructure, and their safe and intelligent operation is essential for delivering high-quality passenger services. To create a smart metro system, IoT technologies (e.g., RFID, wireless, and so on) are configured to create a smart metro system according to the information required. The smart metro system integrates ubiquitous sensing devices, heterogeneous network infrastructure, and advanced computing systems to monitor physical changes within the metro environment, as shown in Fig. 9.3. Detected information is transmitted to the control center via a heterogeneous network, where a high-performance computing system processes and analyzes the data. Then, the control center uses a powerful computing system to process and analyze the collected data and perform urban operations through the feedback component. The metro fire emergency response management system leverages sensor networks to detect and locate hazards, identify personnel (including ignition points, trapped individuals, and firefighters), and facilitate communication between the environment and the control center. HTCPN is developed based on a multi-layer event model. In Petri nets, places represent system activity states, while transitions denote activity processes, which can be assigned execution times. Colored tokens, integrated with IoT technologies, represent specific types of information, enabling Petri net states to be dynamically updated based on real-time sensor data.

9.3.1.3 HTCPN Model Validation

Once the fire emergency HTCPN model of a metro station is constructed, it is crucial to verify the correctness and rationality of its structure. Structural analysis, particularly invariants, is the primary method for evaluating a Petri net model. Structural validity analysis ensures that the Petri net accurately represents the actual execution process. The structural effectiveness of Petri nets consists of three key properties: boundedness, reachability, and liveness. Boundedness prevents resource overflow during the metro fire emergency response. Reachability ensures that all

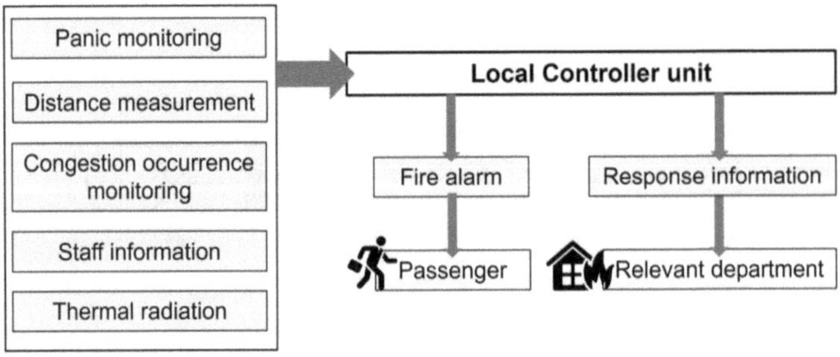

Fig. 9.3 Procedures to perform emergency response under fire hazards in metro stations

necessary states in the emergency response process can be achieved. Liveness guarantees the absence of deadlocks or conflicts, allowing the Petri net to function correctly while maintaining logical consistency and structural integrity.

Define $P = (p_1, p_2, \ldots p_n)$ as a finite set of places, $T = (t_1, t_2, \ldots t_m)$ as a finite set of transitions, and $F \subseteq (P \times T) \cup (T \times P)$ represents the set of directed arcs connecting places and transitions and transitions and places. Based on the structural relationship between each link, a_{ij}^{+} represents the weight of the arc from the input transition t_i to output place p_j, while a_{ij}^{-} represents the weight of the arc from the input place p_j to output transition t_i. When there is no value for the weight on the arc, the weight defaults to 1. Then, the incidence matrix $C = (a_{ij})_{n \times m}$ with n rows and m columns can be calculated and used to infer the structural effectiveness of Petri net. The element a_{ij} can be expressed as Eq. (9.1), in which a_{ij}^{+} and a_{ij}^{-} are given in Eqs. (9.2) and (9.3).

$$a_{ij} = a_{ij}^{+} - a_{ij}^{-}, i \in \{1,2,\ldots,n\}, j \in \{1,2,\ldots,m\} \tag{9.1}$$

$$a_{ij}^{+} = \begin{cases} 1, & \text{if}\left(t_i, p_j\right) \in F \\ 0, & \text{otherwise} \end{cases} \quad i \in \{1,2,\ldots,n\}, j \in \{1,2,\ldots,m\} \tag{9.2}$$

$$a_{ij}^{-} = \begin{cases} 1, & \text{if}\left(p_j, t_i\right) \in F \\ 0, & \text{otherwise} \end{cases} \quad i \in \{1,2,\ldots,n\}, j \in \{1,2,\ldots,m\} \tag{9.3}$$

Given that $X^{(m \times 1)}$ represents a non-negative integer vector that satisfies $C^T \times X = 0$, which is a P_invariant of the Petri net; $Y^{(n \times 1)}$ represents a non-negative integer vector Y that satisfies $C \times Y = 0$, which is a T_invariant of the Petri net. P_invariant and T_invariant satisfy a dual relationship. P_invariant and T_invariant represent the flow status and behavior characteristics of the resources in the Petri net. On this basis, the boundedness and activity of the Petri net structure can be inferred, and the reachable sequence of the transition in the Petri net can be obtained.

9.3.2 *Simulation-Based Rescue Performance Analysis*

Metro stations have extremely high passenger density, especially during peak hours, which poses a huge challenge to efficiently complete emergency responses within limited time and resources. This section defines two key criteria for evaluating rescue performance: emergency response time and the number of firefighters assigned. Given the extensive use of CPN Tools in Petri net modeling, rescue performance can be effectively monitored within a simulation environment equipped with CPN Tools.

CPN Tools was developed by the CPN Group at Aarhus University, Denmark, in collaboration with various software firms. It is a comprehensive software suite for editing, simulation, state-space analysis, and characteristic analysis of Petri nets. CPN Tools not only translates Petri net models into computational language but also accurately and efficiently replicates the original simulated system. Additionally, it evaluates system performance using state-space reports and simulation-derived parameters. The logic and modeling framework in CPN Tools is implemented in Standard Meta-Language (SML), a general-purpose functional programming language. In CPN Tools, tokens in places represent the resources and information required for emergency response, while transitions model the sequence of actions. Time functions are incorporated into transitions to represent action durations. For the time functions in CPN Tools, "@+expTimed(x)" is a timed statement, indicating that the time obeys the exponential distribution of the parameter x, and x represents the average time of transition execution. To collect simulation data on emergency response time and firefighter team size, count monitors are added to key places and transitions before executing the model.

9.3.2.1 Emergency Response Time

Emergency response time is a key optimization objective in fire emergency management. Firefighting and rescue operations are critical components of fire emergency control, and their duration significantly impacts overall effectiveness. Reducing fire emergency response time mitigates infrastructure damage and enhances the safety of trapped passengers. Conversely, delayed fire control can create a cascading effect, leading to greater losses. The emergency response process begins when sensors detect environmental changes and issue a fire warning signal. It concludes when firefighters leave the emergency site. The timeline of the emergency response process is shown in Fig. 9.4.

The time to finish the whole fire emergency response in the HTCPN model can be calculated. $T_e(p)$ denotes the time to start task p if each pre-task is finished. However, during the response process, even if each pre-task can be finished in time, $T_e(p)$ may not be the time to start p because some tasks before p may be delayed. Hence, the actual time to start p, $E(p)$, is calculated as Eq. (9.4), in which the waiting time of p for p_{ex}, $W(p, p_{ex})$, is calculated by Eq. (9.5).

$$E(p) = \begin{cases} 0 & p = p_S \\ T_e(p) + W(p, p_{ex}) & \text{otherwise} \end{cases} \tag{9.4}$$

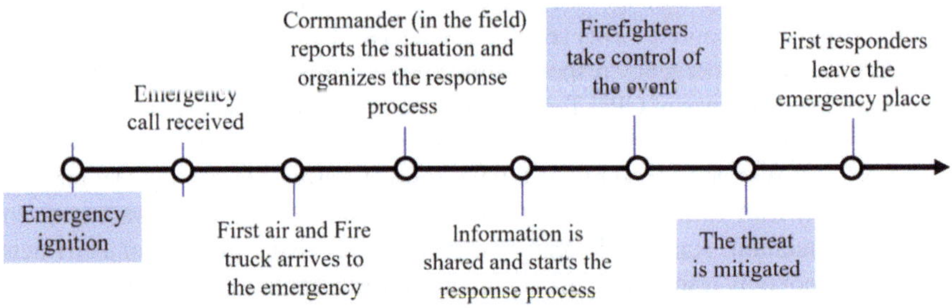

Fig. 9.4 Emergency response timeline

$$W(p,p_{ex}) = \begin{cases} 0 & T_e(p_1)+\alpha(p_1) \le T_e(p) \\ T_e(p_{ex})+\alpha(p_{ex})-T_e(p) & \text{otherwise} \end{cases} \tag{9.5}$$

where p_s represents the start place of an emergency process, p_{ex} is the last pre-task of p, $\alpha(p_{ex})$ is the time to execute the task p_{ex}. Then $E(p)$ is the time to finish the whole fire emergency response process if all tasks p_i before p can be finished in $\alpha(p_i)$.

9.3.2.2 Firefighter Dispatch

The allocation of firefighters during an emergency directly impacts response time and influences both public safety and property damage. Therefore, the number of firefighters deployed during a fire serves as a key optimization objective in emergency response. Traditionally, commanders assigned firefighters based on information gathered from phone reports. However, with the advancement of the IoT, real-time sensor data enables commanders to make informed decisions, enhancing firefighter coordination and optimizing resource utilization. Based on the total available firefighter resources, an optimal number of firefighters should be dispatched. The firefighter allocation is calculated as Eq. (9.6):

$$N = \sum_{n=1}^{N_f} x_n + \sum_{m=1}^{N_r} y_m \quad n \in \{1,2,\ldots,N_f\}, m \in \{1,2,\ldots,N_r\} \tag{9.6}$$

where N_f and N_r are the numbers of the firefighting team and passenger rescue team, respectively; x_n is the number of firefighters in the nth firefighting team; y_m is the number of firefighters in the mth passenger rescue team.

9.3.3 MOO for Rescue Performance Improvement

In this section, we develop an MOO model for enhanced fire emergency response to minimize the number of firefighters and the total emergency response time for a metro fire disaster. The Skyline operator is used to solve the MOO problem.

9.3.3.1 MOO Formulation

The objective function to minimize the number of firefighters and the total emergency response time is given in Eq. (9.7):

$$\begin{cases} \min \ F_1 \\ \min \ F_2 \end{cases} \tag{9.7}$$

where F_1 is the number of firefighters (unit: person); F_2 is the total emergency response time (unit: min). The objective function F_2 is obtained by the HTCPN model simulation in the CPN Tools.

Constraints on the number of firefighting teams are written in Eqs. (9.8) and (9.9).

$$X_{\min} \le x_n \le X_{\max} \quad n \in \{1,2,\ldots,N_f\} \tag{9.8}$$

$$Y_{\min} \le y_m \le Y_{\max} \quad m \in \{1,2,\ldots,N_r\} \tag{9.9}$$

where $X_{\max}$ and $X_{\min}$ are the maximum and minimum capacities of a firefighting team, respectively; $Y_{\max}$ and $Y_{\min}$ are the maximum and minimum capacities of a passenger rescue team, respectively.

9.3.3.2 Skyline Operator

In the MOO problem, one significant concept is "dominate", which represents a relationship between two solutions in the feasible solution space. To be more specific, for two solutions A and B in a minimization problem, when no component value of B is smaller than A and at least one component value of B is strictly greater than A, we consider that "A dominates B". The optimal solution set of an MOO problem comprises a series of "non-dominated solutions", which means that the solution is not dominated by any other solutions in the feasible solution space. The task of an MOO problem is to find out the optimal solution set, and hence the corresponding set formed by the objective function values is called "the Pareto front".

In this chapter, the Skyline operator is utilized to find the Pareto front of the MOO problems. The Skyline operator in the database has become a hotspot in previous research because of its interesting applications. To the best of our knowledge, Stephan [15] first proposed the Skyline operator to solve problems in the database environment. The Skyline operator mainly suggests a baseline for multi-criteria decision-making. Specifically, given a set of n objectives $\{k_1, k_2, \ldots, k_n\}$, the Skyline operator returns all objects k_i such that k_i is not dominated by any other object k_j. The notion of the Skyline is the same as finding the Pareto optimal tuples with respect to the objectives; thus, this is also called the Pareto operator. For the MOO problem constructed in this chapter, the dataset obtained by the CPN Tools prompts us to establish the Skyline operator in Python 3.7 by installing the "paretoset" package to find the Pareto front in the given data. The Skyline operator process is outlined in Algorithm 9.1.

Algorithm 9.1 Skyline operator

```
Input: Data, sense
Output: Pareto front set (mask)
if sense ← max_clos then
    col_Data ← -col_Data
end if
next_point_index ← 0
while next_point_index < length[Data] do
  this_Data ← Data[next_point_index]
  for i←1 to length[Data]
    for j←1 to length [objective]
      if Data[i, j] < this_Data[j]
        current_efficient_points[i] ← True
      end if
    end for
  end for
  current_efficient_points[next_point_index] ← True
  is_efficient ← is_efficient[current_efficient_points]
  Data ← Data[current_efficient_points]
  next_point_index ←next_point_index + 1
end while
is_efficient_mask[is_efficient] ← True
return is_efficient_mask
```

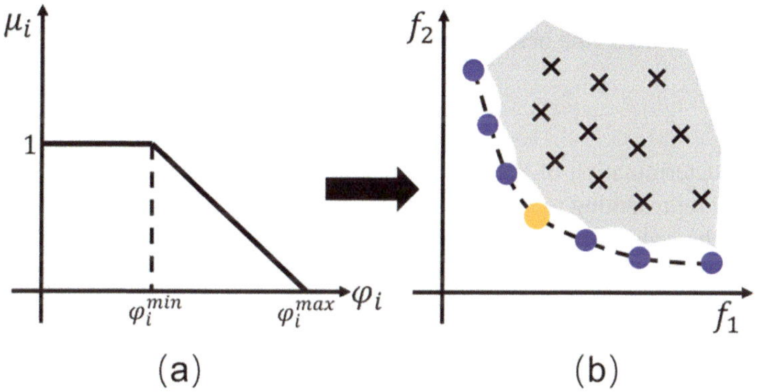

Fig. 9.5 The fuzzy mechanism for the best compromise solution in the Pareto front: (**a**) the membership function; (**b**) the best compromise solution in the Pareto front

9.3.3.3 Selecting the Best Solution

After obtaining the Pareto front using the Skyline operator, this chapter further employs a fuzzy-based mechanism to effectively select one solution as the best compromise solution, as shown in Fig. 9.5. The advantage of the fuzzy-based mechanism is that there are more subjective parameters involved. Therefore, it is more flexible for the managers to determine the importance and expectations of goals based on actual needs and hence make more aggressive or conservative decisions.

By calculating the membership function $\propto_i^k$, each ith objective function among the k-optimal Pareto solution can be expressed in Eq. (9.10):

$$\mu_i^k = \begin{cases} 1 & \varphi_i \le \varphi_i^{\min} \\ \left(\varphi_i^{\max} - \varphi_i\right) / \left(\varphi_i^{\max} - \varphi_i^{\min}\right) & \varphi_i^{\min} \le \varphi_i \le \varphi_i^{\max} \\ 0 & \varphi_i \ge \varphi_i^{\max} \end{cases} \tag{9.10}$$

where $\varphi_i^{\max}$ and $\varphi_i^{\min}$ are the maximum and minimum expected values of ith objective, respectively. For the non-dominated solutions, the normalized membership function μ^k can be optimized by the undermentioned weighted sum technique:

$$\begin{cases} \mu^k = \dfrac{\sum_{i=1}^{N} w_i \mu_i^k}{\sum_{k=1}^{M} \sum_{i=1}^{N} w_i \mu_i^k} \\ \text{subject to}: \sum_{i=1}^{N} w_i = 1 \end{cases} \tag{9.11}$$

where M is the number of non-dominated solutions; w_i is the weight of the ith objective function; N is the total number of optimization targets. Then, the maximum value of μ^k is called "the best compromise solution", written as Eq. (9.12):

$$\mu^{\max} = \max\left(\mu^1, \ldots, \mu^k\right) \tag{9.12}$$

With the determination of the optimal results, a total reduced percentage could be perceived, representing to which extent the optimized results can be achieved compared to the values before optimization. The total reduced percentage (TRP) is based on measuring the reduced percentage of both the emergency time and the number of firefighters, written as Eq. (9.13).

$$\text{TRP} = \frac{K_0 - K}{K_0} + \frac{R_0 - R}{R_0} \tag{9.13}$$

where K_0 and R_0 are the original values of total emergency response time and the number of firefighters; K and R are the total emergency response time and the number of firefighters in the best compromise solution, respectively.

9.4 Experimental Studies

Wuhan Metro is an urban rail transit system serving Wuhan, Hubei Province, China. Its first line, Wuhan Metro Line 1, officially opened in July 2004 and now covers most of the city. Before the COVID-19 pandemic, Wuhan Metro recorded an average daily ridership of 1.0187 million passengers in 2019. As passenger volume continues to grow, effective disaster management has become essential to ensuring metro system safety. In this research, we select the Simenkou metro station of the rail transit line 5 in Wuhan as the studied case to explore the optimization of fire emergency response.

9.4.1 Data and Resources

Simenkou Metro Station is an island-style station with two underground levels. The first basement level (B1) serves as the concourse, while the second basement level (B2) houses the train platform. The total construction area of the station is 15,151.37 m^2. The platform and concourse levels are connected by two sets of passageways, each comprising an escalator and a staircase. Additionally, a vertical elevator is centrally located within the station; however, it ceases operation in the event of a fire. The platform level is linked to external spaces through four entrances and exits, each equipped with both an escalator and a staircase, with effective widths of 1.0 m and 3.86 m, respectively. Ticket gates are positioned near the entrances and exits within the concourse level. The simplified layout of the metro station is shown in Fig. 9.6.

9.4.2 Rescue Model Development

In this section, an HTCPN simulation model of Simenkou Station is developed in CPN Tools to simulate the fire emergency response process. Based on the metro station's fire emergency response plan, an HTCPN-based metro fire emergency response model is constructed, as illustrated in Fig. 9.7. When a fire occurs, sensors immediately transmit data to the relevant departments, triggering the corresponding emergency response level and dispatching emergency signals. These signals are sent to key entities, including rescue teams (including fire brigades and medical teams), the metro dispatching center, the management team (including the ticketing center and power control center), and the metro duty room. Upon receiving the emergency signal, all departments must take immediate action. If fire hazards and evacuations are managed effectively, the central control center will deactivate the emergency response system and initiate preparations to resume metro operations.

Fig. 9.6 Photos of the Simenkou metro station of the rail transit line 5 in Wuhan, China. (**a**) Entrance of the station; (**b**) Station floor; (**c**) Lift and stairs; (**d**) Platform floor; (**e**) Street map and station layout plan; (**f**) Station exit sign

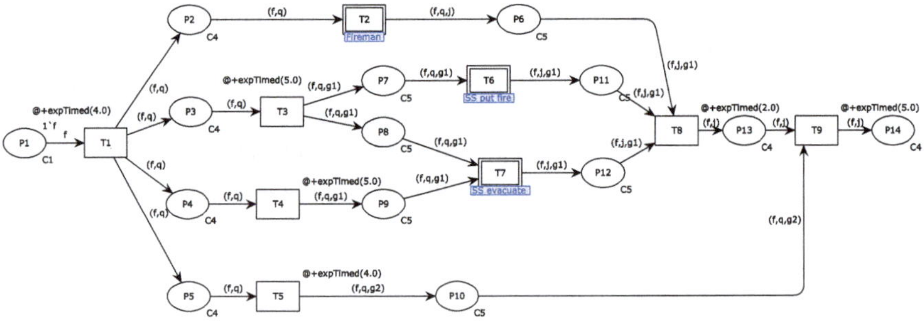

Fig. 9.7 Constructed metro fire emergency response model of main-HTCPN

Finally, the structure of the HTCPN simulation model is qualitatively validated by P_invariant.

Table 9.1 provides explanations for the places and transitions in the main HTCPN model, while Table 9.2 defines the color representations used in the model. Table 9.3 describes the places and transitions in the sub-HTCPN model, and Table 9.4 presents the initial parameter settings of the HTCPN model, incorporating 5% randomness. Among these transitions, T2, T6, and T7 are the most strongly influenced by management decisions and exhibit the highest complexity. Proper configuration of these transitions is expected to significantly enhance emergency response efficiency. In contrast, other transitions have relatively fixed execution times with minor fluctuations. Therefore, in this chapter, within the overall system, three subsystems, T2, T6, and T7, are set up to form a process model, as shown in Fig. 9.8. They are the firefighting team, the passenger evacuation team, and the rescue team, respectively.

Table 9.1 Meanings of places and transitions in the main HTCPN model

Places	Description	Transitions	Description
P1	Fire broke out	T1	Initiate emergency response
P2	The rescue team (119, 120) received the signal	T2	Rescue team firefighting and rescue
P3	Metro duty room receives the signal	T3	PIS receives instructions and gathers metro staff
P4	Metro dispatch center receives the signal	T4	The dispatching center plans passenger evacuation plan
P5	The management team (AFC + SCADA) received the signal	T5	Measures are taken by the management team (AFC + SCADA)
P6	Fire extinguishing is successful and the rescue is over	T6	Staff is putting out the fire
P7	Staff is assembled	T7	The staff is assisting the evacuation of passengers
P8	PIS broadcast information completed	T8	Report results to OCC
P9	Evacuation plan generation	T9	Ready to resume operations
P10	AFC stops ticket sales and SCADA power adjustment completed		
P11	Staff put out the fire		
P12	Passengers' evacuation completed		
P13	OCC confirmed correction		
P14	End emergency response		

Table 9.2 Meanings of colors in the HTCPN model

Color	Value	Description
C1	f	Type of accident: f is a fire accident
C2	q, j	Fire status: fire (q), fire controlled (j)
C3	g1, g2	Emergency organization: response group (g1), management group (g2)
C4	C1 and C2 combined	Combination type
C5	C1, C2, and C3 combined	Combination type
C6	q1 and q2 combined	Combination type
R	C5 and sensor data combined	Combination type
T1	d1	The index number of tokens t1 in the place "team1 num"
T2	d2	The index number of tokens t2 in the place "team2 num"
T3	d3	The index number of tokens t3 in the place "team3 num"
T4	d4	The index number of tokens t4 in the place "team4 num"

1. In transition T2, upon arrival at the scene, the fire brigade of the rescue team is divided into two sub-teams that operate simultaneously. One team is the firefighting team that carries out the firefighting work, and the other is the rescue team that rescues the trapped passengers. During the fire extinguishing process, the heat radiation of the scene changes and depends on the number of firefighters.

Table 9.3 Meanings of places and transitions in the sub-HTCPN model

Places	Description	Transitions	Description
team1	The firefighting teams arrive at the scene	rush1	The rescue team rush to the scene
team2	The passenger rescue teams arrive at the scene	rush2	Staff in charge of firefighting ready to assemble
team3	The staff firefighting teams are assembled	rush3	Staff in charge of evacuation ready to assemble
team4	The staff prevent stampede teams are assembled	team1 ready	The firefighting teams prepare
team5	The staff evacuation team is assembled	team2 ready	The passenger rescue teams prepare
wait1	The firefighting teams wait for action	team3 ready	The staff firefighting teams prepare
wait2	The passenger rescue teams wait for action	team4 ready	The staff prevent stampede teams prepare
wait3	The staff firefighting teams wait for action	start 1	The firefighting team move to target location
wait4	The staff prevent stampede teams wait for action	start 2	The passenger rescue team move to target location
count1	Sensors identify the large ignition points	start 3	The staff firefighting team move to target location
count2	Sensors identify the trapped passenger points	start 4	The staff prevent stampede team move to target location
count3	Sensors identify the small ignition points	stop 1	The firefighting team perform missions
count4	Sensors identify the evacuation congestion points	stop 2	The passenger rescue team perform missions
busy1	The firefighting team is ready	stop 3	The staff firefighting team perform missions
busy2	The passenger rescue team is ready	stop 4	The staff prevent stampede team perform missions
busy3	The staff firefighting team is ready	stop 5	The staff evacuation team perform missions
busy4	The staff prevent stampede team is ready	leave1	The rescue team leave the scene
team1num	The firefighting team is on standby	leave2	Staff in charge of firefighting leave the scene
team2num	The passenger rescue team is on standby	leave3	Staff in charge of evacuation leave the scene
team3num	The staff firefighting team is on standby		
team4num	The staff prevent stampede team is on standby		
finish1	The firefighting teams complete all missions		

Table 9.3 (continued)

Places	Description	Transitions	Description
finish2	The passenger rescue teams complete all missions		
finish3	The staff firefighting teams complete all missions		
finish4	The staff prevent stampede teams complete all missions		
finish5	The staff evacuation team completes all missions		

Table 9.4 Initial color parameter setting of the HTCPN model (5%)

Parameters	Definition	Value	Unit
n1	Number of large ignition points identified by the sensors	/	
n2	The number of trapped passenger points identified by the sensors	/	
T1	Number of extinguishing teams of firefighters	3	Group
T2	Number of rescue teams of firefighters	2	Group
x	Heat radiation of large ignition points	/	kW/m^2
y	Number of dispatched firefighters per group	/	
d	Distance between trapped passengers and firefighters	190–210	m
s1	Speed of firefighters and trapped passengers	52–58	m/min
s2	Speed of firefighters	162–178	m/min
n3	Number of small ignition points recognized by the sensors	5	
T3	Number of extinguishing teams of the metro staff	3	Group
n4	Number of evacuation congestion points identified by sensors	3	
T4	Number of rescue teams of the metro staff	2	Group
q1	Number of passengers on metro train	888–945	
q2	Number of people waiting on the metro platform	57–63	
A1	Escalator passing capacity	160	Person/(min·m)
A2	Stair passing capacity	70	Person/(min·m)
B	Number of escalators	3	
N	The total width of the stairs	2	Meter

Note: The parameter values in Table 9.4 are obtained according to the survey of the metro station

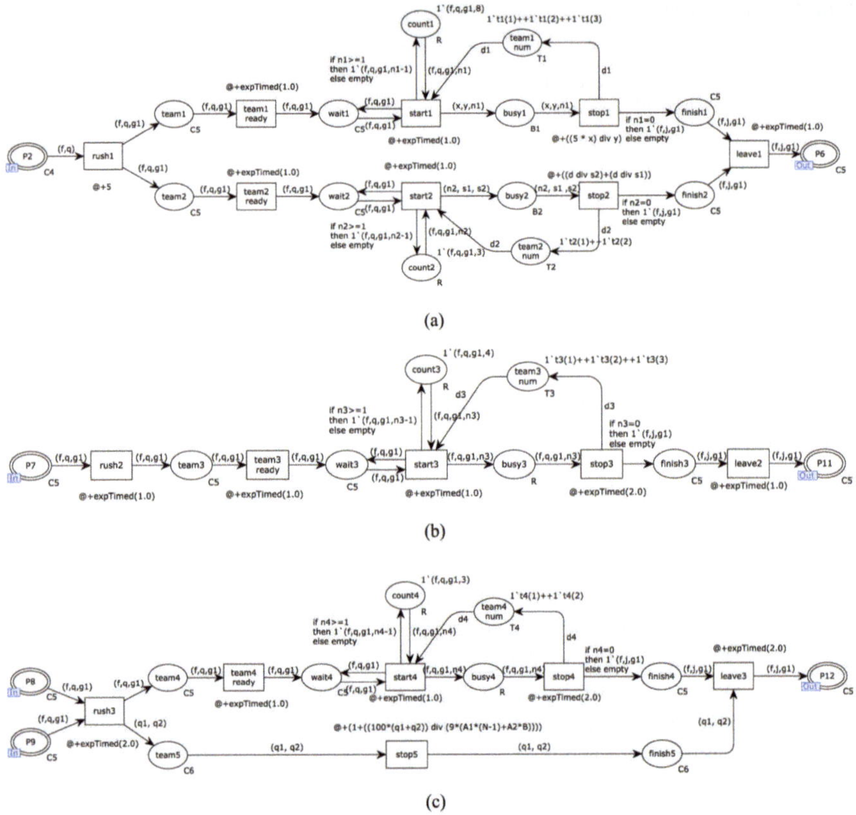

Fig. 9.8 Constructed metro fire emergency response model of sub-HTCPN: (**a**) detailed procedure module of the transition T2; (**b**) detailed procedure module of the transition T6; (**c**) detailed procedure module of the transition T7

Zhou et al. [16] proposed a model to describe the variation of the heat radiation when firefighters were working on extinguishing the fire, written as Eq. (9.14).

$$r(t+\Delta t)=\begin{cases} r(t)-0.2\cdot p_1\cdot\Delta t, & r/2\le p_1 \\ r(t)+0.2\cdot\left(r(t)/2-p_1\right)\cdot\Delta t, & r/2>p_1 \end{cases} \tag{9.14}$$

where $r(t)$ represents the heat radiation at time t; p_1 is the number of firefighters carrying out firefighting work; Δt is the time increment (unit: minutes). This relationship is based on the assumption that a firefighter can extinguish 2 kW/m^2 heat radiation of a fire within 10 min. For the rescue team that rescues the trapped passengers, the rescue time is affected by many factors, such as the number of trapped passengers, distance, and thermal radiation. Eq. (9.15) describes the time it takes for the firefighters to rescue trapped persons, written as:

$$t = \frac{d}{s_1} + \frac{d}{s_2} + \frac{r(t)\cdot(c+1)}{2\cdot p_2} \tag{9.15}$$

where d is the length of the firefighters' route to rescue trapped passengers; s_1 is the moving speed of the firefighters; s_2 is the moving speed of the firefighters when carrying passengers; c is the number of trapped passengers; p_2 is the number of firefighters in the rescue team; $r(t)$ is the radiation at time t.

2. In transition T6, before the fire brigade arrives, on-site staff must take immediate measures to contain less severe fires that remain within their control capacity. The fire extinguishing team is divided into three sub-teams, with fire suppression positions determined based on data collected from thermal radiation sensors. In transition T7, staff members are primarily responsible for guiding passengers in an orderly evacuation. Upon receiving a congestion signal from sensors, they immediately proceed to congested locations and assist passengers until the evacuation is complete. According to the calculation method outlined in GB 50157–2013: Code for Design of Metro, the evacuation time for the metro platform level T, can be computed as Eq. (9.16)

$$T = 1 + \frac{Q_1 + Q_2}{0.9\left[A_1(N-1) + A_2 B\right]} \tag{9.16}$$

where Q_1 is the number of passengers on the train; Q_2 is the sum of the number of people waiting on the platform and the number of staff on the platform; A_1 is the escalator passing capacity (unit: person/(min·m)); A_2 is the pedestrian staircase passing capacity (unit: person/(min·m)); N is the number of escalators; B is the total width of the pedestrian stairs. This type is mainly used in metro stations with wide platforms and multiple stairs and elevators.

3. The model built is correct, reasonable, and effective since it meets the requirements of boundedness, activity, and accessibility. According to the incidence matrix C in Eq. (9.17), the P_invariant can be derived from $C^T \times X = 0$. Then, we have $X_1^T = (1,1,0,0,0,1,0,0,0,0,0,0,1,1)$, $X_2^T = (1,0,0,1,0,0,0,1,0,0,1,0,1,1)$, $X_3^T = (1,0,0,0,1,0,1,0,0,0,0,0,0,1)$, $X_4^T = (1,0,1,0,0,0,0,0,1,0,0,1,1,1)$, and $X_5^T = (1,0,1,0,0,0,0,0,0,1,1,0,1,1)$. The correlation matrix and P-invariant analysis confirm that the model exhibits boundedness, ensuring smooth connections between activities in the emergency response process. Additionally, the model incorporates both sequential and parallel structures, eliminating the possibility of dead tasks. Based on the definition of accessibility, all states in the model are reachable, indicating the absence of deadlocks. Consequently, the Petri net accurately represents the metro station fire emergency response process and can serve as a practical guide for real-world emergency management.

$$C = \begin{bmatrix} -1 & 0 & 0 & 0 & 0 & 0 & 0 & 0 & 0 \\ 1 & -1 & 0 & 0 & 0 & 0 & 0 & 0 & 0 \\ 1 & 0 & -1 & 0 & 0 & 0 & 0 & 0 & 0 \\ 1 & 0 & 0 & -1 & 0 & 0 & 0 & 0 & 0 \\ 1 & 0 & 0 & 0 & -1 & 0 & 0 & 0 & 0 \\ 0 & 1 & 0 & 0 & 0 & 0 & 0 & -1 & 0 \\ 0 & 0 & 0 & 0 & 1 & 0 & 0 & 0 & -1 \\ 0 & 0 & 0 & 1 & 0 & -1 & 0 & 0 & 0 \\ 0 & 0 & 1 & 0 & 0 & 0 & -1 & 0 & 0 \\ 0 & 0 & 1 & 0 & 0 & -1 & 0 & 0 & 0 \\ 0 & 0 & 0 & 0 & 0 & 1 & 0 & -1 & 0 \\ 0 & 0 & 0 & 0 & 0 & 0 & 1 & -1 & 0 \\ 0 & 0 & 0 & 0 & 0 & 0 & 0 & 1 & -1 \\ 0 & 0 & 0 & 0 & 0 & 0 & 0 & 0 & 1 \end{bmatrix} \tag{9.17}$$

9.4.3 Analysis of Results

In this section, the proposed MOO model is applied to the fire emergency response at Simenkou Metro Station. To evaluate its effectiveness, the optimization problem is simulated under four different scenarios: (1) Low thermal radiation and few ignition points; (2) High thermal radiation and few ignition points; (3) Low thermal radiation and many ignition points; (4) High thermal radiation and many ignition points. Thermal radiation exceeding 6 kW/m^2 is classified as high, while 6 kW/m^2 or lower is considered low. Similarly, scenarios with more than five ignition points are categorized as many, whereas those with five or fewer are considered few. To assess optimization performance, five test cases are generated within each scenario, varying in thermal radiation intensity and number of ignition points, as illustrated in Fig. 9.9. The HTCPN-based emergency response process for each test case is executed 2000 times in CPN Tools to account for randomness in the simulation. The overall Pareto front is then derived from the thousand-run-time Pareto points. To determine the best compromise solution, a fuzzy-based mechanism is applied. Finally, optimization results for 20 different test cases are obtained.

To perform the optimization, several assumptions are made: (1) According to the sensor feedback data, the fire intensity is defined as three levels: large (Scenarios I), medium (Scenarios II and Scenarios III), and small (Scenarios IV). (2) For each of the fire intensities, the number of firefighters is distributed equally in each group based on experience and is used as a baseline to compare with the optimized results. For firefighting teams, eight firefighters in each group for large fires; six firefighters in each group when the fire is medium; four firefighters in each group for small fires.

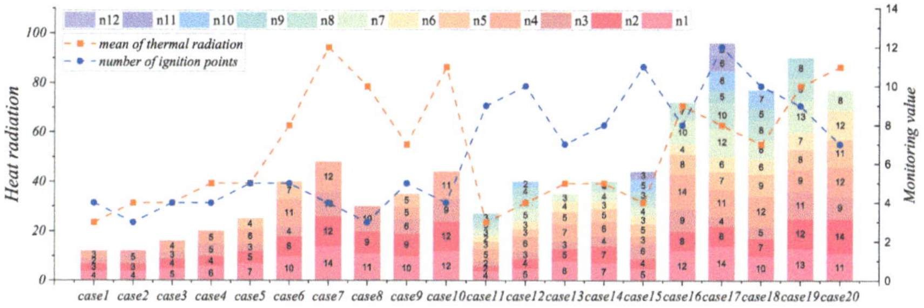

Fig. 9.9 Heat radiation and the number of ignition points for the 20 test cases

For the rescue team, three firefighters are fixed. Then, the baseline total numbers of firefighters for Scenarios I–IV are 18, 24, 24, and 30, respectively. (3) After the firefighters arrive at the scene, they will be assigned to five teams. Three firefighting teams are responsible for extinguishing fires, and two rescue teams are responsible for rescuing trapped passengers. (4) After the firefighters finish the current task, the team settings remain unchanged and continue to execute the next task. (5) After the firefighters of the rescue team have performed their tasks, they do not consider aiding the firefighters who are still putting out the fire (and vice versa). (6) The emergency response time is expected to be within 40 min, and the total number of firefighters should preferably be fewer than 30.

After optimization, the optimal results of the four scenarios are listed in Table 9.5. Figure 9.10 shows the Pareto front, in which the red stars indicate the selected optimal solutions from Scenario I to Scenario IV. Figure 9.11 gives the compromise solutions between the original results and optimized results of the 20 test cases in the four scenarios. Figure 9.12 presents the comparisons between the optimal dispatch and experience dispatch of firefighters. The discussions are as follows.

1. MOO is an effective approach for addressing the complex and conflicting relationship between emergency response time and firefighter deployment. As shown in Fig. 9.10, within the set of non-dominated solutions, an increase in the number of firefighters generally results in a shorter emergency response time. However, due to resource constraints, the number of assigned firefighters cannot be increased arbitrarily, while insufficient firefighter deployment may lead to excessive response times. Therefore, it is essential to allocate the minimum feasible number of firefighters while maintaining an acceptable response time. In Scenarios I–IV, the fuzzy-based mechanism selects an optimal solution from the Pareto front, ensuring that emergency response is completed within 40 min. The fuzzy-based mechanism provides an effective method for identifying the best compromise solution in emergency response optimization.
2. In small- and medium-scale fires, the optimized emergency response time is effectively controlled, enabling a rational allocation of firefighters based on the actual fire conditions. As shown in Fig. 9.11, all 15 test cases in Scenarios I, II, and III exhibit significant optimization, with both emergency response time and

Table 9.5 Emergency response time and the number of firefighters before and after the optimization

Scenarios	Emergency response time (min)			Number of firefighters (person)			TRP
	Before optimization	After optimization	Reduced percentage	Before optimization	After optimization	Reduced percentage	
Scenario I	31.9	20.8	34.8%	18	16.2	10.0%	44.8%
Scenario II	35.3	23.9	32.3%	24	21.6	10.0%	42.0%
Scenario III	34.3	25.6	25.4%	24	20.2	15.8%	41.3%
Scenario IV	42.6	28.6	32.9%	30	31.8	−6.0%	26.8%
Avg. value	36.0	24.7	31.4%	24	22.5	6.3%	38.7%

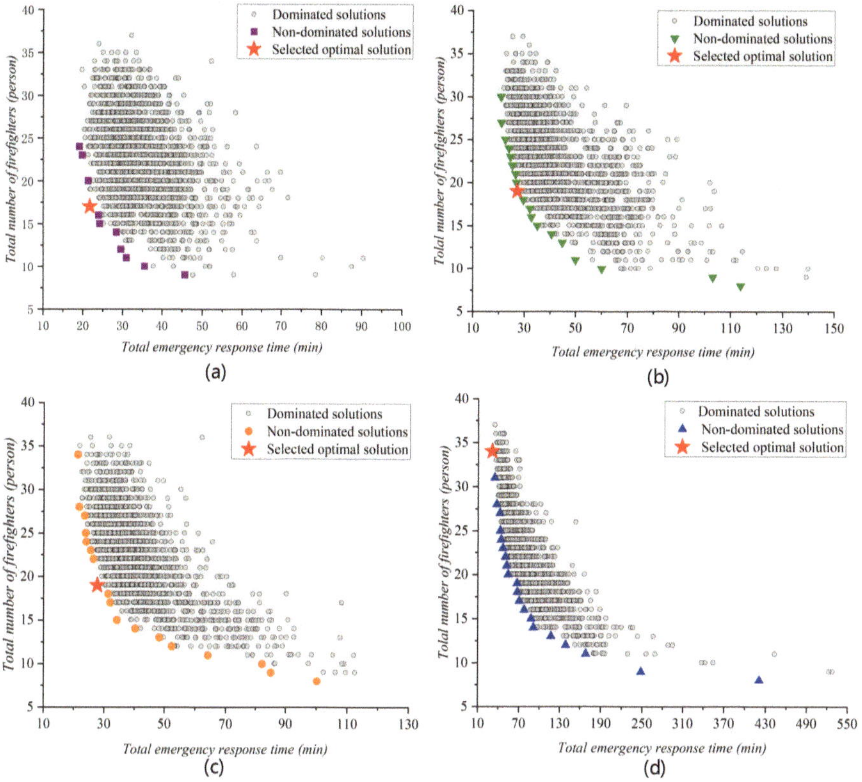

Fig. 9.10 Pareto front and the selected optimal solutions under the four scenarios: (**a**) case3 of Scenario I; (**b**) case6 of Scenario II; (**c**) case12 of Scenario III; (**d**) case17 of Scenario IV

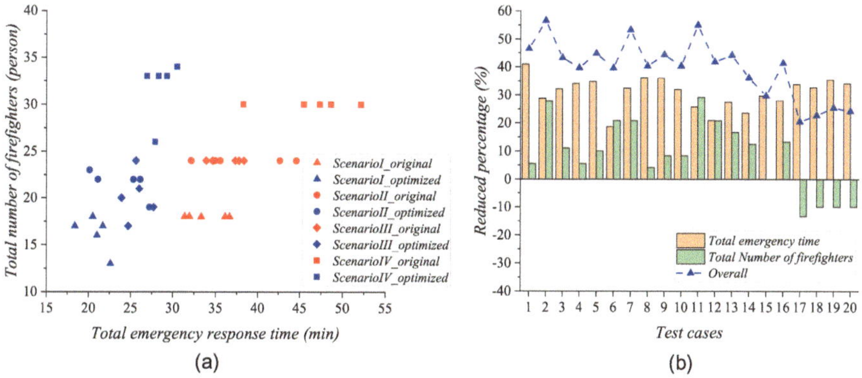

Fig. 9.11 Comparisons of the emergency response time and the number of firefighters between the original and optimized results and the total reduced percentage: (**a**) the emergency response time and the number of firefighters comparison; (**b**) the total reduced percentage

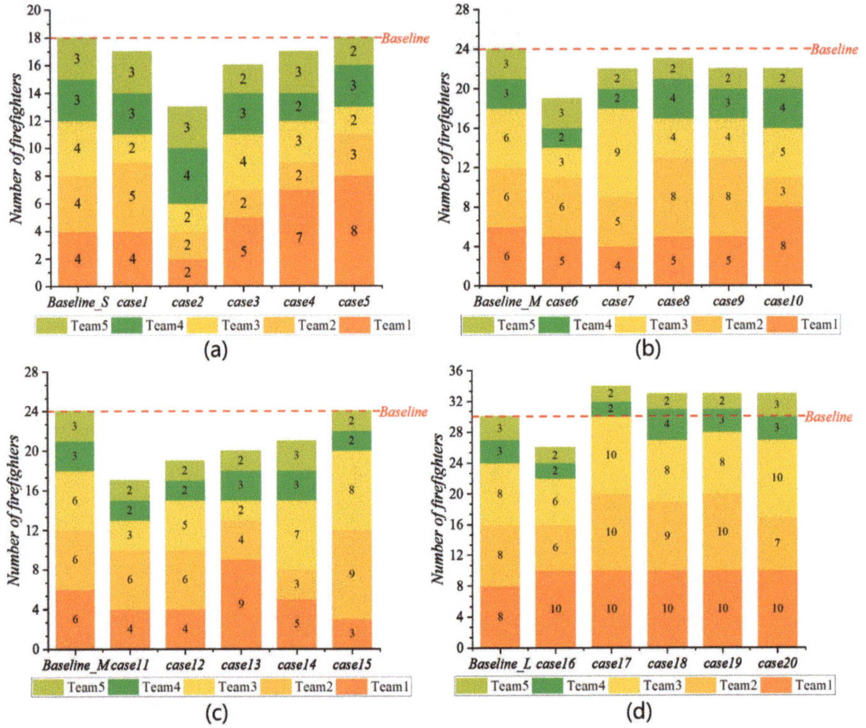

Fig. 9.12 Comparisons between the optimal and experience dispatch of firefighters in the 20 test cases: (**a**) Scenario I; (**b**) Scenario II; (**c**) Scenario III; (**d**) Scenario IV

firefighter deployment reduced compared to pre-optimization levels. As detailed in Table 9.5, TRP achieved for Scenarios I, II, and III are 44.8%, 42.0%, and 41.3%, respectively. The reduction in emergency response time is greater than the optimization in firefighter deployment. In Scenarios I and II, a 10% reduction in the number of firefighters results in an emergency response time decrease of 34.8% and 32.3%, respectively. In Scenario III, the firefighter reduction reaches 15.8%, exceeding that of Scenarios I and II. However, since firefighter availability and response time are inherently conflicting, the emergency response time is only reduced by 25.4%, as fewer firefighters are available.

3. In large-scale fires, completing the metro fire emergency response within a reasonable time often requires an increased number of firefighters, thereby preventing significant economic losses. As shown in Fig. 9.10, to ensure that the emergency response is completed within 40 min, the Pareto frontier selects the leftmost point, representing the shortest response time. As detailed in Table 9.5, the emergency response time in Scenario IV was reduced by 32.9%, but the number of firefighters increased by 6%. Among the five test cases in Scenario IV, at least one team reached ten members, as illustrated in Fig. 9.12d. This finding suggests that experience-based allocation is unreliable, and the firefighter group

size limit of ten personnel, which applies to small- and medium-scale fires, is not suitable for large fires. When facing a large fire, relying on experience-based allocation often results in excessively long response times due to an insufficient number of firefighters. Under these circumstances, additional personnel must be deployed to ensure the emergency response is completed within an acceptable timeframe.

4. Even when the total number of firefighters remains the same, different team allocations can significantly impact emergency response time. The Skyline operator optimizes firefighter distribution among teams to minimize response time while reducing overall firefighter deployment. As shown in Fig. 9.12, compared to experience allocation, the number of firefighters in each group will be different in optimization results. In Scenarios I, II, and III, except for case 5 and case 15, the total number of firefighters in other cases is reduced, which saves the human resource of firefighters. In case 2 and case 12, when the heat radiation of each ignition point is not very different, the distribution of the firefighting team is relatively even. However, in cases 5, 9, and 13, one or two teams have a larger number of firefighters to suppress large fires, while the remaining teams have fewer personnel assigned to extinguish smaller fires. Therefore, the reasonable number of firefighters in each team can be determined by the MOO approach, which can appropriately reduce the total number of assigned firefighters and improve the efficiency of emergency response.

9.5 Merits Analysis

As demonstrated above, the simulation-based MOO approach effectively enhances emergency response efficiency while optimizing firefighter resource allocation. Simulation results indicate that firefighting teams generally require more time to complete their tasks compared to rescue teams. Therefore, the interaction between firefighting and rescue teams, where rescue teams assist in fire suppression, plays a crucial role in improving response effectiveness. To evaluate the optimization results of emergency response when considering the interaction of the firefighting teams, this section reconstructs the HTCPN model in the interactive situation (Fig. 9.13) and evaluates the emergency responses in the four scenarios. Table 9.6 lists the optimal results of the four scenarios after optimization. Figure 9.14 gives the comparisons between the original results and interaction-based optimized results of the 20 test cases in the four scenarios. Figure 9.15 presents the comparisons between the optimal dispatch and experience dispatch of firefighters considering the interaction. Detailed discussions are presented as follows.

1. After accounting for interaction effects, the average total reduction percentage across 20 test cases improves, enabling a faster fire emergency response with fewer firefighters. As shown in Table 9.6, the average TRP reaches 44.5%, compared to 38.7% in the optimization without interaction effects (Table 9.5).

Table 9.6 Emergency response time and the number of firefighters before and after the optimization considering the interaction

Scenarios	Emergency response time (min)			Number of firefighters (person)			TRP
	Before optimization	After optimization	Reduced percentage	Before optimization	After optimization	Reduced percentage	
Scenario I	31.9	21.4	32.9%	18	15.4	14.4%	47.4%
Scenario II	35.3	25.3	28.3%	24	19	20.8%	49.2%
Scenario III	34.3	26.9	21.6%	24	18	25.0%	46.6%
Scenario IV	42.6	31.4	26.3%	30	27.4	8.7%	35.0%
Avg. value	36.0	26.3	27.3%	24.0	20.0	17.2%	44.5%

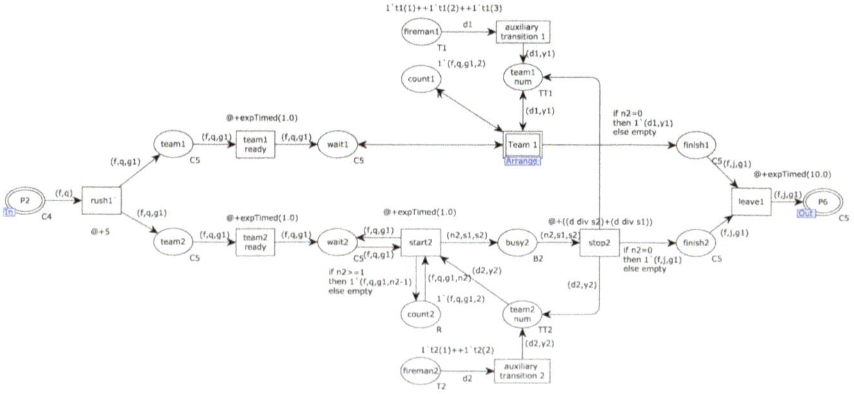

Fig. 9.13 HTCPN model in the interactive situation

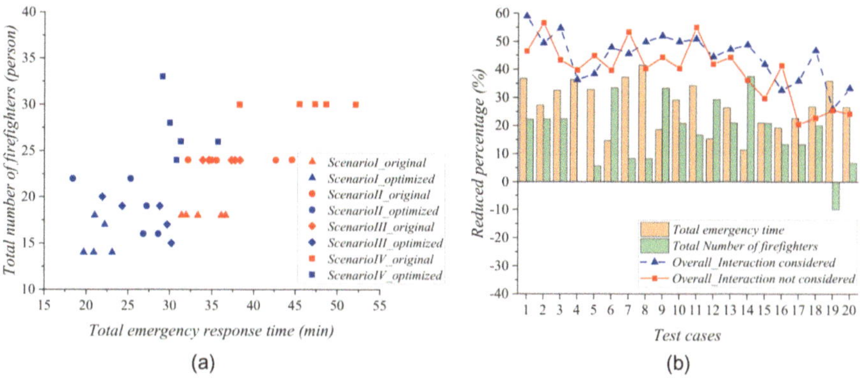

Fig. 9.14 Comparisons of the emergency response time and the number of firefighters between the original and interaction-based optimized results and the total reduced percentage: (**a**) the emergency response time and the number of firefighters comparison; (**b**) the total reduced percentage

Specifically, emergency response time improves by 27.3%, while the required number of firefighters decreases by 17.2% on average. These findings demonstrate that incorporating interaction effects in simulation and optimization enhances the effectiveness of emergency response planning.

2. The interactive effects vary across the four scenarios. In large-scale fires (Scenario IV), optimization results incorporating interaction effects show a significant reduction. In contrast, the impact of interaction in smaller fires (Scenarios I, II, and III) is less pronounced. As shown in Tables 9.5 and 9.6, the TRP increases by 2.6%, 7.2%, 5.3%, and 8.2% in Scenarios I, II, III, and IV, respectively. In Scenario I, the TRP increase is only 2.6% due to the limited number of fire points and lower heat radiation. Consequently, the time difference between firefighting and rescue operations is minimal. By the time the rescue team has completed its mission, the fire extinguishing team has nearly finished its task,

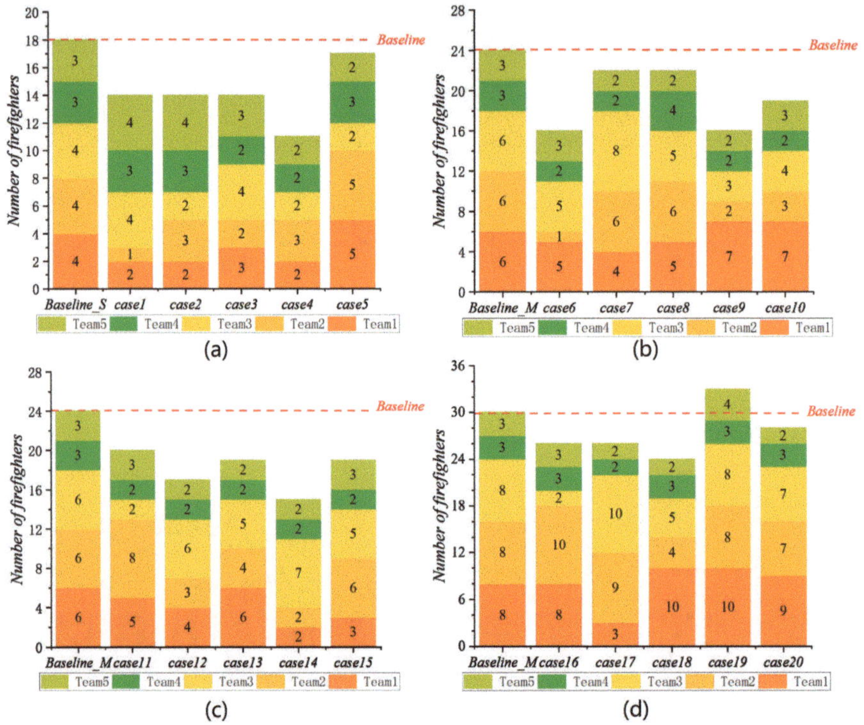

Fig. 9.15 Comparisons between the optimal and experience dispatch of firefighters in the 20 test cases considering the interaction: (**a**) Scenario I; (**b**) Scenario II; (**c**) Scenario III; (**d**) Scenario IV

resulting in negligible interaction effects. However, in Scenario IV, the TRP increases by 8.2%, indicating that interaction effects contribute significantly to fire emergency control. Considering interaction effects allows emergency response efficiency to be achieved without increasing the number of firefighters. Specifically, without interaction effects, firefighter deployment must increase by 6% to achieve a 32.9% reduction in emergency response time. In contrast, with interaction effects, the optimized emergency response time is reduced by 26.3% (<32.9%), but the required number of firefighters decreases by 8.7%. These findings suggest that incorporating interaction effects positively impacts emergency response optimization, particularly in large-scale fires.

3. After accounting for interaction effects, the optimal firefighter allocation decreases compared to scenarios without interaction effects. This suggests that the rescue team's assistance alleviates the workload of the firefighting team. As shown in Fig. 9.15a–c, in small- and medium-sized fires, each fire extinguishing team requires one to two fewer firefighters. In large-scale fires, firefighter teams are no longer at full capacity. In Cases 16, 17, and 18, smaller teams are strategically positioned to await rescue team support, optimizing firefighter deployment. However, in Cases 19 and 20, where no low-heat fire points exist, maximizing firefighter deployment is necessary to contain the fire as quickly as possible.

9.6 Chapter Summary

This chapter proposes an approach integrating the simulation techniques and the MOO methods for the simultaneous optimization of both the emergency response time and the number of firefighters allocated. The HTCPN model is constructed to simulate the process of emergency response at the metro station, and the Skyline operator is employed to find out the Pareto front of the MOO problem based on the simulation dataset. Finally, the optimal compromise solution is selected through a fuzzy-based mechanism. The novelty of this research lies in that: (a) Through Petri nets combined with the emergency response process of metro stations, an HTCPN model is constructed, which appropriately represents the relationship between the relevant parts; (b) The optimal emergency response strategy is given from a MOO perspective based on the simulated dataset, considering response time and allocation of firefighters.

A case study of fire emergency response at Simenkou metro station in Wuhan, China, is conducted to assess the feasibility and effectiveness of the proposed approach. The findings indicate that: (1) the HTCPN model of the Simenkou station, which is established and simulated in the CPN Tools, is correct, reasonable, and effective since it meets the requirements of boundedness, activity, and accessibility. (2) MOO can solve the complex and conflicting relationship between emergency response time and the number of firefighters in different situations. For all four scenarios, the proposed optimization method can reduce the emergency response time and the number of assigned firefighters, with a 38.7% total reduced percentage on average. (3) Incorporating interaction effects in simulation and optimization further improves performance, resulting in an average total reduction percentage of 44.5%. This enhancement enables faster fire emergency responses with fewer firefighters.

The proposed approach fills the research gap that few current studies investigate the simultaneous optimization of time control and resource allocation of the fire emergency response at metro stations. Consequently, it provides scientific guidance for real-time decision-making. Additionally, this approach serves as a decision-support tool for emergency command departments, aiding in the development of effective optimization plans that account for the complexity and multi-objective nature of fire emergency response events.

References

1. Zhang L, Liu M, Wu X, AbouRizk SM (2016) Simulation-based route planning for pedestrian evacuation in metro stations: a case study. Autom Constr 71:430–442. https://doi.org/10.1016/j.autcon.2016.08.031
2. Liu Q, He R, Zhang L (2022) Simulation-based multi-objective optimization for enhanced safety of fire emergency response in metro stations. Reliab Eng Syst Saf 228:108820. https://doi.org/10.1016/j.ress.2022.108820

3. Zhao H, Niu C, Dou X, Liang J (2024) Urban multipoint fire disaster emergency simulation based on web information. Int J Disaster Risk Reduct 101:104223. https://doi.org/10.1016/j.ijdrr.2023.104223
4. Liu Q, Guo K, Wu X, Xiao Z, Zhang L (2024) Simulation-based rescue plan modeling and performance assessment towards resilient metro systems under emergency. Reliab Eng Syst Saf 241:109642. https://doi.org/10.1016/j.ress.2023.109642
5. Ben Attia H, Kahloul L, Benhazrallah S, Bourekkache S (2020) Using hierarchical timed coloured petri nets in the formal study of TRBAC security policies. Int J Inf Secur 19:163–187. https://doi.org/10.1007/s10207-019-00448-9
6. Scheller DS, Reglen D (2021) Improving fire department turnout times: training versus sanctions in a high public service motivation environment. Int Public Manag J 24:973–996. https://doi.org/10.1080/10967494.2019.1635236
7. Wang L, Zhao X, Wu P (2024) Resource-constrained emergency scheduling for Forest fires via artificial bee Colony and variable neighborhood search combined algorithm. IEEE Trans Intell Transp Syst 25:5791–5806. https://doi.org/10.1109/TITS.2023.3338017
8. Özyer T, Zhang M, Alhajj R (2011) Integrating multi-objective genetic algorithm based clustering and data partitioning for skyline computation. Appl Intell 35:110–122. https://doi.org/10.1007/s10489-009-0206-7
9. Zhu R, Becerik-Gerber B, Lin J, Li N (2023) Behavioral, data-driven, agent-based evacuation simulation for building safety design using machine learning and discrete choice models. Adv Eng Inform 55:101827. https://doi.org/10.1016/j.aei.2022.101827
10. Zhou Z, Zhou Y, Pu Z, Qi Y, Xu Y (2019) An integrated cellular automata approach for spatial evacuation simulation on metro platforms with smoke spreading. Transp Res Rec 2673:851–864. https://doi.org/10.1177/0361198119836985
11. Li X, Chen W, Wang C, Kassem MA (2022) Study on evacuation behavior of urban underground complex in fire emergency based on system dynamics. Sustainability 14:1343. https://doi.org/10.3390/su14031343
12. Liu H, Hua C, Lei C (2021) Planning for time-varying volunteer firefighter systems under probabilistic service disruptions. Transp Res Part E Logist Transp Rev 154:102459. https://doi.org/10.1016/j.tre.2021.102459
13. Li Q, Deng Y, Liu C, Zeng Q, Lu Y (2016) Modeling and analysis of subway fire emergency response: an empirical study. Saf Sci 84:171–180. https://doi.org/10.1016/j.ssci.2015.12.003
14. Zhou J, Reniers G (2022) Petri-net based cooperation modeling and time analysis of emergency response in the context of domino effect prevention in process industries. Reliab Eng Syst Saf 223:108505. https://doi.org/10.1016/j.ress.2022.108505
15. Borzsony S, Kossmann D, Stocker K (2001) The Skyline operator. In: Proceedings 17th international conference on data engineering, pp 421–430. https://doi.org/10.1109/ICDE.2001.914855
16. Zhou J, Reniers G (2016) Petri-net based simulation analysis for emergency response to multiple simultaneous large-scale fires. J Loss Prev Process Ind 40:554–562. https://doi.org/10.1016/j.jlp.2016.01.026

Chapter 10
Resource Preparedness Planning

10.1 Introduction

In regard to enhancing resilience in metro systems during disaster events, one of the key measures is the protection of metro stations through the deployment of emergency resources. This involves mobilizing resources from nearby warehouses to the affected stations. A typical scenario is the transportation of flood control resources, such as sandbags and water pumps, from resource warehouses to flooded stations during a storm flood. This allows the protected stations to continue operating during the flood, thereby fulfilling the essential travel needs of citizens. If total resources are limited, those stations that are unable to receive emergency supplies will have to be closed [1]. Therefore, during a flood, an optimal scheme for transporting resources from warehouses to stations will determine which stations can be protected and thus continue operating, thereby determining the service performance of the metro network.

Efficient resource transportation requires reasonable resource preparedness planning, which typically refers to location and inventory planning of resource warehouses. However, most urban flood control warehouses are not specifically designed for metro systems, resulting in several limitations in both warehouse placement and inventory adequacy for station protection. To specify, certain critical metro stations, such as those with high passenger volume or major interchange hubs, may lack nearby warehouses, or the nearby warehouses may not have sufficient resources to provide effective protection. Conversely, large warehouses with ample supplies may be located near stations that are relatively safe from flooding. This mismatch in resource distribution can severely affect metro network performance during flood events and even pose risks to public safety. Thus, establishing flood control resource warehouses dedicated to metro systems is essential, and to this end, it is significant to conduct careful planning for the warehouse location and inventory, which requires optimization techniques to achieve reliable solutions. Although there has been

L. Zhang et al., *Resilience in Urban Transit Systems*, Advances in 21st Century Human Settlements, https://doi.org/10.1007/978-981-95-4322-9_10

substantial research focused on optimizing the location-inventory management of emergency facilities, no research has specifically addressed the location-inventory optimization of resource warehouses for metro floods.

Therefore, this chapter proposes a multi-condition bilevel optimization (MBO) model. As preliminaries for the optimization model, we employ the time-weighted metro network proposed in Chap. 8 to simulate station closures within the metro system and the Average Passenger Utility (APU) is utilized to evaluate the performance of the metro network. We then utilize GIS to perform spatiotemporal storm flood simulations, enabling the calculations of the required resources for each flooded station under a specific flood condition. Subsequently, the MBO model is formulated. The upper-level optimization seeks to maximize the weighted average APU of the metro network across multiple flood conditions with different return periods, while the lower-level optimization focuses on maximizing the total passenger flow of protected stations under a given flood condition. To validate the applicability and effectiveness of the proposed optimization model, we carry out a case study on the Shanghai metro system. Overall, the proposed MBO model incorporates various flood conditions into the bilevel optimization framework, thereby achieving robust and efficient flood control resource positioning.

10.2 Resource Preparedness Planning of Emergency Facilities

The research object in this chapter, flood control resource warehouses, falls under the category of emergency facilities, which are designed to respond to sudden disasters or emergencies. Location and inventory of emergency facilities are considered as strategic decision variables in disaster management. This section reviews related studies in the areas of location optimization and inventory optimization, thereby identifying research gaps that this chapter aims to address.

1. **Location optimization of emergency facilities.** In emergency facility location optimization, most current research concentrates on discrete location problems due to computational constraints, where facilities can only be established at a limited number of candidate locations [2]. Broadly, discrete location problems can be divided into three types: (1) the p-center problem, (2) the p-median problem, and (3) the coverage problem. The first two problem types were introduced by Hakimi [3], Hakimi [4] to minimize either the maximum or weighted average service distance between facilities and demand nodes. These models have been developed and applied in the location optimization of emergency facilities, such as shelters and rescue stations [5–7]. The coverage problem, which is the most widely studied of the three [2], is typically approached from two perspectives. The first minimizes resource use, usually the number of emergency facilities, while meeting demand [8, 9]. The second seeks to maximize the coverage of people and assets within limited resources [10, 11]. Recent research has also

explored probabilistic and stochastic location problems. For example, Rawls and Turnquist [12] introduced a stochastic optimization approach that situates facilities to ensure coverage across various disaster intensities, adapting to unpredictable demand surges in emergency logistics. Similarly, Jia et al. [13] proposed a new two-stage stochastic optimization model to determine the locations for building stations of restoration teams before disasters, considering their routing of conducting restoration tasks after disasters.

2. **Inventory optimization of emergency facilities.** Inventory optimization is also a critical topic for the planning of emergency facilities, which can ensure that resources are available when and where needed. Traditional models, like the Newsvendor model [14], are often applied to supply chain and sales issues; however, they have also been adapted to emergency management to include multiple disaster scenarios and flexible inventory levels [15]. Like location optimization, inventory optimization must account for various uncertainties. In this regard, Pan et al. [16] proposed a medical resource inventory model that addresses uncertain demand and stochastic occurrence time, optimizing costs while ensuring service quality in emergency scenarios. Recent studies have also employed risk-averse models to minimize shortage costs. For example, Noyan [17] proposed a two-stage risk-averse inventory model that balances inventory availability with cost management under uncertain conditions. Such inventory optimization approaches significantly improve resource allocation across locations, enhancing the applicability and effectiveness of emergency facilities.

Despite extensive research, many studies optimize location or inventory in isolation, which risks undermining overall effectiveness during disasters. Integrated location–inventory optimization has emerged to address this limitation. For example, Liu et al. [18] advanced distributionally robust ambulance allocation models, while Wang and Nie [19] introduced a path-based location–inventory–routing model for last-mile supply distribution. However, these models primarily serve isolated demand points requiring resources. For interconnected metro stations, the objective function needs to be refined from simply maximizing the number of covered demand points to enhancing the overall performance of the metro network. Additionally, when planning for location and inventory of flood control resource warehouses, it is crucial to account for various flood conditions to ensure that the solution remains robust across different conditions. To this end, the MBO model proposed in this paper incorporates multiple flood conditions with different return periods, optimizing the weighted average metro network performance across various flood conditions. In this way, this model establishes a more robust framework for resource warehouse planning, tailored to the urban metro systems that demand high responsiveness during flood events.

10.3 Simulation-Based Resource Preparedness Planning Framework

As shown in Fig. 10.1, the overall workflow of the proposed model comprises three main steps. Firstly, we create a network model of the metro system and employ the APU to evaluate the metro system's performance, which will also be used as the objective in the optimization model. Secondly, we employ a Geographic Information System (GIS) to carry out the spatiotemporal flood simulation, which enables the calculation of required resources for flooded metro stations. Thirdly, we establish a multi-condition bilevel optimization (MBO) model. In this model, the upper level optimizes the location-inventory plan to maximize the weighted average APU across different flood conditions, and the lower level optimizes the resource transportation scheme to maximize the protected passengers. The MBO problem is solved by a genetic-algorithm-based (GA-based) nested method. The detailed methodology of each step is elaborated in the following sections.

10.3.1 Metro Network Modelling and Evaluation

As the preliminary of the optimization model, the time-weighted metro network proposed in Chap. 8 that accounts for passengers' travel time is employed, as shown in Fig. 10.2a. Each metro station, represented by a station node S_i, may contain either a single platform (for non-transfer stations) or multiple platforms (for transfer stations), and each platform is represented by a platform node $P_{i,j}$, where the subscripts i and j indicate the station number and the metro line number, respectively. A weighted undirected edge is established between every pair of adjacent platform nodes, with the weight corresponding to the travel time along the edge. The edge connecting platforms P_{i_1,j_1} and P_{i_2,j_2} is denoted as $E_{i_1,j_1}^{i_2,j_2}$, and its weight is denoted as $T_{i_1,j_1}^{i_2,j_2}$.

Using the established metro network model, we can simulate the closures of stations in the metro system. When a station is closed prior to a flood, trains will bypass the closed station, traveling directly from the previous station to the next without stopping. As a result, passengers cannot enter, exit, or transfer at the closed station. In the network model, this station closure is illustrated in Fig. 10.2b. During the flood, metro stations that are not protected by flood control resources will be closed, and our warehouse location and inventory planning aim to optimize the metro network performance under station closures.

To evaluate the performance of a metro network with certain stations closed, we here employ the Average Passenger Utility (APU) in Chap. 8 to assess network performance based on overall passenger satisfaction, which will also serve as the objective of the multi-condition bilevel optimization. The APU, defined as the average of all passengers' utility values, is expressed in Eq. (10.1):

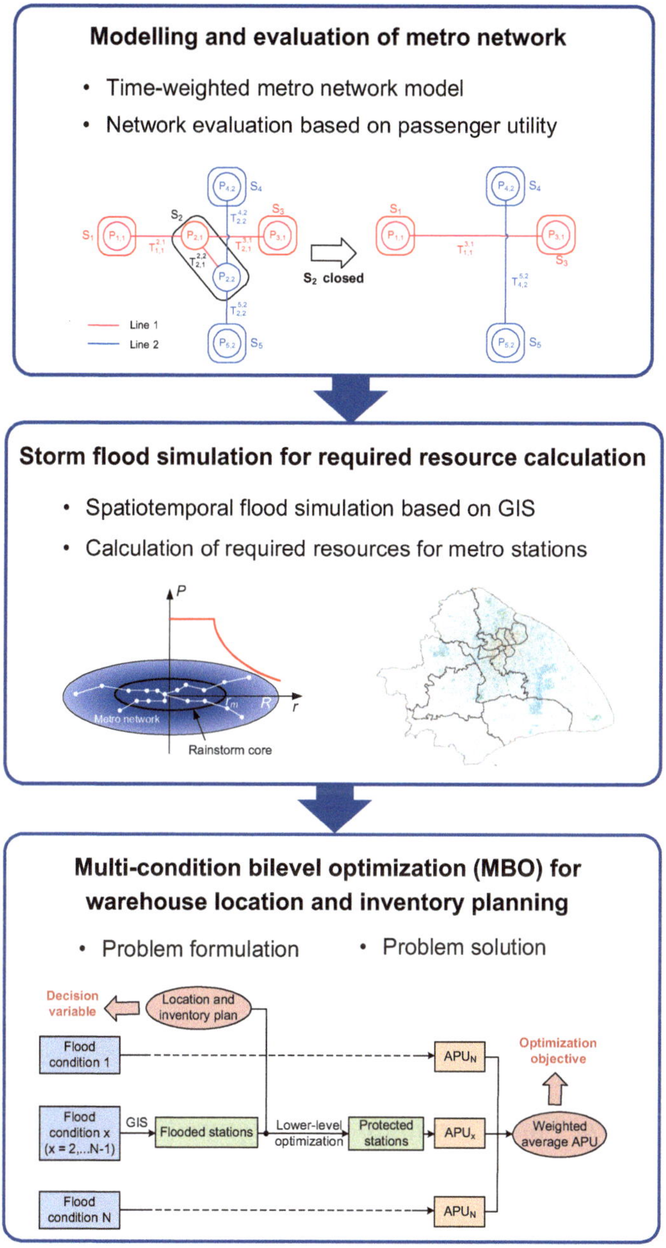

Fig. 10.1 General workflow of the methodology

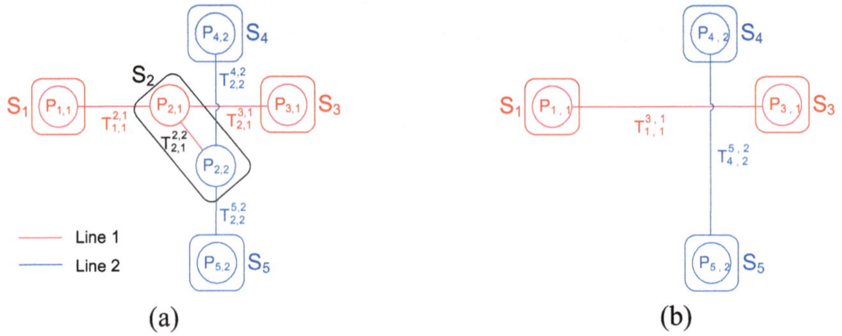

Fig. 10.2 Station closure in a time-weighted metro network: (**a**) normal metro network; (**b**) network with S_2 closed

$$\mathrm{APU} = \frac{\sum_{i_1=1}^{N_S}\sum_{i_2=1, i_2\neq i_1}^{N_S}\left(V_{i_1}^{i_2}\bullet U_{i_1}^{i_2}\right)}{\sum_{i_1=1}^{N_S}\sum_{i_2=1, i_2\neq i_1}^{N_S}V_{i_1}^{i_2}} \tag{10.1}$$

where $U_{i_1}^{i_2} = \begin{cases} -\dfrac{1}{t_e}\Delta ST_{i_1}^{i_2} + 1, & 0 \le \Delta ST_{i_1}^{i_2} \le t_e \\ 0, & \Delta ST_{i_1}^{i_2} > t_e \end{cases}$ denotes the utility for passengers traveling from S_{i_1} to S_{i_2}, in which $\Delta ST_{i_1}^{i_2} = \mathrm{ST}_{i_1}^{\prime i_2} - \mathrm{ST}_{i_1}^{i_2}$ represents the time delay caused by station closures—that is, the difference between the shortest travel time from S_{i_1} to S_{i_2} under partial station closures ($\mathrm{ST}_{i_1}^{\prime i_2}$) and the shortest travel time in the normal network ($\mathrm{ST}_{i_1}^{i_2}$); $V_{i_1}^{i_2}$ represents the daily average passenger flow from S_{i_1} to S_{i_2}, and N_S is the total number of stations in the metro network under normal conditions. With the network structure in normal conditions being given, the APU solely depends on the set of closed stations $\mathcal{S}^c$. Therefore, we can also express the APU as a function of $\mathcal{S}^c$, denoted as $\mathrm{APU}\left(\mathcal{S}^c\right)$.

10.3.2 Storm Flood Simulation for Required Resource Calculation

Strom flood simulation can generate a series of flood conditions, detailing the inundation depths and the required resources of each station under each flood condition. Therefore, it is an important preliminary task for warehouse location-inventory optimization. In this section, we propose a GIS-based technique for spatiotemporal simulation of storm flood, and the flood depth at each metro station entrance is extracted to assess the extent of inundation and used to calculate the required flood control resources.

10.3.2.1 Spatiotemporal Flood Simulation Based on GIS

Before conducting a spatiotemporal flood simulation, it is essential to describe the spatiotemporal characteristics of the rainstorm. For the spatial distribution of rainstorms, we characterize the gradual decrease of rainfall intensity from the core area outward using the areal reduction factor (ARF), which is defined as the ratio of the average rainfall over the studied area to the point rainfall at a specific location. To represent the most adverse rainfall scenario for the metro system, a rainstorm core with the maximum rainfall intensity is set in the area of the city where metro stations are most densely concentrated. The rainfall within the core is maximal and constant, while outside the core, the rainfall gradually decreases based on ARF, as shown in Fig. 10.3. The rainfall at a distance r from the center of the rainstorm core, $P_s(r)$, is expressed by Eq. (10.2):

$$P_s(r) = \begin{cases} P_c,\ r < r_m \\ P_c\left[f\left(\pi(r-r_m)^2\right) + \pi(r-r_m)^2 f'\left(\pi(r-r_m)^2\right)\right],\ r \geq r_m \end{cases} \tag{10.2}$$

where P_c is the rainfall in the core area, r_m is the radius of the core area, and f is a known function derived from local rainfall records.

A rainstorm is also non-uniform in terms of temporal scale, and the Chicago hyetograph has been widely used in describing the temporal characteristics of a rainstorm. Therefore, this section employs the Chicago hyetograph to calculate the rainstorm time series, which is expressed as Eq. (10.3):

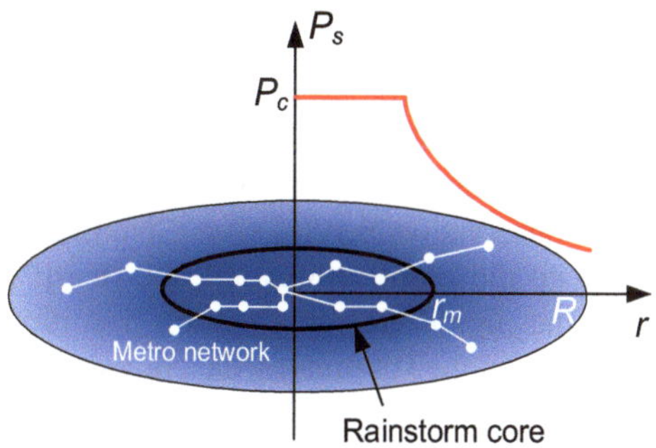

Fig. 10.3 Rainstorm core covering the metro-dense region and the pattern of rainfall variation

$$i(t)=\begin{cases}\dfrac{a\left[\dfrac{t_p-t}{r}(1-n)+b\right]}{\left(\dfrac{t_p-t}{r}+b\right)^{1+n}}, & 0<t\le t_p\\[2ex] \dfrac{a\left[\dfrac{t-t_p}{1-r}(1-n)+b\right]}{\left(\dfrac{t-t_p}{1-r}+b\right)^{1+n}}, & t_p<t\le t_d\end{cases} \tag{10.3}$$

where $i(t)$ (mm/min) is the instantaneous rainfall intensity at time t; t_p and t_d are the time of peak rainfall and rainstorm duration, respectively; r is the peak coefficient; a, b, n are parameters related to the return period, derived from fitting historical storm data.

With the rainstorm spatiotemporal pattern obtained, we can calculate the rainfall at any location within the studied region for any given minute during a storm with a specified return period and duration. Specifically, based on the design storm's return period, duration, and the local values of the parameters a, b, n, r, the minute-by-minute rainfall time series can be computed using the integral of $i(t)$, which represents the rainfall at the rainstorm core. Subsequently, Eq. (10.2) can be applied to determine the rainfall at other locations, written as Eq. (10.4):

$$P(r,k)=\begin{cases}\int_{k-1}^{k} i(t)dt, \ r<r_m\\ \int_{k-1}^{k} i(t)dt\left[f\left(\pi(r-r_m)^2\right)+\pi(r-r_m)^2 f'\left(\pi(r-r_m)^2\right)\right], \ r\ge r_m\end{cases} \tag{10.4}$$

where $P(r,k)$ (mm) is the rainfall at a distance of r from the rainstorm core center in the kth minute. If the location is represented using the Cartesian coordinate system, we can set $r=\sqrt{x^2+y^2}$ and then $P_{xy}(x,y,k)=P\left(\sqrt{x^2+y^2},k\right)$.

Then, the hydrology toolset in a Geographic Information System (GIS) can be employed to implement the rainfall-induced flood simulation. The methodology primarily follows the framework proposed by Zhang and Pan [20]. Specifically, the Soil Conservation Service Curve Number (SCS-CN) method [21] is first used to calculate the runoff; afterward, the runoff is filled into the depressions extracted from the digital elevation model. Specifically, for a given location with rainfall $P_{xy}(x,y,k)$ in the kth minute, the corresponding runoff is calculated using Eq. (10.5) according to the SCS-CN method:

$$q(x,y,k)=\begin{cases}0, & P_{xy}(x,y,k)<I_a(x,y)\\ \dfrac{\left(P_{xy}(x,y,k)-I_a(x,y)\right)^2}{P_{xy}(x,y,k)-I_a(x,y)+S(x,y)}, & P_{xy}(x,y,k)\ge I_a(x,y)\end{cases} \tag{10.5}$$

where $q(x,y,k)$ (mm) is the runoff at location (x,y) in the kth minute; $S(x,y)=\frac{25400}{\mathrm{CN}(x,y)}-254$ (mm) is the potential maximum retention at location (x,y) where $\mathrm{CN}(x,y)$ is a comprehensive parameter reflecting the permeability of the underlying surface at (x,y). $I_a(x,y)$ (mm) is the initial abstraction, and $I_a(x,y)=0.05S(x,y)$.

For the calculated runoff, we should further subtract the portion that is discharged by the drainage system, thus acquiring the final runoff at (x,y) in the kth minute that will converge on the surface, $Q(x,y,k)$, as expressed in Eq. (10.6). By summing $Q(x,y,k)$ over the entire duration t_d, we obtain the maximum accumulated runoff for each location (x,y) during the storm in Eq. (10.7).

$$Q(x,y,k)=\max\left\{0,q(x,y,k)-q_s\right\} \tag{10.6}$$

$$Q_{\mathrm{accu}}(x,y)=\sum_{k=1}^{t_d}Q(x,y,k) \tag{10.7}$$

where $Q_{\mathrm{accu}}(x,y)$ (mm) is the maximum accumulated runoff at (x,y); q_s (mm) is the amount of rainwater that the urban drainage system can discharge per minute.

The basic principle of using GIS for flood simulation involves filling the depressions in the studied region with the accumulated runoff, Q_{accu}. Pits in the DEM should be filled using GIS hydrology tools to identify depression areas from the regional digital elevation model (DEM). Then, depressions can be generated by subtracting the original DEM from the filled DEM. A D8 algorithm is then used to calculate flow direction, based on which the flow accumulation is calculated, and a threshold is set for flow accumulation to extract the stream. On this basis, the depressions can be classified into two categories: (1) isolated depressions and (2) streamside depressions. The isolated depressions are not located on the extracted stream. For each isolated depression, the water that accumulates within it comes solely from the runoff within the depression's boundary. The streamside depressions are located on the extracted streams and are connected by the streams. All these depressions are defined as the sub-catchment outlet node, and the accumulated water of a streamside depression is from both its sub-catchment's runoff and upstream. We can then calculate the accumulated water volume, W_{accu}, for a depression as follows:

$$W_{\mathrm{accu}}=\begin{cases}\int_{y_l}^{y_r}\int_{x_l}^{x_r}Q_{\mathrm{accu}}(x,y)\,dxdy, & \text{for isolated depressions}\\ W_{\mathrm{up}}+\int_{Y_l}^{Y_r}\int_{X_l}^{X_r}Q_{\mathrm{accu}}(x,y)\,\mathrm{d}xdy, & \text{for streamside depressions}\end{cases} \tag{10.8}$$

where $[x_l, x_r]$ and $[y_l, y_r]$ define the intervals of an isolated depression; $[X_l, X_r]$ and $[Y_l, Y_r]$ constitute the boundary of the sub-catchment corresponding to the streamside depression. W_{up} represents the water volume that flows from the upstream depression into the current streamside depression.

Subsequently, using the depression's storage curve $\mathcal{S}$, the water surface elevation in the depression is calculated by Eq. (10.9). The inundation depth at each location within the depression is then obtained by subtracting the elevation of that location from the water surface elevation, written as Eq. (10.10).

$$E_W = \mathcal{S}^{-1}\left(W_{\text{accu}}\right) \tag{10.9}$$

$$I_d\left(x,y\right) = E_W - E_s\left(x,y\right) \tag{10.10}$$

where $\mathcal{S}^{-1}$ is the inverse function of the storage curve $\mathcal{S}$; E_W is the water surface elevation in the depression; $E_s(x, y)$ is the ground surface elevation at (x, y); $I_d(x, y)$ is the inundation depth at (x, y).

For an isolated depression, if the volume of accumulated water exceeds the depression's maximum storage volume $S_{\max}$, the remaining runoff in that depression area, $W_{\text{accu}} - S_{\max}$, will flow into the outlet node of the sub-catchment where it is located. For a streamside depression, the remaining accumulated water will flow downstream into the next streamside depression. Thus, after calculating the inundation depths of all isolated depressions, the inundation depths of all streamside depressions can then be computed according to the streamflow direction, from upstream to downstream. This GIS-based approach to flood simulation can better account for "high elevation depression" conditions and consider depression connectivity [22]. In the study of Zhang and Pan [20], the validity of this approach has already been verified in the Nangang District of Harbin, China.

For a metro station S_i, we can obtain the ground inundation depth at its entrance and then subtract the step height to determine the inundation depth at the entrance, represented by Eq. (10.11):

$$D_i = I_d\left(x_i, y_i\right) - h_{\text{step}} \tag{10.11}$$

where D_i refers to the inundation depth at the entrance of the station S_i; (x_i, y_i) denotes the coordinates of the entrance; h_{step} represents the height of the station's step. According to the research conducted by Lyu et al. [23], we have $h_{\text{step}} = 0.2$ m.

10.3.2.2 Calculation of Required Resources for Metro Stations

Once we have obtained the inundation depths at the entrances of the metro stations, we can proceed to calculate the flood control resource needs for each station. The resources allocated to metro stations include sandbags and pumps. In this study, we relate the quantity of flood control resources required by a metro station to two main factors: (1) the inundation depth at the station entrance and (2) the number of platforms within the station. A higher water depth at the station entrance indicates a greater volume of floodwater entering the metro station, thereby necessitating more flood control resources. Additionally, a larger number of platforms results in a wider distribution of floodwater within the station, which also demands more resources for effective flood control. Note that it is important to differentiate between underground and aboveground platforms in terms of their required resources. Specifically, underground platforms are at higher risk due to the tendency of water to accumulate in lower areas, thus requiring more resources than aboveground platforms.

To simplify our calculations, we assume that the quantity of resources needed for a metro station is directly proportional to both the inundation depth and the number of platforms. Additionally, we assume that for the same inundation depth, an underground platform requires double the resources compared to an aboveground platform. Consequently, we introduce Eq. (10.12) to calculate the resources required for each metro station under specified flood conditions:

$$p_i = k_p \cdot D_i \cdot \left(P_{a,i} + 2P_{u,i}\right) \tag{10.12}$$

where p_i is the quantity of required resources for Station i; D_i (m) is the flood depth at the entrance of Station i; $P_{a,\,i}$ and $P_{u,\,i}$ denote the number of aboveground and underground platforms in Station i, respectively; k_p is the coefficient indicating the resources needed for an aboveground platform per unit of water depth. Through expert elicitation, we know that for a metro station with an aboveground platform, an inundation depth of 0.1 m typically requires approximately 1 high-capacity mobile water pump and 50 water-swelling sandbags for effective flood protection. This configuration is thus defined as 1 unit of resources. Hence, if p_i is expressed in resource units, k_p can be inversely calculated as: $k_p = \dfrac{p_i}{D_i \cdot \left(P_{a,i} + 2P_{u,i}\right)} = \dfrac{1}{0.1 \times (1 + 2 \times 0)} = 10$, indicating that each additional meter of flood depth at an aboveground platform corresponds to a demand of 10 units of resources.

10.3.3 *Multi-Condition Bilevel Optimization (MBO)*

10.3.3.1 Problem Formulation

This section introduces a multi-condition bilevel optimization (MBO) model for the location and inventory planning of resource warehouses. On one hand, it accounts for metro station inundations under various flood conditions, ensuring that the proposed plan demonstrates robust performance across storm floods with different return periods. On the other hand, it follows the general framework of bilevel optimization, considering the interaction between the upper and lower levels. In this study, the upper-level decision-making involves the urban planning department devising a warehouse location and inventory plan, while the lower-level decision-making involves the emergency department transporting resources from warehouses to flooded stations under each specific flood condition and a given location-inventory plan. The upper-level optimization aims to maximize the weighted average APU of the metro network across multiple flood conditions, while the lower-level optimization targets maximizing the total passenger flow of protected stations.

In this chapter, a discrete location problem is considered, where the location of resource warehouses is selected from a finite set of C_L candidate locations. For general discrete location problems, the decision variable is a binary variable x_i $(i = 1,2,\ldots,C_L)$, where $x_i = 1$ and $x_i = 0$ indicate that candidate location i is selected and not selected, respectively. Here, in order to incorporate inventory information, we allow x_i to be a non-negative integer. When $x_i = 0$, it indicates that the inventory at candidate location i is zero, i.e., it is not selected as a warehouse location. When $x_i \neq 0$, it indicates that the location is selected, and the resource inventory is x_i. Consequently, a decision array $X = \left[x_1,x_2,\ldots,x_{C_L}\right]$ is constructed to represent both the location and inventory information, serving as the upper-level decision variable.

Under a specific flood condition, resources need to be transported from warehouses to metro stations. We classify the metro stations under a given flood condition into three categories based on the inundation depth at the station entrance, D_i:

1. $D_i = 0$. These stations are not inundated and are referred to as "safe stations".
2. $0 < D_i \leq D_m$. Here D_m is the threshold for distinguishing between severe and non-severe inundation. If the inundation does not exceed D_m, although the station will experience some water ingress, it is considered manageable and can be protected. The stations satisfying $0 < D_i \leq D_m$ are referred to as "flooded stations", which are the service targets of resource warehouses.
3. $D_i > D_m$. These stations experience significant inundations. Typically, limited flood control resources cannot meet the needs of all inundated stations, and protecting these stations whose $D_i > D_m$ would consume substantial flood control resources, so it is best for these stations to be closed. Therefore, they are termed "closed stations."

Let S_j^s, S_j^f, and S_j^c represent the sets of safe stations, flooded stations, and closed stations, respectively, under a storm flood condition with a return period of T_j. The lower-level decision variable is the resource transportation scheme Y_j from warehouses to flooded stations under this flood condition. Y_j is a matrix of size $C_L \times \left|S_j^f\right|$, where the element $y_{j,ik}$ in the ith row and kth column denotes the quantities of resources transported from the ith warehouse candidate location to the kth flooded station under the storm flood with return period T_j. With the Y_j matrix, we can determine which flooded stations are protected and which are not, defined as S_j^p and S_j^{np}, respectively, as expressed by Eqs. (10.13) and (10.14):

$$S_j^p = \{k|\sum_{i=1}^{C_L} y_{j,ik} \geq p_{j,k}\} \tag{10.13}$$

$$S_j^{np} = S_j^f - S_j^p \tag{10.14}$$

where $p_{j,k}$ is the resource requirement of the kth flooded station under the T_j-year flood condition. From Eqs. (10.13) and (10.14), it is evident that a flooded station is considered protected only when the resources it receives meet its requirements.

Protected stations can continue normal operations, whereas unprotected stations are deemed to be in emergency closure.

1. Upper-level optimization

   In the upper-level location-inventory planning, we consider m flood conditions with different rainstorm return periods T_1, T_2,..., T_m, where the annual occurrence probabilities over a longer time span are $\frac{1}{T_1}, \frac{1}{T_2}, \ldots, \frac{1}{T_m}$, respectively. Hence, the upper level maximizes the weighted average APU of the metro network across different flood conditions, using the occurrence probabilities as weights, formulated as Eq. (10.15). The upper-level optimization is subject to the total resources constraint, written as Eq. (10.16).

$$\max_{X} \frac{\sum_{j=1}^{n} \mathrm{APU}_j\left(T_j, Y_j, X\right)\frac{1}{T_j}}{\sum_{j=1}^{n}\frac{1}{T_j}} \tag{10.15}$$

$$\text{s.t. } \sum_{i=1}^{C_L} x_i \le p_T \tag{10.16}$$

where APU_j is the metro network APU under the T_j-year flood condition; p_T is the total resources available.

   APU_j depends on T_j, the location-inventory plan X, and the resource transportation scheme Y_j. Specifically, we need to identify the corresponding safe station set S_j^s, flooded station set S_j^f, and closed station set S_j^c for the given T_j-year flood condition. For the flooded stations S_j^f, further, based on X and Y_j, we determine the protected station set S_j^p and the unprotected stations S_j^{np}, thereby allowing us to calculate APU_j for the metro network. In Sect. 10.3.1, APU is regarded as a function of all closed stations in the metro network, i.e., $\mathrm{APU}(S^c)$. Hence, $\mathrm{APU}_j(T_j, Y_j, X)$ can be transformed into an equivalent form in Eq. (10.17):

$$\mathrm{APU}_j\left(T_j, Y_j, X\right) \equiv \mathrm{APU}_j\left(S_j^c\left(T_j\right) \cup S_j^{np}\left(T_j, Y_j, X\right)\right) \tag{10.17}$$

2. Lower-level optimization

   The Y_j in Eqs. (10.15) should be acquired by solving the lower-level optimization expressed as Eq. (10.18), in which V_j is the total passenger flow of protected stations under T_j-year flood condition, written as Eq. (10.19). The lower-level optimization is subject to the constraints in Eqs. (10.20)–(10.22):

$$Y_j = \arg\max V_j\left(Y_j | , T_j | , X\right) \tag{10.18}$$

$$V_j = \sum_{k=1}^{\left|S_j^f\right|} \frac{\sum_{i=1}^{F_W} y_{j,ik}}{p_{j,k}} v_k \tag{10.19}$$

$$\text{s.t. } y_{j,ik} = 0 \ \ if \ d_{ik} > d_s \tag{10.20}$$

$$\sum_{k=1}^{|S_j^f|} y_{j,ik} \le x_i \tag{10.21}$$

$$\sum_{i=1}^{F_W} y_{j,ik} = \begin{cases} p_{j,k}, & \text{Station } k \text{ protected} \\ 0, & \text{Station } k \text{ unprotected} \end{cases} \tag{10.22}$$

where v_k is the daily average passenger flow of the kth flooded station, i.e., the average number of passengers entering, exiting, or transferring (if applicable) at the kth flooded station each day, which can be easily obtained from AFC data; d_{ik} is the distance between the ith warehouse candidate location to the kth flooded station; d_s is the maximum distance from which a warehouse can serve a flooded station.

In the lower-level optimization, Eq. (10.18) indicates that we need to optimize the resource transportation scheme Y_j given T_j and X in order to maximize V_j, the total passenger flow of the protected stations under the flood condition. This is a simple, effective, and reasonable station protection strategy that is highly feasible in practice. Equation (10.20) represents the service distance constraint, which means that resources cannot be transported if the distance between a warehouse and a flooded station exceeds d_s. Equation (10.21) is the inventory constraint, stating that the total quantities of resources dispatched from each warehouse cannot exceed its inventory. Equation (10.22) is the stations' resource receipt constraint, indicating that the resources received by a flooded station must either equal its requirement (meaning the station is protected) or be zero (meaning the station is not protected). This constraint ensures that resources are not wasted. Based on Eq. (10.22), we know that $\frac{\sum_{i=1}^{F_W} y_{j,ik}}{p_{j,k}} = 1$ for a protected station and $\frac{\sum_{i=1}^{F_W} y_{j,ik}}{p_{j,k}} = 0$ for an unprotected station.

Therefore, $\frac{\sum_{i=1}^{F_W} y_{j,ik}}{p_{j,k}}$ can be used as a 0–1 variable to represent the protection status of a flooded station. This is also why it is used as the coefficient of the passenger flow v_k in Eq. (10.19) to calculate the total protected passenger flow.

The framework of MBO is illustrated in Fig. 10.4.

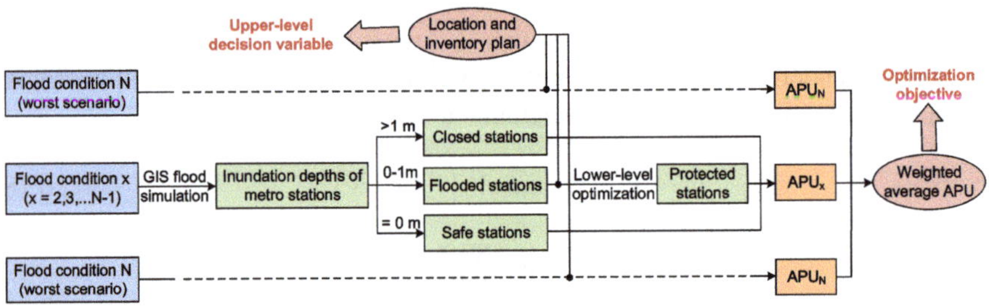

Fig. 10.4 Framework of MBO

10.3.3.2 Problem Solution

For a bilevel optimization problem, when the lower-level problem is convex and sufficiently regular, it can be transformed into a single-level optimization using the Karush-Kuhn-Tucker (KKT) conditions. However, in this study, although the lower-level problem is a linear programming problem, the upper-level problem is highly nonlinear. Therefore, a nested evolutionary algorithm is a suitable choice for solving this problem, where the lower-level problem is solved for each upper-level individual [24]. In this section, we develop a genetic-algorithm-based (GA-based) nested method for solving the proposed problem. Specifically, the upper level employs a GA to optimize the location-inventory plan, while the lower level uses a linear programming method to determine the optimal resource transportation scheme. The detailed steps of the algorithm are as follows:

Step 1: Initialization. Randomly generate an initial population of size N_P, where each individual is represented by a $1 \times C_L$ array, representing the location-inventory plan.

Step 2: Individual correction. Each individual should satisfy the resource constraint. To maximize resource utilization, we correct each individual so that the sum of its elements equals the total available resources. Specifically, when the sum of all elements of an individual does not equal the total available resources, we adjust each element by dividing it by the current sum of elements and then multiplying by total available resources.

Step 3: Fitness calculation. For each individual, perform the lower-level linear programming to obtain the optimal resource transportation scheme, then compute the upper-level objective function, i.e., the weighted average APU in Eq. (10.15), to determine the individual's fitness.

Step 4: Generation of the next generation population. Select individuals from the current population using the tournament selection method. Generate N_P offspring individuals through uniform crossover, uniform mutation, and constraint-based correction. Compute the fitness of the offspring using the method in Step 3. Employ the elite strategy to merge the parent and offspring populations and select the N_P individuals with the highest fitness to form the next generation population.

Step 5: Repeat Step 4 until the fitness of the best individual in the population no longer changes over a certain number of generations. The best individual, in terms of fitness, is taken as the optimal solution for the upper-level optimization, i.e., the optimal location-inventory plan.

10.4 Experimental Studies

10.4.1 Data and Resources

To verify the applicability and validity of the proposed model, this section examines the Shanghai metro system in China as a case study. Similar to Chap. 8, we obtained the metro system map from its official website and the travel time between platforms from the Gaode Map App. To calculate the network APU, we used AFC data from April 2015 provided by the 2015 Shanghai Open Data Application. Since the Shanghai Open Data Application only released AFC data for that particular month, we rely solely on this data for our study. Therefore, the constructed metro network is also based on April 2015 to align with the available AFC data.

Building on this foundation, the case study on flood control warehouse planning can be conducted for the Shanghai metro system. At first, to implement the methodology described in Sect. 10.3.2.1, the pink area in Fig. 10.5 (center: 121°28′49″E, 31°12′54″N; radius: 6694.27 m) is identified as the rainstorm center covering the metro-dense area. For this rainstorm center, we set five storm flood conditions with return periods of 2, 3, 5, 10, and 20 years, each lasting 120 min. According to the Design Rainstorm Distribution for Shanghai [25], the parameters a, b and n in Eq. (10.3) are given as $a = 9.581(1 + 0.846 \log P)$, $b = 7.0$, $n = 0.656$. Additionally, according to this standard, within the design return periods of 2–100 years and rainfall durations of <180 min, a uniform peak coefficient of 0.405 can be applied. Therefore, the instantaneous rainfall intensity based on the Chicago hyetograph can be expressed as Eq. (10.23):

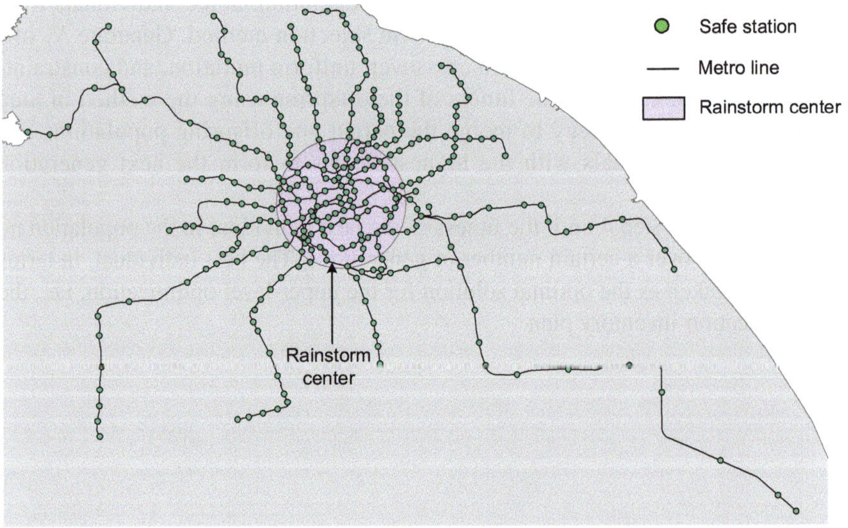

Fig. 10.5 Rainstorm center covering the metro-dense region in Shanghai

$$i(t)=\begin{cases}\dfrac{9.581(1+0.846\log P_R)\left[\dfrac{120\times 0.405-t}{0.405}(1-0.656)+7.0\right]}{\left(\dfrac{t_p-t}{0.405}+7.0\right)^{1+0.656}}, & 0<t\le t_p\\[2ex] \dfrac{9.581(1+0.846\log P_R)\left[\dfrac{t-120\times 0.405}{0.405}(1-0.656)+7.0\right]}{\left(\dfrac{t-t_p}{1-0.405}+7.0\right)^{1+0.656}}, & t_p<t\le t_d\end{cases} \tag{10.23}$$

where $i(t)$ (mm/min) is the instantaneous rainfall intensity at time t; P_R (year) is the rainstorm return period.

Using Eq. (10.23) and Eq. (10.4), the rainstorm time series for the center area and the area outside the center can be calculated. For Shanghai, the function f in Eq. (10.2) is $f = -0.058 \ln \pi r^2 + 1.0529$ [26]. In terms of drainage capacity, the design standard in Shanghai typically adopts a 1-year return period for rainfall in general areas and a 3- to 5-year return period for critical areas [27]. In this study, due to the unavailability of a detailed drainage parameter map, we uniformly adopt the 1-year return period rainfall as a conservative estimate. Based on the Standard of Rainstorm Intensity Formula for Shanghai [25], this corresponds to a drainage capacity of 36 mm/h.; hence, q_s in Eq. (10.8) is 0.6 mm. Then we obtained the DEM of Shanghai from Geospatial Data Cloud (http://www.gscloud.cn/) hosted by Chinese Academy of Sciences. Based on this, the flood simulation is conducted using GIS, and the flood depths at the metro station entrances are extracted. As mentioned in Sect. 10.3.3.1, under a specific flood condition, metro stations are categorized based on the flood depth at the entrances as safe stations, closed stations, and flooded stations. Here the threshold D_m is set as 1 m. Figure 10.6 shows the inundation of the Shanghai area and metro stations under a 5-year flood event.

10.4.2 Candidate Location of Warehouses

The proposed MBO model requires selecting several warehouse locations from a given set of candidate locations and then managing inventory. Due to the lack of information on the candidate locations, this section introduces a method based on Delaunay triangulation and set covering to generate candidate warehouse locations.

First, based on the results of station inundations, we compile all flooded stations under the five flood conditions, as shown in Fig. 10.7a. These 72 flooded stations represent potential service targets of the warehouses, while the other stations remain either safe or closed under the five flood conditions and do not require any flood control resources.

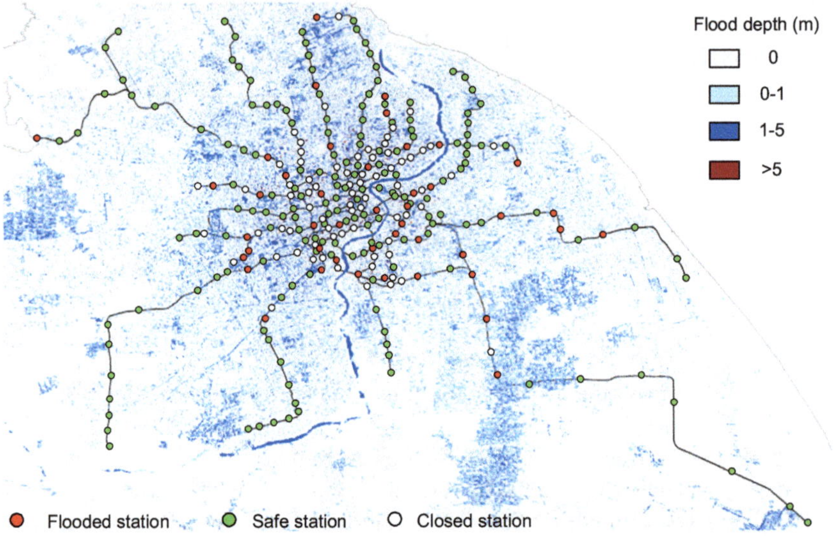

Fig. 10.6 Inundation of the Shanghai area and metro stations under a 5-year flood event

Next, we construct a Delaunay triangulation mesh for these flooded stations, as illustrated in Fig. 10.7b, and generate preliminary candidate locations using the following rules:

1. For each triangle, calculate the distances from the centroid to its three vertices. If all distances are less than d_s, the centroid is considered a preliminary candidate location. If not, proceed to step 2.
2. Calculate the distance from the midpoint of the shortest side of the triangle to its two endpoints. If the distance is less than d_s, the midpoint and the vertex not on the shortest side are considered preliminary candidate locations. Otherwise, the three vertices of the triangle are selected as preliminary candidate locations.

This rule-based approach ensures that every flooded station has at least one warehouse within a distance of d_s. In this study, d_s is set to 5 km, resulting in 146 preliminary candidate locations, as shown in Fig. 10.7b. To minimize the number of candidate locations while ensuring that all flooded stations are covered, we applied a greedy algorithm to identify the minimum number of candidate locations necessary to cover all flooded stations. The final candidate locations with their number and coverage are illustrated in Fig. 10.7c. It is observed that only 15 candidate locations are sufficient to cover the 72 flooded stations fully.

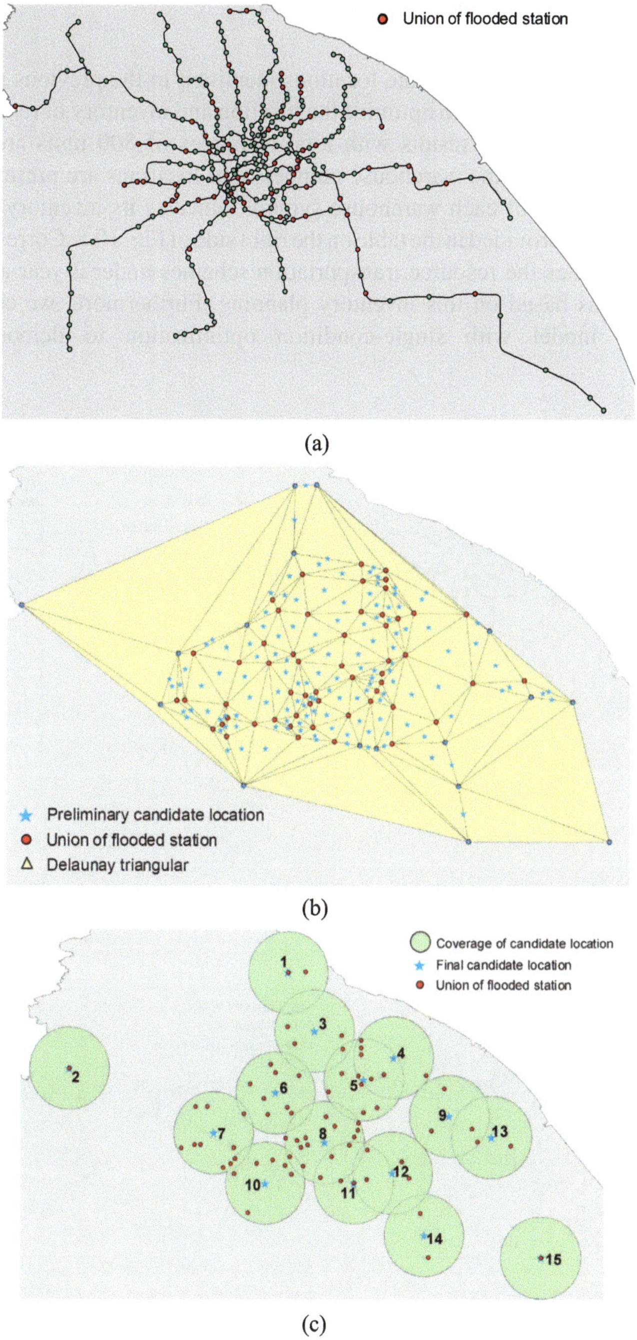

Fig. 10.7 Generation of candidate locations: (**a**) the union of flooded stations in the five flood conditions; (**b**) Delaunay triangulation mesh and preliminary candidate locations; (**c**) 15 final candidate locations with their numbers and coverage

10.4.3 Optimization Results Analysis

Based on the warehouse candidate locations identified in the previous section, the MBO model can be applied to optimize the location and inventory of resource warehouses. The optimization results with total resources of 300 units are shown in Fig. 10.8. In Fig. 10.8, the warehouse numbers and locations are presented on the map, where the size of each warehouse symbol indicates its inventory level, with specific quantities provided in the table on the right side of Fig. 10.8. Correspondingly, Fig. 10.9 illustrates the resource transportation schemes under 2-year and 20-year flood conditions based on this inventory planning. Furthermore, we compare the optimization model with single-condition optimization to demonstrate the

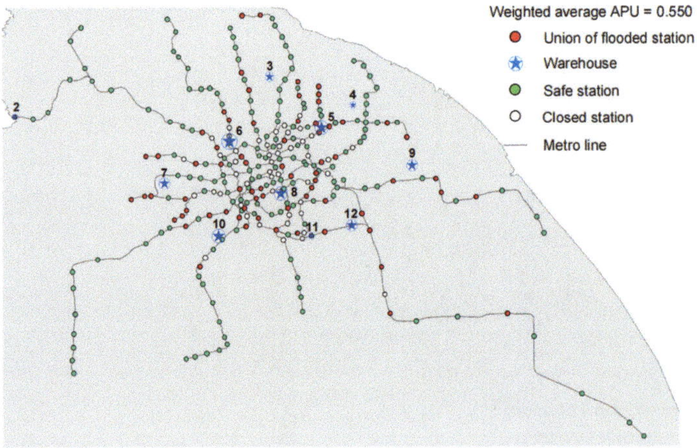

Inventory Table:

Warehouse No.	Inventory (units)
2	1
3	8
4	1
5	54
6	73
7	19
8	57
9	19
10	49
11	1
12	18

Fig. 10.8 Location and inventory planning of resource warehouses with 300 units of total resources

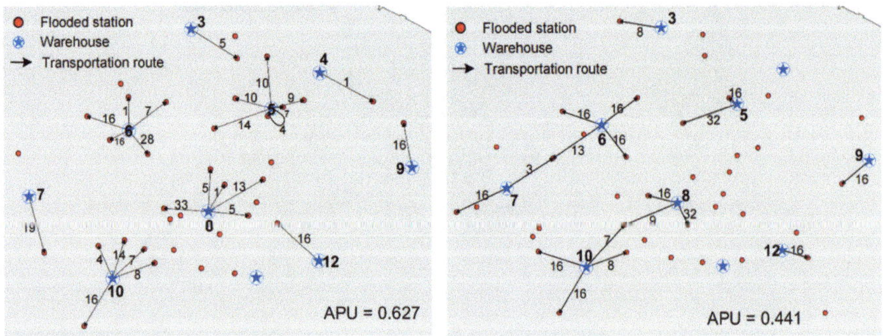

(a) (b)

Fig. 10.9 Resource transportation scheme from warehouses to flood stations: (**a**) 2-year flood condition; (**b**) 20-year flood condition robustness of the multi-condition optimization, and the results are shown in Fig. 10.10 and Table 10.1. The detailed analysis is as follows:

1. The MBO model effectively determines warehouse locations and inventory to achieve the best metro network performance under floods. As shown in Fig. 10.8, the selected warehouse locations are distributed across different areas of Shanghai. Warehouses located in metro-dense areas have significantly higher inventory levels compared to those in peripheral regions. Specifically, Warehouses 5, 6, and 8 have the most resources, with 54, 73, and 57 units, respectively. According to Fig. 10.7c, they also cover a large number of metro stations, specifically 12, 9, and 16 stations, respectively. Based on AFC data, the total average daily passenger flow of the stations covered by Warehouses 5, 6, and 8 is also quite high, at 249331, 349850, and 637,286 person-time, respectively. This configuration effectively protects metro-dense areas and ensures that as many people as possible can travel during floods, thus contributing to a satisfactory weighted average APU of 0.550. Additionally, it can be observed that Warehouses 1 and 11, which have an inventory of 1 and are co-located with flooded stations, allow these limited resources to be directly stored within the metro stations, thus reducing warehouse construction costs. Furthermore, due to the limited total resources

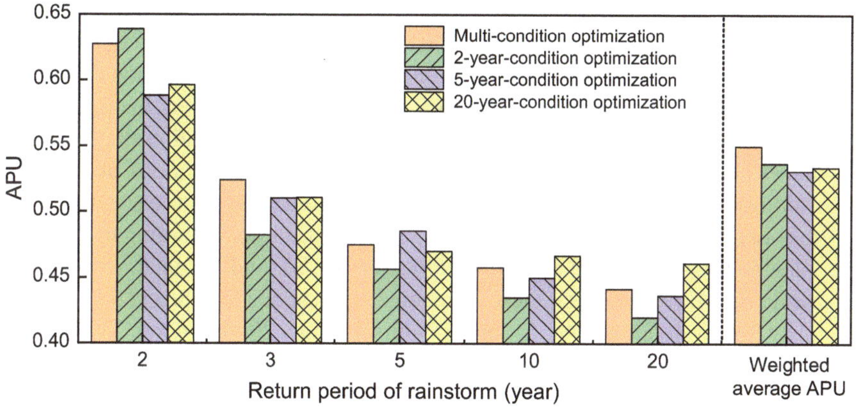

Fig. 10.10 APU of various optimization methods under different flood conditions

Table 10.1 APU of various optimization methods under different flood conditions

Return Period	Multi-condition optimization	Single-condition optimization		
		2-year	5-year	20-year
2	0.627	**0.639**[a]	0.588	0.596
3	**0.524**	0.482	0.510	0.511
5	0.475	0.456	**0.485**	0.470
10	0.458	0.435	0.450	**0.467**
20	0.441	0.420	0.437	**0.461**
Weighted average APU	**0.550**	0.537	0.531	0.534

[a]The boldface entries indicate the highest APU under each flood condition

and varying importance of stations, not all flooded stations can be covered by warehouses. By comparing the candidate locations in Fig. 10.7c with the final selected locations in Fig. 10.8, it is evident that candidate locations 1, 13, 14, and 15 were not chosen as final locations. Consequently, the flooded stations covered solely by these candidate locations are not covered by any warehouse. This is likely due to their relatively low passenger traffic and the fact that they are not major transfer stations, offering limited improvement to APU. Therefore, even though the inundation depth of these flooded stations under the corresponding flood conditions is less than 1 m, they should still be closed.

2. Lower-level optimization effectively transports resources from warehouses to protected stations. As shown in Fig. 10.9, the transportation scheme derived from the lower-level optimization allocates resources from warehouses to their nearby metro station. By calculating the distance between warehouses and protected stations, we found that under 2-year and 20-year flood conditions, the average transportation distance per unit of resource is 3.48 km and 3.42 km, respectively. This suggests that rapid deployment of resources is feasible during floods using the proposed method. Additionally, warehouses can allocate resources differently based on station demand. For example, under a 2-year flood, Warehouse 8 transported 33, 5, 1, 13, and 5 units of resources to five of its nearby metro stations. This is determined by constraint Eq. (10.22), which aims to protect flooded stations without wasting resources. Furthermore, Fig. 10.9b indicates that a single flooded station is not necessarily served by only one warehouse. Under a 20-year flood condition, Warehouse 10 and Warehouse 8 supply 7 and 9 units of resources to a flooded station located between them; a similar situation occurs with Warehouses 6 and 7. This flexible allocation strategy maximizes the use of resources to protect as many flooded stations as possible. Additionally, the total quantity of resources dispatched from each warehouse accounts for a large portion of its inventory. For instance, according to Fig. 10.8, Warehouses 5, 6, and 8 have inventories of 54, 73, and 57 units, respectively, while Fig. 10.9 shows that these warehouses dispatched 54, 68, and 57 units during a 2-year flood, and 48, 61, and 57 units during a 20-year flood. This demonstrates that the stored resources are effectively utilized to protect the flooded stations.
3. The multi-condition optimization method yields robust metro network APU performance across various flood conditions compared with single-condition optimizations. In a single-condition optimization, we extract flooded stations (with inundation depth between 0 and 1 m) under a specific flood condition and generate a Delaunay triangulation mesh to identify candidate locations, then use a GA-based nested method to optimize inventory levels. Note that in this case, the upper-level optimization objective becomes the APU under the specific flood condition, while the lower-level optimization objective is the resource transportation scheme under that flood condition. Figure 10.10 compares the APU obtained by multi-condition optimization and single-condition optimizations based on 2-year, 5-year, and 20-year floods. The left side of the figure shows the comparison of APU achieved by the optimization methods under five different

flood conditions, while the far right displays the weighted average APU across the five flood conditions. The specific values in Fig. 10.10 are presented in Table 10.1. As illustrated, the APU obtained from multi-condition optimization ranks either first or second across different flood conditions, and its weighted average APU is superior to that of any single-condition optimization. Specifically, under the 2-year, 3-year, 5-year, 10-year, and 20-year storm floods, the multi-condition optimization achieves the APU of 0.639, 0.524, 0.485, 0.467, and 0.461, respectively, performing best in the 3-year flood and second best under other flood conditions. As for the weighted average APU, the multi-condition optimization achieves a value of 0.550, outperforming the best single-condition optimization, which is based on a 2-year flood by 1.3 percentage points. This demonstrates that the multi-condition optimization integrates network performance under different flood conditions into its objective function, thus providing a comprehensive consideration of these conditions. By contrast, in single-condition optimization, although it achieves the best APU in its respective designed condition, its performance becomes unsatisfactory under other conditions.

10.5 Merits Analysis

The MBO model involves two key parameters: the flood depth threshold D_m and the maximum service distance d_s. This section conducts a sensitivity analysis on these two parameters. First, the flood depth threshold D_m determines which stations are categorized as "flooded stations," which become the service targets of warehouses and may receive resource protection. A lower D_m reflects a more conservative approach, allowing protection only for stations with minor water ingress, whereas a higher D_m means that more severely inundated stations are included in the set of serviceable flooded stations. In the experimental study, we set $D_m = 1$ m to generate candidate warehouses and optimize their locations and inventories. In this section, to explore the impact of D_m, we calculate the APU under various D_m values and different flood conditions, as shown in Fig. 10.11. For each threshold D_m, we extract the respective set of risky stations, construct the Delaunay triangulation mesh to identify candidate warehouse locations, and then optimize the inventory levels using the GA-based nested method. It can be observed that as D_m increases, both the maximized weighted average APU and the APU under different flood conditions also increase. The most significant improvement occurs when D_m increases from 0.6 m to 0.8 m, as a large number of stations have inundation depths in this range. Therefore, when D_m rises from 0.6 m to 0.8 m, many flooded stations—some of which are high-passenger-flow or interchange stations that strongly contribute to APU—are included as serviceable targets, leading to a noticeable increase in APU.

The maximum service distance d_s refers to the upper limit of distance within which a warehouse can rapidly transport resources to a flooded station during a flood event. If this distance is exceeded, the warehouse is considered unable to deliver resources to the corresponding station. Figure 10.12 present the APU values

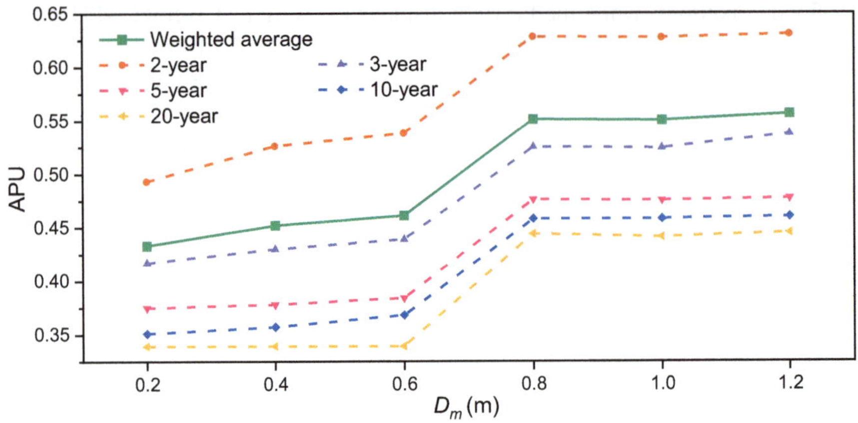

Fig. 10.11 APU under different D_m values and flood conditions

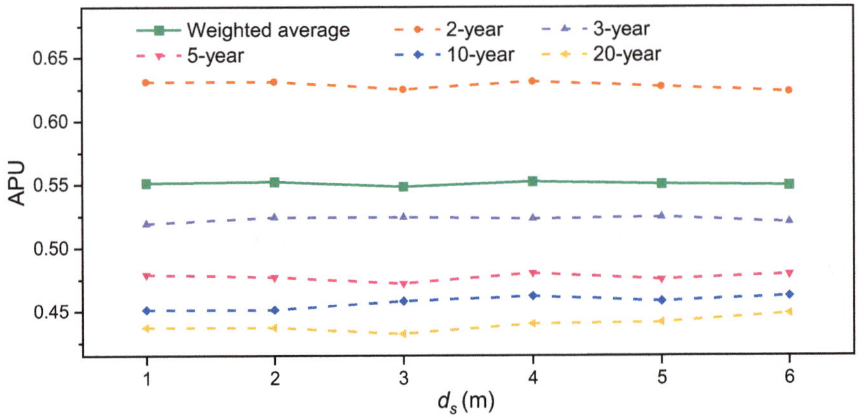

Fig. 10.12 APU under different d_s values and flood conditions

under different d_s settings and flood conditions. It is worth noting that for each d_s, we reconstruct the Delaunay triangulation mesh to generate candidate warehouse locations, followed by inventory optimization using the GA-based nested method. It is seen that the APU is relatively insensitive to changes in the maximum service distance d_s. As illustrated, the maximum weighted average APU fluctuates within a narrow range (0.548–0.552) as d_s varies from 1 km to 6 km, and no consistent increasing trend is observed with increasing d_s. This can be attributed to two main reasons. First, although a larger d_s theoretically allows warehouses to serve more flooded stations, the total available resources remain fixed. Expanding the service range does not increase the quantity of resources, but rather enlarges the competition among stations, which may dilute the protection for high-priority stations. Second, for each d_s, the candidate warehouse locations are re-generated using a

Delaunay triangulation-based rule. In other words, changes in d_s not only modify the service coverage but also alter the spatial structure of the warehouse configuration. This leads to a situation where, when d_s is smaller, although the service range of each warehouse is reduced, the candidate warehouse locations become more numerous and densely distributed (to ensure coverage of all risky stations), potentially offsetting the disadvantage of a smaller coverage radius.

10.6 Chapter Summary

This chapter proposes an MBO model for integrated planning of the location and inventory of resource warehouses for urban metro systems. Initially, a time weighted metro network model is created, based on which the APU is employed to measure network performance. Then, GIS-based flood spatiotemporal modeling is utilized to calculate flood depths for metro station entrances, thereby determining the required flood control resources for each station. On these bases, the MBO model is developed, in which the upper-level optimization aims to maximize the weighted average APU of the metro network across multiple flood conditions, while the lower-level optimization focuses on maximizing the total passenger flow of protected stations under each flood condition. A case study on the Shanghai metro system is conducted to demonstrate the effectiveness of the proposed model.

Based on the case study, several key conclusions can be summarized: (1) The MBO model effectively determines warehouse locations, inventory, and resource transportation schemes during floods, thereby optimizing metro network performance under multiple flood conditions. (2) Compared to single-condition optimizations, the multi-condition optimization provides robust and top-tier metro network performance across multiple flood conditions. (3) In comparison to baseline inventory strategies, the MBO model consistently delivers the highest APU performance for the metro network under all flood conditions. (4) The maximum weighted average APU increases with the flood depth threshold but is not sensitive to changes in the maximum service distance.

Compared to existing research on location and inventory optimization for emergency facilities, this study makes several contributions. First, it is among the first to optimize flood control resource warehouses for metro systems, a novel real-world scenario that is becoming increasingly important due to climate change. In developing the optimization model for this problem, we incorporate network topology to model the metro system and station closures, and use GIS for spatiotemporal simulation of storm floods. Second, this study contributes methodologically. The proposed MBO model not only achieves integrated planning of both location and inventory but also uses the average performance of the metro network under multiple flood conditions as the objective in a bilevel optimization. Additionally, the bilevel framework effectively captures the interdependence between resource positioning (upper level) and resource transportation (lower level), making the MBO model more aligned with the practical needs of metro flood emergency management.

References

1. He R, Zhang L, Tiong RL (2025) Integrated location-inventory planning of resource warehouses for metro floods: a multi-condition bilevel optimization model. Reliab Eng Syst Saf 265:111498
2. Wang W, Wu S, Wang S, Zhen L, Qu X (2021) Emergency facility location problems in logistics: status and perspectives. Transp Res Part E Logist Transp Rev 154:102465
3. Hakimi SL (1964) Optimum locations of switching centers and the absolute centers and medians of a graph. Oper Res 12:450–459
4. Hakimi SL (1965) Optimum distribution of switching centers in a communication network and some related graph theoretic problems. Oper Res 13:462–475
5. Demange M, Gabrel V, Haddad M, Murat CJ (2020) A robust p-Center problem under pressure to locate shelters in wildfire context. EURO J Comput Optim 8:103–139
6. Wang W, Yang K, Yang L, Gao Z (2022) Tractable approximations for the distributionally robust conditional vertex p-center problem: application to the location of high-speed railway emergency rescue stations. J Oper Res Soc 73:525–539
7. İbrahim Miraç E, Eren Ö (2020) P-median and maximum coverage models for optimization of distribution plans: a case of United Nations Humanitarian response depots. In: Smart and sustainable supply chain and logistics – trends, challenges, methods and best practices. Springer, pp 225–246
8. Toregas C, Swain R, ReVelle C, Bergman L (1971) The location of emergency service facilities. Oper Res 19:1363–1373
9. Schreuder JAM (1981) Application of a location model to fire stations in Rotterdam. Eur J Oper Res 6:212–219
10. Aktaş E, Özaydın Ö, Bozkaya B, Ülengin F, Önsel Ş (2013) Optimizing fire station locations for the Istanbul metropolitan municipality. Interfaces 43:240–255
11. Su Q, Luo Q, Huang SH (2015) Cost-effective analyses for emergency medical services deployment: a case study in Shanghai. Int J Prod Econ 163:112–123
12. Rawls CG, Turnquist MA (2012) Pre-positioning and dynamic delivery planning for short-term response following a natural disaster. Soc-Econ Plan Sci 46:46–54
13. Jia C, Zhang C, Li Y-F, Li Q-L (2023) Joint pre- and post-disaster planning to enhance the resilience of critical infrastructures. Reliab Eng Syst Saf 231:109023
14. Dana JJD, Petruzzi NC (2001) Note: the newsvendor model with endogenous demand. Manag Sci 47:1488–1497
15. Duran S, Gutierrez MA, Keskinocak PJI (2011) Pre-positioning of emergency items for CARE international. Interfaces 41:223–237
16. Pan W, Guo Y, Jin L, Liao S (2017) Medical resource inventory model for emergency preparation with uncertain demand and stochastic occurrence time under considering different risk preferences at the airport. PLoS One 12:e0183472
17. Noyan N (2012) Risk-averse two-stage stochastic programming with an application to disaster management. Comput Oper Res 39:541–559
18. Liu K, Li Q, Zhang Z-H (2019) Distributionally robust optimization of an emergency medical service station location and sizing problem with joint chance constraints. Transp Res B Methodol 119:79–101
19. Wang Q, Nie X (2023) A location-inventory-routing model for distributing emergency sup plies. Transp Res Part E Logist Transp Rev 175:103156
20. Zhang S, Pan B (2014) An urban storm-inundation simulation method based on GIS. J Hydrol 517:260–268
21. Mishra SK, Singh VP (2003) SCS-CN method. In: Mishra SK, Singh VP (eds) Soil conservation service curve number (SCS-CN) methodology. Springer Netherlands, Dordrecht, pp 84–146
22. Yin Z, Xu S, Yin J, Wang J (2010) Small-scale based scenario modeling and disaster risk assessment of urban rainstorm water-logging. Acta Geogr Sin 65:553–562

23. Lyu HM, Shen SL, Yang J, Yin ZY (2019) Inundation analysis of metro systems with the storm water management model incorporated into a geographical information system: a case study in Shanghai. Hydrol Earth Syst Sci 23:4293–4307
24. Sinha A, Malo P, Frantsev A, Deb K (2014) Finding optimal strategies in a multi-period multi-leader–follower Stackelberg game using an evolutionary algorithm. Comput Oper Res 41:374–385
25. Shanghai Quality and Technical Supervision Bureau (SQTSB) (2017) Standard of rainstorm intensity formula and design rainstorm distribution (in Chinese)
26. Tao X, Zhang X, Wang X (2019) Analysis on characteristics of rainfall distribution in Shanghai. Water Purif Technol 38:25–28. (in Chinese)
27. Shanghai Water Affairs Bureau (SWAB) (2020) Urban rainwater drainage planning in Shanghai (in Chinese)

Chapter 11
System Repair Planning

11.1 Introduction

The resilience of urban transit systems (UTS) encompasses four key attributes: preparedness, robustness, recoverability, and adaptation [1]. Among these, the first two attributes can be enhanced through actions taken during the preparedness and mitigation phases. However, this requires a deep understanding of the system and its surrounding environment. Given the uncertainties in the timing, pathways, and intensity of future natural disasters, accurate prediction remains challenging. Moreover, due to limited resources, preparedness and robustness alone are insufficient to address all the threats that natural disasters may pose to UTS. In such cases, the complementary attribute—recoverability—becomes particularly crucial, and it needs to be strengthened through proactive system repair planning.

Recent studies have increasingly focused on modeling the recovery/repair processes of UTS following service disruptions caused by natural disasters. Early approaches largely relied on static topological features, such as node centralities, to determine the priority of station or link repairs [2]. With the growing recognition of the role of service demand, later studies began incorporating operational characteristics, such as passenger flow distributions and service efficiencies, into repair strategies [3]. Moreover, various optimization models have been developed to identify efficient repair sequences, taking into account factors like limited repair resources and passenger behavior dynamics [4, 5]. Simulation-based frameworks and multi-objective optimization methods have been employed to assess the effectiveness of different repair strategies under varying disruption scenarios [6]. In summary, the studies in this field have progressively evolved from purely topology-based analyses toward more comprehensive recovery modeling that considers network structure, service demand, operational constraints, and available resources.

The ultimate purpose of UTS in operation is to satisfy passenger flow demand efficiently and safely. Its service quality relies on accurate monitoring of passenger

L. Zhang et al., *Resilience in Urban Transit Systems*, Advances in 21st Century Human Settlements, https://doi.org/10.1007/978-981-95-4322-9_11

flow. To estimate the trajectories of passengers, Yang et al. [7] developed a space-time-state hyper network for passenger flow assignment. Most studies that include passenger flow analysis assess the impacts of frequent operational incidents on limited stations, while few consider the scarce, almost simultaneous, large-scale effect on the whole system when multiple stations are destroyed in natural disasters. When the UTS service is partially affected, passengers may choose to wait or shift from rail travel to bus travel near the incident stations. Mo et al. [8] defined two metrics, i.e., stay ratio and delay time, to assess the impacts of disruptions. These service-focused models can effectively evaluate the disruption under existing, static infrastructure. However, they cannot simulate the changes with intervention strategies on infrastructures.

Modeled as graphs with nodes representing stations and links representing connections between stations, the evolving mechanisms of UTS can be viewed from a system perspective. System resilience can be quantified based on system response, which is aggregated from node measures. The node measures can be purely topological (e.g., degree, betweenness, clustering coefficient) or functionality-based (e.g., load, efficiency, and utility). The most critical stations can be identified by node measures or by optimizing the interdiction models from the attacker side, whose results are method-dependent and flow-modulated. Prioritizing the service quality, Shao et al. formulated a cost optimization model for the UTS network with the link-level flow, where cost was assumed to be travel time and travel time was estimated by some shape function [9]. It is hard to tune the parameters of the analytical shape function so that it fits the real-world situation. Travel time was divided into intervals for queueing, boarding, on-board, alighting, and exiting, which were logically correct but hard to estimate in practice and may further affect the time-based resilience assessment.

Instead, this chapter uses a purely data-driven method to estimate link-level passenger flow and further investigate system performance under large-scale disruptive events and possible recovery strategies [10]. To do so, a two-layer complex network is first constructed. It explicitly distinguishes the infrastructure layer and service layer of UTS. Mathematical formulations of the two layers are given. Secondly, based on smart card data, origin-destination passenger flow is mapped to each link level for better localization purposes. Then, multiple attack protocols are applied to the constructed network to investigate the system resilience in the absorptive dimension. Finally, a recovery strategy based on localized functionality is proposed and compared to a state-of-the-art recovery strategy in terms of system resilience in the adaptive and/or recovery dimension.

11.2 System Repair Planning Under Disruptions

Recent years have witnessed significant progress in research on repair planning for UTS after disruptions (including natural disasters, human-induced damage, and equipment failures). These studies aim to enhance the rapid recovery capability of UTS. This section reviews the existing literature from the following perspectives:

1. **Repair planning based on node importance**. Many studies in this area primarily relied on complex network theory, utilizing static topological properties of networks (such as node degree, betweenness centrality, closeness centrality, etc.) to identify critical stations or sections and accordingly develop repair sequences. For instance, Yadav et al. [4] tested five repair strategies for London's rail system, including ones based on betweenness centrality and a greedy algorithm. They found that centrality-based strategies could match or outperform greedy heuristics, particularly in complex, interconnected networks. However, considering only the topology-based node importance is insufficient to capture the actual operational state of the system; hence, some researchers have incorporated operational characteristics into system repair planning. For example, Bi and MacAskill [3] integrated structural importance (e.g., betweenness centrality) and operational importance (e.g., passenger flows) to assess recovery paths for the London metro system under various flood scenarios. They evaluated recovery performance using both global efficiency and the number of satisfied OD-pair journeys, capturing network-wide functionality. These studies represent a shift from purely topological analyses toward performance-centered modeling frameworks that emphasize user-level impacts during the recovery phase.
2. **Repair planning based on optimization models**. Besides the node importance-based repair strategies, optimization methods represent another significant category of approaches for UTS restoration, as they can systematically handle complex system constraints and multiple objectives. Scholars have introduced optimization-based models that consider system constraints and real-world operations. For example. Zheng et al. [5] proposed a bi-level programming model combining passenger path choice and infrastructure restoration to maximize overall recovery efficiency. Tan et al. [11] introduced a three-level model framework, in which the middle and lower levels jointly identified the most vulnerable station combinations, while the upper level optimized the restoration sequence. Zhu et al. [6] employed a multi-objective evolutionary algorithm (NSGA-II) to optimize the recovery sequence under limited repair resources, providing a practical framework for decision-making under constraints. Taking both budget limitations and repair time uncertainty into account, Zhang et al. [12] optimized the restoration sequence scheme for maximizing the global average efficiency under the condition that the network accessibility meets given resilience requirements. Therefore, the optimization-based approaches not only enrich the methodological toolbox for metro system repair planning but also provide decision-makers with more realistic and flexible strategies to balance efficiency, resilience, and resource constraints in post-disaster recovery of UTS.

While significant progress has been made in modeling the recovery of urban rail transit networks, most existing studies primarily rely on static topological metrics or simplified operational proxies to design repair strategies. Methods such as those based on node centralities (e.g., betweenness or degree) or greedy heuristics often assume homogeneous network importance and overlook the dynamic, spatially heterogeneous nature of passenger flows. Additionally, few studies fully incorporate fine-grained, real-world service data into the recovery modeling process. Consequently, current approaches may not adequately capture the localized functional losses or the actual operational priorities that emerge during real disruptive events. To address these gaps, this chapter proposes a novel two-layer network framework that explicitly distinguishes the infrastructure and service layers of an urban rail transit system. A new dynamic link flow (LF) measure, derived from smart card data, is introduced to represent localized functionality with fine spatial resolution, on which basis a link-flow-guided repair strategy is developed and tested against traditional recovery cost-based strategies. It is expected that this chapter will offer a practical and data-driven pathway to enhance the UTS resilience.

11.3 Simulation-Based Optimization for Repair Planning in UTS

The proposed method can provide a data-driven, systematic, and comprehensive simulation of service disruption and restoration for the UTS under attacks and with recovery considering dynamic flow demand. Resilience in the dimensions of absorption and adaptation is assessed accordingly. Resilience assessment results from the constructed model help to (1) quantify the impacts of given attacks in disaster preparedness and mitigation stages, and (2) prioritize the stations to be repaired in disaster response and recovery stages. A framework of the proposed method is shown in Fig. 11.1. The four modules and their possible integration with the development of sensing technology and disaster risk prediction are described below.

11.3.1 Two-Layer Graph Model of UTS

The infrastructure of UTS is relatively stable while its service demand, i.e., passenger flow, varies between weekdays and weekends, and between peak hours and non-peak hours. To distinguish the static attributes of infrastructure and dynamic ones of service in UTS operation, a two-layer graph, including the infrastructure layer and service layer, is constructed to model the structure and functionality of UTS, respectively. Built on top of the infrastructure layer, connections in the service layer, which are empirically constructed from passenger flow data, display a large

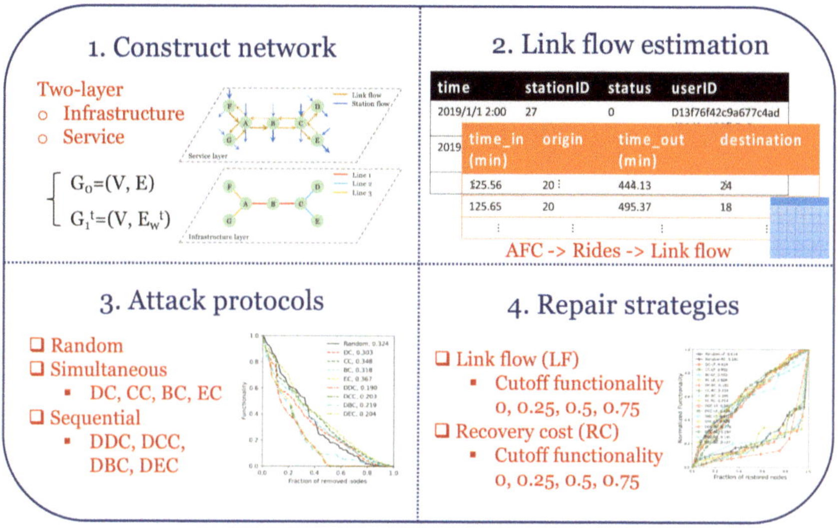

Fig. 11.1 A framework of the proposed method for service disruption and recovery of UTS in natural disaster shocks

heterogeneity in load intensity [13] and are thus modeled by a weighted graph rather than the binary states as in the infrastructure layer. Mathematically, the infrastructure layer representing the stations and their connections is modeled via an undirected graph, denoted as G_0. Parallelly, the service layer representing the passenger flow from place A to place B, whose value is inherently time-varying, is modeled via a directed graph with a time index, denoted as G_1^t. The two-layer graph model is expressed as shown in Eq. (11.1). An illustration of the two-layer graph model on a simple metro transit system is shown in Fig. 11.2.

$$\begin{cases} G_0 = (V,E) \\ G_1^t = (V,E_w^t) \end{cases} \tag{11.1}$$

where the node set $V = \{1,2,\ldots,n\}$ represents stations, the link set $E = \{(i,j)\}$ represents connections between adjacent stations i and j, and E_w^t is the weighted link set at time t. The existence of links in E is indicated by an adjacency matrix $A = [a_{11},\ldots,a_{ij},\ldots,a_{nn}]$, whose element a_{ij} is determined according to Eq. (11.2). Similarly, the intensity of connections in E_w^t is indicated by a weighted adjacency matrix $A_w^t = \left[a_{11}^t,\ldots,a_{ij}^t,\ldots,a_{nn}^t\right]$, whose element a_{ij}^t is determined according to Eq. (11.3).

$$a_{ij} = \begin{cases} 1, & \text{if } i \text{ is adjacent to } j \\ 0, & \text{otherwise} \end{cases} \tag{11.2}$$

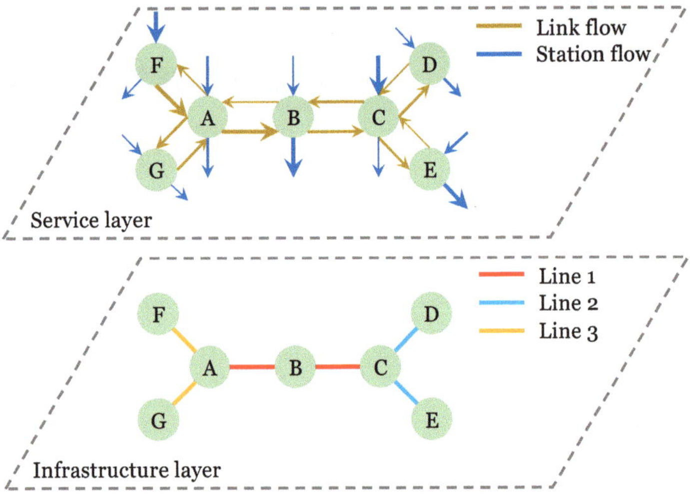

Fig. 11.2 Illustration of the two-layer graph model on a simple metro transit system

$$a_{ij}^{t}=\begin{cases} w_{ij}^{t}, & \text{if } i \text{ is adjacent to } j \\ 0, & \text{otherwise} \end{cases} \tag{11.3}$$

where w_{ij}^{t} is the intensity of connection from node i to node j at time t.

In this chapter, we propose a real travel time-based fuzzy set mapping method for link flow (LF) estimation, which incorporates the uncertainty from travel time deviation and relative distance to stations. The estimated LF with time index is then utilized as the weight in the service layer of the UTS network. The fuzzy LF estimation method is elaborated on in the next section.

11.3.2 Passenger Flow Estimation

Estimation of passenger flow is essential in the operation of UTS as the passenger volume is closely related to its functionality. In this section, we map every single ride based on the real travel time information from a fuzzy perspective, after which the time and space information of all the rides are utilized to estimate the distributed flow, i.e., LF. To this end, smart card records, also known as AFC data, are used to derive two types of passenger flow data: (1) total ridership in the UTS networks, which can be accurately derived; and (2) its distribution over the links, which is estimated with uncertainty. Related quantities and the proposed LF estimation method are introduced as follows.

The AFC system can record ride-related information with a standard format, where each record includes a digital user identity document (ID), binary behavior indicators (i.e., in/out, equivalently, enter/exit), the time instance when activity is detected, and station ID. Based on AFC records, entering/exiting behavior can be paired with a complete ride record. Each ride is defined by the inbound/outbound time and OD locations information, as shown in Eq. (11.4). For each ride, its shortest path refers to the one passing the minimum stations, which is denoted as a sequence of its passing stations, as shown in Eq. (11.5). Rides aggregated during a certain period are interpreted as the total passenger flow in the network. The aggregation operator of the ridership data is controlled by a parameter called time step T, which is multiple times of the sampling rate of the AFC system. Based on historical ride records R and given time step T, passenger flow at discrete time instance t can be accurately derived, as shown in Eq. (11.6). The passenger flow can be distributed into OD pairs, as expressed in Eq. (11.7).

$$r = \left[t_{\text{in}}, s_{\text{in}}, t_{\text{out}}, s_{\text{out}}\right] \tag{11.4}$$

where r denotes a ride, t_{in} denotes the time entering the station s_{in}, and t_{out} is the time exiting the station s_{out}. The stations s_{in} and s_{out} are also known as the origin and destination of the ride, respectively.

$$\text{path} = \left[s_{\text{in}}, \ldots, u, v, \ldots, s_{\text{out}}\right] \tag{11.5}$$

where u, v are the intermediate stations in the path. Two adjacent stations in the path define a travel segment, which is known as an atomic interval (i.e., link), such as $u \rightarrow v$ here.

$$f(t) = \text{count}\left(R\left[t_{\text{in}} \leq t + T \text{ and } t_{\text{out}} \geq t\right]\right) \tag{11.6}$$

where $f(t)$ is the aggregated rides from time t to $t + T$, count(•) is the count function, and $t = nT$, n is a positive integer.

$$f(t) = \sum\nolimits_{i,j \in V} f_{i \rightarrow j}^{t} \tag{11.7}$$

where $f_{i \rightarrow j}^{t}$ denotes the passenger flow from origin i to destination j at time instance t, which is an element of the OD matrix, and V is the set representing metro stations. As a ride refers to distinct OD stations, the diagonal terms follow that $f_{i \rightarrow i}^{t} = 0$.

To obtain LF values, a real travel time-based fuzzy set mapping is customized to assign each ride to a specific link location along its path (Fig. 11.3a). As a result, each ride is represented as a fuzzy set indicating the beliefs over each link in its path and the uncertainty element. It has the form shown in Eq. (11.8). The estimation process consists of three steps: determination of the shortest path, fuzzy set assignment, and link flow aggregation. Firstly, the path most likely to be given the origin and destination is set to the shortest path by default for simplicity, with an introduced uncertainty component reflecting the deviation of travel time. Secondly, the stations along the shortest path and the AFC time record (i.e., the inbound and outbound time) are used for fuzzy set assignment. The uncertain element in the fuzzy

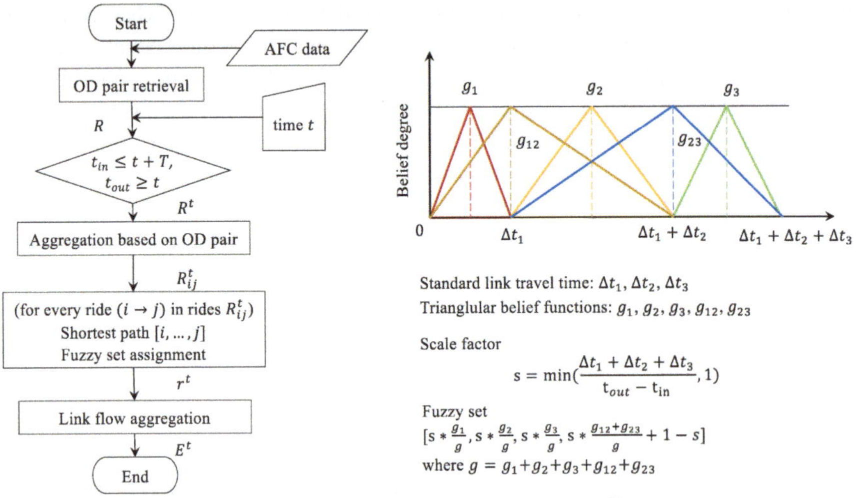

(a) (b)

Fig. 11.3 Real travel time-based fuzzy set mapping method for link flow estimation. (**a**) Flowchart of link flow aggregation from AFC data; (**b**) An example of real travel time based fuzzy set assignment using triangular belief functions, which is suitable for riders that pass four stations in the shortest path (i.e., three atomic intervals, thus have 3+1 elements in the fuzzy set)

set explicitly captures two parts: the deviation between the real travel time and the standard travel time, which corresponds to path choice; and the relative location to the nearest stations, which corresponds to location mapping.

An example of constructing a fuzzy set using triangular belief functions is illustrated in Fig. 11.3b. It specifies a linear change in beliefs between the most uncertain situation in location mapping (i.e., at the stations) and the least uncertain condition (i.e., in the middle of links). Finally, LF at time t is acquired by taking the sum of the belief degrees over each link for all the rides iterating over each OD pair $f^t_{i\to j}$. For flow conservation, the uncertainty belief u^t is equally distributed to the previous $r^t_{u\to v}$ elements. Meanwhile, the average $u^t_{i\to j}$ for all rides with the same OD pair is kept as an indicator of the uncertainty level in LF estimation from ride records. Rearranging the elements, the estimated LF and its uncertainty can be expressed in Eq. (11.9).

$$r^t = \left[\cdots, r^t_{u\to v}, \cdots, u^t\right] \tag{11.8}$$

where r^t is a fuzzy set corresponding to a ride record in $f^t_{i\to j}$, $r^t_{u\to v}$ is the belief that the rider is at the link $u \to v$, and u^t is the uncertainty degree in the fuzzy set assignment. The fuzzy values sum to 1, that is, $\sum_{u,v\in \text{path}} r^t_{u\to v} + u^t = 1$.

$$L^t = \begin{bmatrix} e^t_{1\to1} & \cdots & e^t_{1\to v} & \cdots & e^t_{1\to n} \\ \vdots & \ddots & \vdots & \ddots & \vdots \\ e^t_{u\to1} & \cdots & e^t_{u\to v} & \cdots & e^t_{u\to n} \\ \vdots & \ddots & \vdots & \ddots & \vdots \\ e^t_{n\to1} & \cdots & e^t_{n\to v} & \cdots & e^t_{n\to n} \end{bmatrix} U^t = \begin{bmatrix} u^t_{1\to1} & \cdots & u^t_{1\to j} & \cdots & u^t_{1\to n} \\ \vdots & \ddots & \vdots & \ddots & \vdots \\ u^t_{i\to1} & \cdots & u^t_{i\to j} & \cdots & u^t_{i\to n} \\ \vdots & \ddots & \vdots & \ddots & \vdots \\ u^t_{n\to1} & \cdots & u^t_{n\to j} & \cdots & u^t_{n\to n} \end{bmatrix} \quad (11.9)$$

where L^t denotes the LF matrix at time t, $e^t_{u\to v}$ is the LF value aggregated at the link $u \to v$ at time t, U^t denotes the uncertainty in flow estimation from rides, and $u^t_{i\to j} \in (0,1)$ is the uncertainty level of LF estimation from rides with path $[i,\ldots,j]$. LF matrix L^t is sparser than the uncertainty matrix U^t as a link is a specific case of paths that concern two stations only.

11.3.3 Attack Protocols for Simulating Natural Disaster Shocks

In simulating the consequent functionality degradation of the UTS system in natural disaster shocks, this chapter focuses on node removal attacks, which are attacks on stations. A new functionality measure, the directly affected LF, is used to indicate the impacts of the attacks on the serviceability of the UTS. The separation in physical locations of stations and links is not considered in the network model. However, as stations at the end of the link are part of the link, it is assumed that when station i is attacked, all the links it connects would be a complete loss in functionality. This situation can be considered the severest condition under attack, compared to a partial loss of local functionality. Attacks made on the infrastructure layer will subsequently affect the functionality of UTS as modeled by the service layer.

A normalized version of the remaining functionality is thus formed as shown in Eq. (11.10). Considering all the attacked nodes, the thus affected LF-based functionality is expressed in Eq. (11.11). The functionality of the network versus δ from 0 to 1 is often plotted, known as degradation curve, which simulates a gradually increasing intensity of attack as δ increases. The accumulated performance of the UTS from slight to severe attack under certain attack protocols can be represented as the area under the degradation curve, denoted as SA, as shown in Eq. (11.12). The smaller the SA, the larger the impact can be caused by the attack. SA changes with time t as LF changes with t.

$$F^t(\delta) = 1 - \frac{f^t(\delta)}{f(t)} \quad (11.10)$$

where $F^t(\delta) \in [0, 1]$ represents the normalized functionality of UTS at time t as a function of the fraction of removed nodes δ, $f^t(\delta)$ is directly affected by LF at time t due to the attacks, and $f(t)$ is the normal passenger flow at time t.

$$f^t(\delta) = \sum_{i \in V_A} \text{sum}\left(L^t[i,:]\right) + \text{sum}\left(L^t[:,i]\right) \tag{11.11}$$

where $V_A \subseteq V$ is the set representing attacked nodes, $\delta = |V_A| / |V|$, sum($L^t[i,:]$) is the flow from station i to one of its adjacent stations at time t, and sum($L^t[:,i]$) is the flow from one of its adjacent stations to station i at time t.

$$\text{SA} = \int_0^1 F^t(\delta)\, d\delta \tag{11.12}$$

where SA denotes the score under attacks, which is a function of time t in the context of dynamic flow.

Node attack protocols specify the number of nodes to be attacked and their attack sequence, usually denoted by a ranked list of nodes. There are random attacks and targeted attacks. For targeted attacks, two common schemes exist in central attacks: simultaneous attacks, where the ranked list of nodes is determined before the attack according to the existing topological properties of the graph; and sequential attacks, where the ranked list of nodes is updated based on the topological properties of the evolving graphs with every node removed [14]. Here, both simultaneous and sequential attacks are extensively investigated with δ varying from 0 to 1, representing the slightest and severest attacks, respectively.

Centrality is an important topological metric of networks to quantify the disintegration of giant components and connectivity patterns, and it is highly relevant to network functionality during attack and protection [14]. Four centrality measures, i.e., DC [15], CC [15], BC [16], and EC [15], are investigated for target selection under simultaneous and sequential attacks. As random node attacks do not depend on centrality measure, overall, nine attack protocols will be tested, belonging to three sets: (1) Random; (2) simultaneous: DC, CC, BC, EC; and (3) sequential: DDC, DCC, DBC, DEC. Different attack protocols will generate different attack sequences of stations and induce varying consequences in terms of service disruption.

11.3.4 Repair Strategies with Varying Initialization

Cascading failure is a well-documented phenomenon/consequence in complex systems due to the gradual onset of dysfunctionality, such as aging, where the failure of one component can lead to the failure of its neighboring components due to, for example, demand redistribution and thus cause point overload [17]. However, in the occurrence of hazards, the infrastructure system becomes paralyzed in a very short time, and self-organization for functionality may not form in the attacked network. Such instantaneously occurring multiple-point failures cannot be repaired at the same time with a limited crew and resources. Sequential recovery is a problem to determine the optimal sequence for repair, which is essentially a tradeoff between the cost (i.e., user cost, repair cost) and resources (i.e., crew availability [18, 19].

A dynamic LF-based repair strategy is introduced to determine the node repair sequences. A cutoff value, i.e., the critical functionality, is used to indicate the beginning of the restoration of attacked nodes, as denoted in Eq. (11.13). The proposed repair strategy sorts out the sequence of the nodes to be repaired according to LF. The sum of the bidirectional LF is used as the recovery cost function, as shown in Eq. (11.14). This repair strategy is compared with the recovery cost (RC) metric [20] to investigate the effect of the repair strategies on the recovery of UTS functionality. The dynamic version of the RC metric is formulated as shown in Eq. (11.15). Different recovery cost functions lead to different repair priorities for nodes to be restored, as denoted by repair sequences.

$$V_A = \{s_1, s_2, \ldots, s_l\} \text{ with } F^t(\delta) \le f_c \tag{11.13}$$

where $V_A \subseteq V$ is a set representing the nodes to be attacked, so that the functionality of the UTS will fall below the critical value f_c.

$$\text{LF}^t(i,j) = \frac{L_{ij}^t + L_{ji}^t}{C} \tag{11.14}$$

where L_{ij}^t is the LF value from node i to node j at time t, and C is a constant considering equal cost per link repair.

$$\text{RC}^t(i,j) = \frac{f_{ij}^t + f_{ji}^t}{C_{ij}} \tag{11.15}$$

where f_{ij}^t is the OD flow from origin i to destination j at time t, and C_{ij} is the least number of nodes that need to be restored to make the OD pair (i,j) connected.

Adaptive/recovery resilience based on a recovery curve can be defined based on algebraic functions, such as the ratio of recovery and loss or integral functions such as recovery area in the reverted triangle/trapezoid resilience model [21, 22]. The algebraic definition depends on two static points (i.e., the initial point and the endpoint) to capture the relative change, while the integral definition requires more data points and reflects the cumulative performance of the dynamic system. Here, the process of recovery is represented by a recovery curve. The recovery curve depicts the trend of functionality versus the fraction of restored nodes. The overall performance of the repair strategy is evaluated as the integral of the normalized restored functionality over the fraction of restored nodes, as shown in Eq. (11.16).

$$\text{SR} = \int_0^1 \hat{F}^t(c)\,dc \tag{11.16}$$

where SR denotes the score of repair performance, $\hat{F}^t(c)$ is the recovery curve, and c is the fraction of repaired nodes.

The workflow of simulating an attack-repair scenario is shown in Fig. 11.4. The key difference between the repair strategy relying on the LF metric and that relying on the RC metric is that the LF metric optimizes the restoration based on LF, while

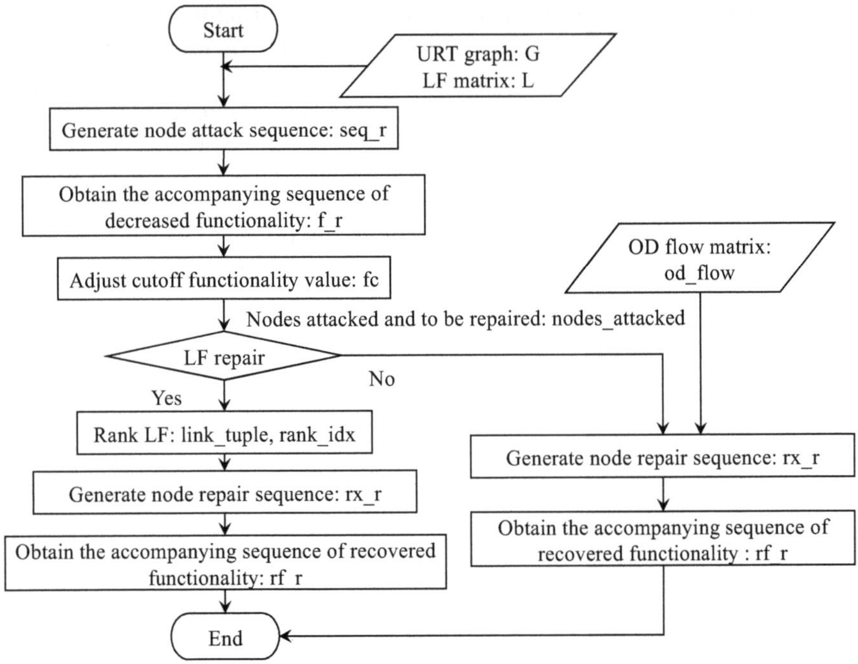

Fig. 11.4 Flowchart for attack-repair simulations

the RC metric optimizes the restoration based on discounted OD flow. As there is usually more than one station for every OD pair travel, which cannot be distinguished in prioritizing the repair, the LF metric has better localization ability than the RC metric, i.e., one link versus one path.

11.4 Experimental Studies

11.4.1 Data and Resources

The proposed method is applied to the Hangzhou Metro system in China. The AFC data from 1st Jan 2019 to 25th Jan 2019 was prepared by Alibaba Cloud and is publicly accessible on the Tianchi competition website. Specifically, the metro system covers three main lines, including Line 1, Line 2, and Line 4, which are composed of 81 stations. Checking the data, one transit station has double labels, "10" corresponding to Line 1 and "51" corresponding to Line 2. Other stations have unique labels. The metro networks are visualized as a graph, as shown in Fig. 11.5.

To facilitate an effective analysis of the passenger flow, only distinct OD pairs were retrieved from the AFC data. Aggregating all the records on 1st Jan 2019, a total of 900,443 rides were obtained from the 2,539,592 pieces of AFC records. The

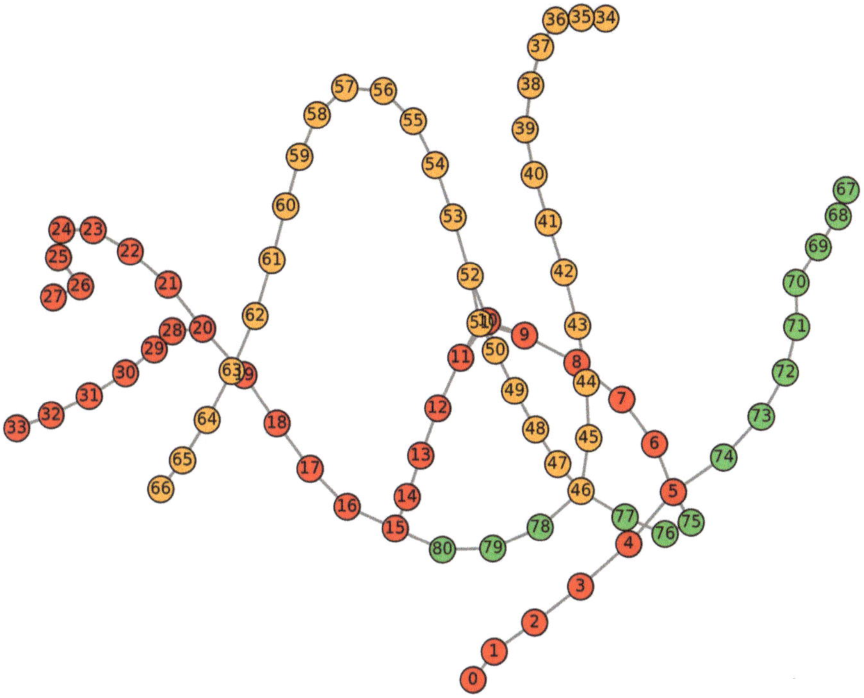

Fig. 11.5 A graph representing Hangzhou metro networks. Red: Line 1; Orange: Line 2; and Green: Line 4

removed data corresponds to frequent in and out behaviors, which can be identified as, for example, maintenance workers and metro staff. Set the time step to 10 min, the passenger flow can then be acquired from ride records according to Eq. (11.6). The obtained flow with a discrete-time indicator exhibits a steady increasing trend until 2 pm, with a peak value of 55k, and then a steady decline till midnight.

11.4.2 Link Flow Estimation

In the travel time-based fuzzy mapping for LF from OD rides (Fig. 11.3), firstly, the standard travel time of each atomic interval, i.e., travel between two adjacent stations, is obtained from the historical data. Specifically, daily rides between adjacent stations are retrieved from the daily OD matrix. For every single link, the median of the real travel time for all rides is then used as the standard atomic travel time for that link. Iterated over all the links, the number of rides for links, and thus estimated standard atomic interval travel time are obtained as shown in Fig. 11.6.

The standard atomic interval travel time is then used to construct the triangular fuzzy belief functions for the fuzzy set representing location beliefs and uncertainty.

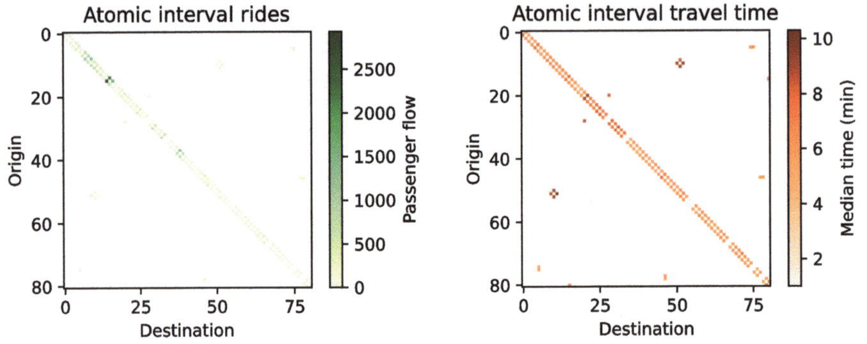

Fig. 11.6 Link travel time estimation is based on (**a**) aggregated daily rides for links, where the standard travel time for each link is taken as (**b**) the median travel time for all rides passing the link

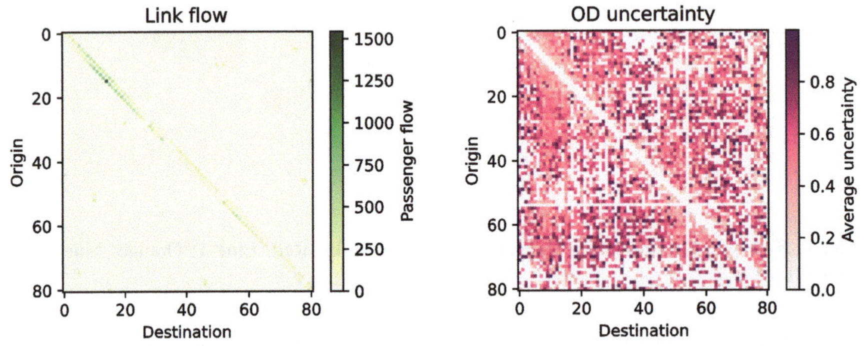

Fig. 11.7 Link flow mapping results from OD rides: (**a**) Estimated link flow at time instance 10:00 am; (**b**) Uncertainty due to travel time deviation and relative location to stations

The uncertainty components for the two situations are explained below: (1) For a ride whose shortest path concerns two stations, the rider is certainly located on the link connecting the two stations, and the uncertainty arises only from extremely long travel time; and (2) for a ride whose shortest path concerns more than two stations, there is an additional uncertain component besides the long travel time, that is, the relative location to each station on the shortest path. The relative location to each station, inferred based on the travel time at time t and the standard travel time, is of the largest uncertainty when the ride is going to pass one station, as there is a large possibility that the rider can be on the link before passing the station or another one after passing the station. The thus derived LF matrix with uncertainty is shown in Fig. 11.7. OD rides with a short path, such as the link rides corresponding to the situation (1), have a much lower uncertainty, as evidenced by the near diagonal items on the map (Fig. 11.7b).

11.4.3 Analysis of Results

Two extreme conditions are considered in the attack and repair simulation: (1) the severest consequence of the attack, that is, attacked nodes will cause full loss of service functionality in their connecting links due to the high impacts of natural disasters; and (2) the earliest repair, that is, once the functionality falls below the critical values, restoration starts instantly even extra attack of nodes may not cause further loss of functionality (i.e., as indicated by a horizontal line segment in the degradation curve). Condition (2) means prompt action after failure detection. For the nine attack protocols, the gradual loss of functionality from normal to complete failure under node attacks, as captured by the degradation curve, is shown in Fig. 11.8. In the repair process, given the cutoff points, which specify the severity of the attack where restoration comes in, the recovery process under the two repair strategies is characterized by the recovery curve.

Repair sequences under LF and RC strategies for one attack-repair scenario are given in Fig. 11.9, depicting the varying priority in node repair relying on the two strategies. A comprehensive simulation of the attack-repair scenarios of the UTS network under varying degrees of attack severity (i.e., critical functionality being 0, 0.25, 0.5, and 0.75), each considering the full combinations from the nine attack protocols and the two repair strategies (i.e., totally 18 cases), is shown in Fig. 11.10. The SR scores for each case are shown in Table 11.1. A statistic summary of the SR scores is shown in Table 11.2. The effect of attack protocols and repair strategies on the UTS functionality is elaborated below.

1. The response of the UTS network under the nine attack protocols forms a regime between the degradation curves of random attack/simultaneous attacks and

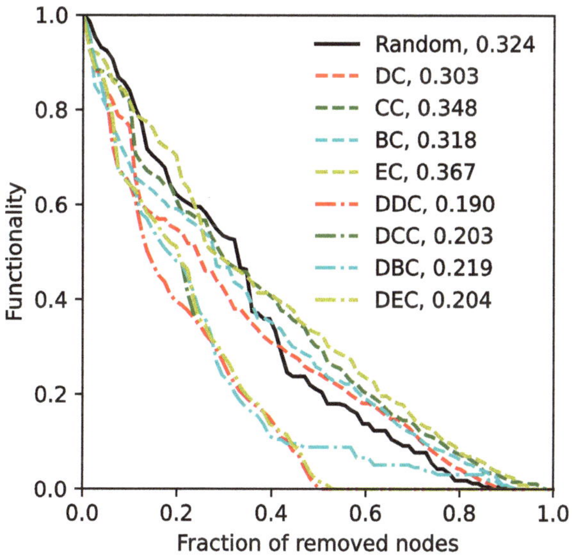

Fig. 11.8 Degradation curve under three sets of node attacks: random attacks; simultaneous attacks (i.e., DC, CC, BC, EC); and sequential attacks (i.e., DDC, DCC, DBC, DEC). SA value is appended for each attack scenario

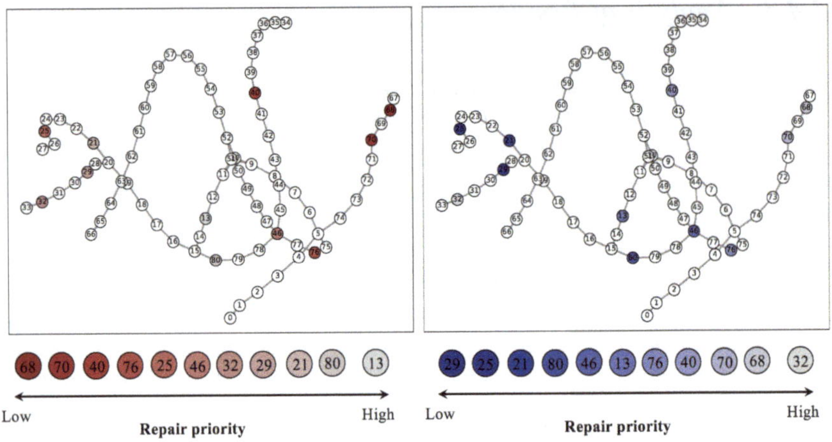

(a) (b)

Fig. 11.9 Repair sequences under the two repair strategies for random attack with $f_c = 0.75$. (**a**) Node repair sequences based on LF; (**b**) Node repair sequences based on RC

sequential attacks, which serves as a portfolio of the possible consequences under natural hazard attacks with different onsets. Referring to Fig. 11.8, all the attacks exhibit a concave-down trend, which means that the more important nodes are identified at an earlier stage than the less important ones. Besides, the concave-down degree varies, indicating varying power of the attacks. Based on the degradation curves and their corresponding SA values, it can be concluded that sequential attacks (i.e., DDC, DCC, DBC, and DEC) are the most powerful, while simultaneous attacks (i.e., DC, CC, BC, EC) and random attacks do not exhibit an obvious difference. Degradation of functionality is not sensitive to centrality measures as the SA scores are similar for the four centrality-related metrics, regardless of the attack schemes (i.e., simultaneous or sequential). The simulation does not consider geological vulnerability to natural disasters from historical events, while previous studies have shown the suitability of using attack protocols to simulate real attack situations in natural disaster shocks. For example, the degradation in UTS functionality caused by flooding, which was obtained from the long-term flood risk map, is similar to that of a random attack and less severe than that of a targeted attack [4].

2. With the same number of nodes repaired, the proposed LF repair strategy restores the functionality of the UTS 2~3 times faster than the existing RC strategy, as tested on multiple attack scenarios and with varying severity of the attacks. To demonstrate the attack-repair process, a specific example depicting a random attack scenario that causes the UTS to fail below 0.75 in functionality (i.e., critical functionality in repair is 0.75) is given. The two metrics, LF and RC, have different priorities over nodes to be repaired, as indicated in the graphs in Fig. 11.9a and Fig. 11.9b, respectively. Specifically, LF gives higher priority over nodes involved in loops as links are assumed to have equal repair cost,

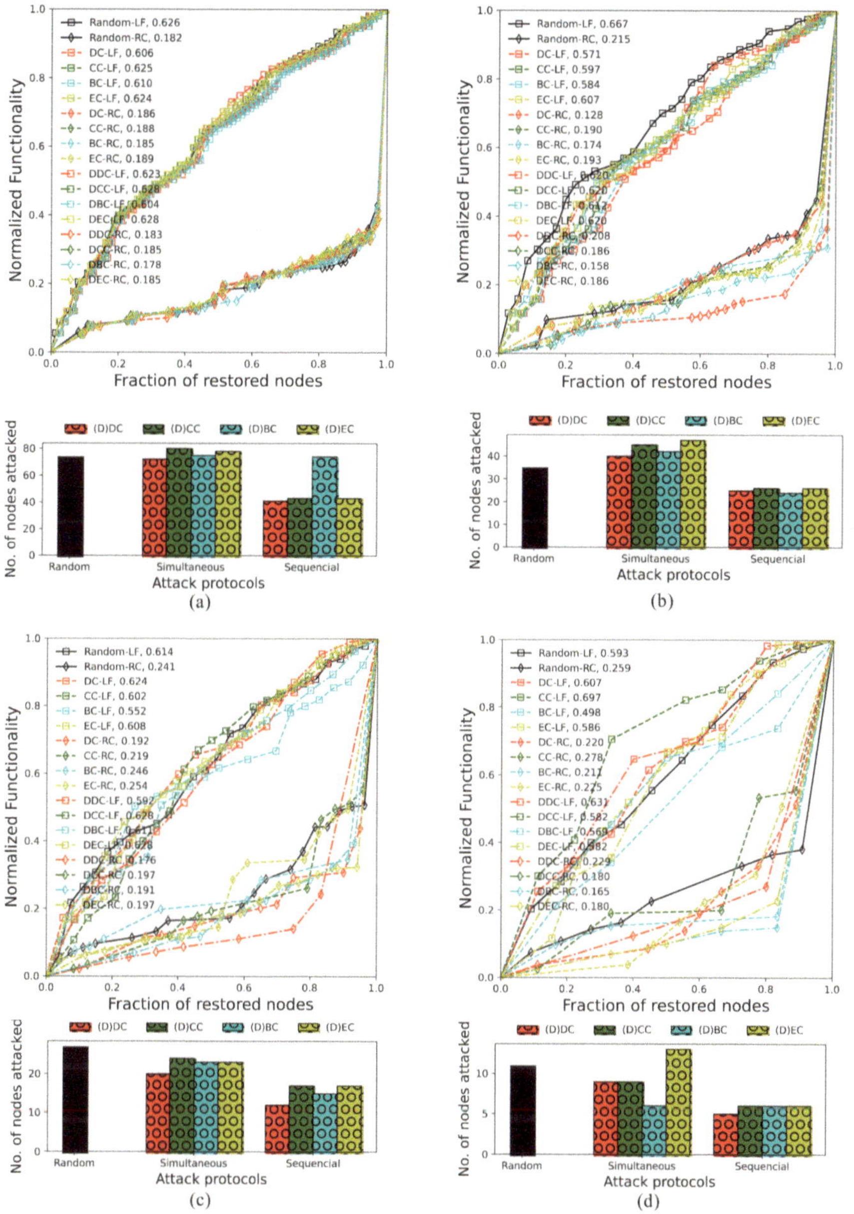

Fig. 11.10 Recovery curves since critical functionality: (**a**) $f_c = 0$, indicating paralysis of the UTS network; (**b**) $f_c = 0.25$, indicating extremely severe loss of functionality; (**c**) $f_c = 0.50$, indicating half loss of functionality; and (**d**) $f_c = 0.75$, indicating a quarter loss of functionality. SR value is appended for each attack-repair scenario

Table 11.1 SR scores of repairs under multiple attack-repair cases and varying severity of attacks

Attack-repair scenario	Critical functionality f_c			
	0	0.25	0.50	0.75
Random-LF	0.626	0.667	0.614	0.593
Random-RC	0.182	0.215	0.241	0.259
DC-LF	0.606	0.571	0.624	0.607
CC-LF	0.625	0.597	0.602	0.697
BC-LF	0.610	0.584	0.552	0.498
EC-LF	0.624	0.607	0.608	0.586
DC-RC	0.186	0.128	0.192	0.220
CC-RC	0.188	0.190	0.219	0.278
BC-RC	0.185	0.174	0.246	0.211
EC-RC	0.189	0.193	0.254	0.225
DDC-LF	0.623	0.620	0.592	0.631
DCC-LF	0.628	0.620	0.628	0.582
DBC-LF	0.604	0.612	0.611	0.569
DEC-LF	0.628	0.620	0.628	0.582
DDC-RC	0.183	0.208	0.176	0.229
DCC-RC	0.185	0.186	0.197	0.180
DBC-RC	0.178	0.158	0.191	0.165
DEC-RC	0.185	0.186	0.197	0.180

Table 11.2 Statistic summary of SR scores over multiple attack scenarios

Repair strategy	Critical functionality f_c	First 4-quantiles	Second 4-quantiles	Third 4-quantiles	Mean	Std Dev	Std Dev/ Mean
LF	0	0.610	0.624	0.626	0.619	0.009	0.015
	0.25	0.597	0.612	0.620	0.611	0.026	0.042
	0.50	0.602	0.611	0.624	0.607	0.023	0.037
	0.75	0.582	0.586	0.607	0.594	0.050	0.084
RC	0	0.183	0.185	0.186	0.185	0.003	0.017
	0.25	0.174	0.186	0.193	0.182	0.025	0.136
	0.50	0.192	0.197	0.241	0.213	0.027	0.125
	0.75	0.180	0.220	0.229	0.216	0.035	0.162

whereas RC prioritizes nodes near the end of lines, as OD travel concerning nodes in the loop may have a larger discount effect on repair cost due to a larger number of nodes to be repaired. Given the cutoff functionality, attacked nodes are determined, which are then repaired according to the repair sequence till recovery. The value of the cutoff functionality can be interpreted as a combined effect of the severity of attacks and the response delay from the start of the attacks to the start of repairs. The lower the value, the more powerful the attack and/or the slower the response.

The recovery curves and the number of attacked nodes for the 18 scenarios (constructed based on the nine attacks and two repair strategies) with the cutoff values being 0, 0.25, 0.5, and 0.75 are shown in Fig. 11.10. The recovery process based on the LF repair strategy is more rapid than that of the RC strategy for all the cases. The SR scores for each scenario are shown in Table 11.1. The SR scores of repairs corresponding to LF repair are larger than those for RC. Therefore, the repair strategy LF is more effective than RC regardless of the attack protocols and critical functionality values. As the path from each OD pair (i.e., whose length is larger or equal to 2) is generally longer than that of the link (i.e., whose length has a fixed value of 2), the LF strategy can identify a precise location (i.e., a link concerning only two stations) compared to the RC strategy (i.e., a path concerning a series of stations), therefore LF is more effective in identifying the critical stations and thus performs better than RC in the repair process.

3. Under severe attacks, the LF repair strategy has a more robust performance than RC for the tested attack protocols, as evidenced by a smaller normalized standard deviation in repair performance. In the simulated scenarios representing natural disaster shocks, even the slightest attack condition with a critical functionality of 0.75 will induce 5–10 nodes to be attacked (Fig. 11.10d), which is much more serious than single-node attacks. Therefore, the conditions with critical functionality being 0, 0.25, 0.5, and 0.75 are termed paralyzed, extremely severe, very severe, and severe, respectively, for convenience. The smaller the cutoff value, the more severe the disruption. For both the LF and RC strategies, the recovery curves exhibit larger dispersion among attack-repair scenarios with varying attack protocols as the cutoff value increases (Fig. 11.10). For the four cutoff values, the recovery curves with the same repair strategy in paralyzed failure almost overlap each other, while recovery curves under the LF repair strategy in severe failure exhibit a large dispersion, and the cases for extremely severe failure and very severe failure lie in between. A statistical summary of the SR scores is shown in Table 11.2. The standard deviation (Std Dev) and mean-normalized Std Dev of the SR scores over different attack protocols increase from paralyzed, to extremely severe, to very severe, and to severe failure. The normalized Std Dev of the SR scores based on the LF repair strategy is smaller than that based on RC for all the failure degrees (i.e., paralyzed, extremely severe, very severe, and severe). The results indicate that the LF repair strategy has a more robust performance than the RC strategy in recovery performance from severe attacks. The difference is more obvious in non-paralyzed networks than in paralyzed ones.

11.5 Merits Analysis

To further investigate the feasibility of representing UTS network functionality by LF, another three quantities, including the number of concave-up points, the number of concave-down points, and the number of inflection points, are used to evaluate the system performance under attacks in addition to the SA score. Dynamic service load from historical records is incorporated. From the results in Sect. 11.3.3, it is the repair strategy rather than the attack protocol that dominates the recovery performance and absorptive resilience. Thus, one attack under each of the three attack schemes is applied to investigate service disruption and absorptive resilience with dynamic service load. Other attacks with similar effects are not presented. The results of one day's passenger flow and its corresponding system response under attacks are shown in Fig. 11.11. An extended temporal investigation period, i.e., 2 weeks, and its characteristics are further analyzed. The results are shown in Fig. 11.12.

Those three quantities can be obtained by differentiating the degraded functionality versus the fraction of attacked numbers. Their meaning is shortly depicted as follows: (1) Concave-up points are points where the slopes of the connective segments are increasing. In the context of degradation curves, each concave-up point indicates that the newly attacked node successfully catches the station so that the maximum possible loss of functionality will be induced. Intuitively, the larger the number of concave-up points, the severe the attack is. (2) Concave-down points are points where the slopes of the connective segments are decreasing. Each concave-down point indicates that the previously attacked node does not induce the maximum possible loss of functionality. Intuitively, the smaller the number of concave-down points, the severe the attack is. (3) An inflection point is a point where the double derivative equals zero. An inflection point indicates equal importance for three subsequent attacked nodes to cause functionality loss. The number of the inflection point is neutrally related to the impacts of the attack, while it indicates a change of concavity. The three indices reflect local features of the degradation curves, whereas SA score reflects the integral performance of attack protocols and can be considered as a global index. All four indicators are used to evaluate the UTS's performance under attacks with dynamic LF.

Relatively stable system performance under all the tested attacks is observed during the daily operation of UTS. Since the opening of the UTS at 6:00 am on the day 1st Jan 2019, the passenger flow in the Hangzhou Metro system gradually increased to a peak value of around 55k at 2:10 pm, then gradually decreased till the closing hour at midnight (Fig. 11.11a). Within a short period (i.e., around 1 h) since its opening, the passenger flow in the UTS increased to 20k (i.e., around 1/3 of the peak value), after which a higher level of flow was maintained until late in its operation hours (i.e., the last 2 h of its operation). During the main daytime when the flow is over 1/3 of the peak value, the overall flow exhibits a dynamic pattern, however, a relatively stable spatial distribution of flow in the UTS is enabled by the sufficient passengers, as evidenced by the relatively stable values of SA between 7 am to

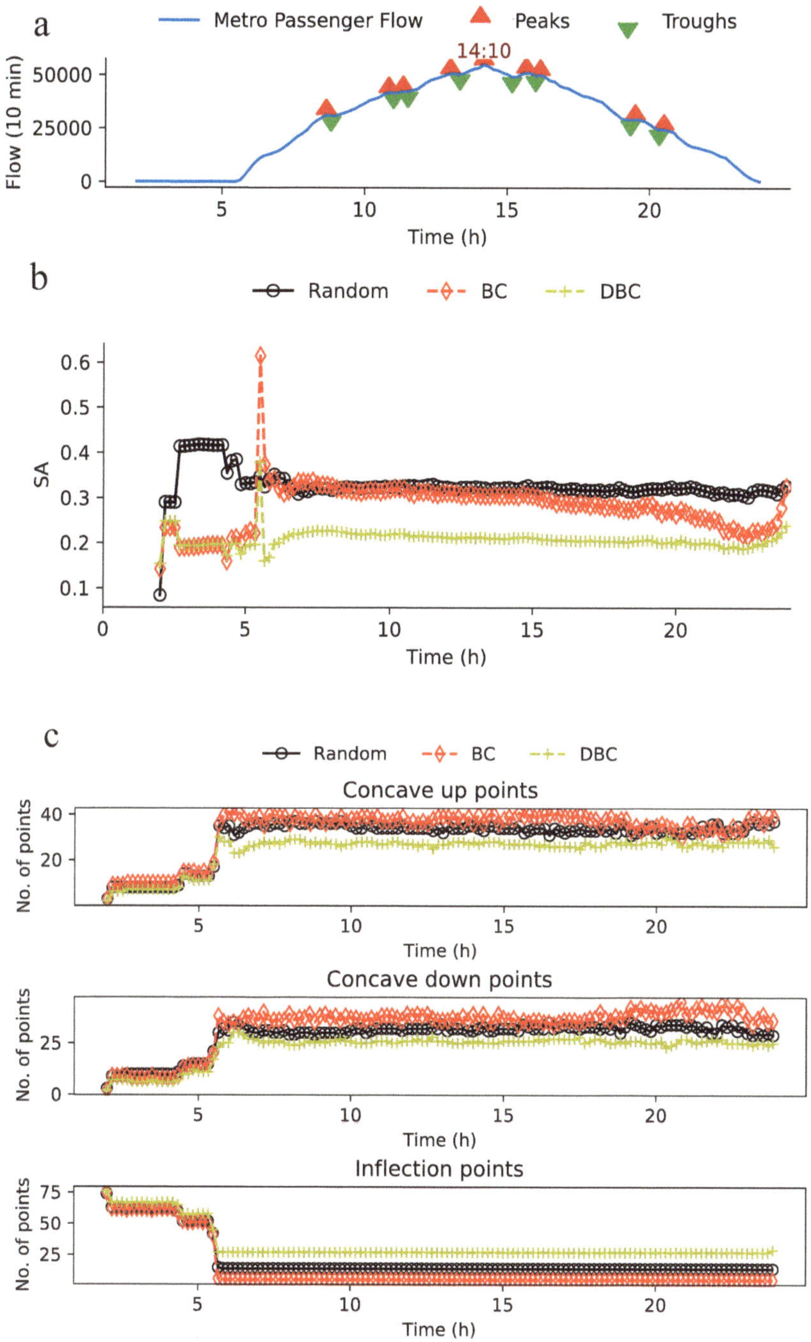

Fig. 11.11 Daily variation in the flow and system absorptive resilience. (**a**) Flow in the UTS network on the day of 1st Jan 2019. (**b**) SA scores under attacks. (**c**) No. of points belonging to the concave-up group (positively related to the impacts of the attack), the concave-down group (negatively related to the impacts of the attack), and the inflection group (neutrally related to the impacts of the attack)

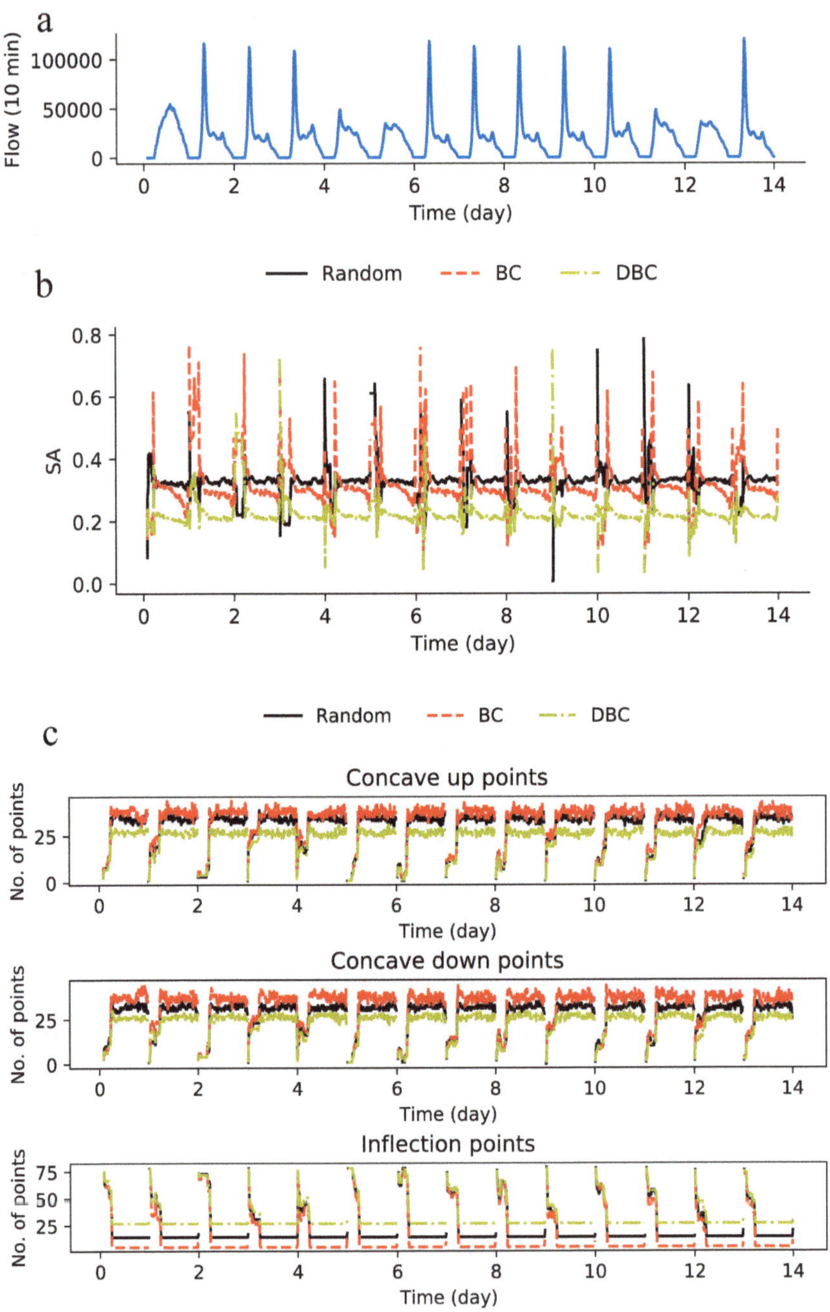

Fig. 11.12 A weekly pattern of flow and corresponding system response under attacks. (**a**) Flow in the UTS network from 1st Jan 2019 to 14th Jan 2019. (**b**) SA scores under attacks. (**c**) No. of points belonging to the concave-up group (positively related to the impacts of the attack), concave-down group (negatively related to the impacts of the attack), and inflection group (neutrally related to the impacts of the attack)

10 pm (Fig. 11.11b). The stable trend is consistently captured by the three indicators, i.e., the number of concave-up points, the number of concave-down points, and the number of inflection points (Fig. 11.11c). The results indicate that high passenger flow (i.e., over 1/3 of the peak value), which lasts the majority of its operation hours except the earliest and latest hours, may enable heterogeneous traveling choices and thus stabilize the spatial distribution of passenger flow. Therefore, one strategy should work well enough to protect against natural disaster attacks during the daily operation of Hangzhou Metro. Attention and resources are suggested to prioritize an optimized strategy rather than being equally spent on time-specific ones that may have inferior performance.

Based on AFC data over two consecutive weeks, a shorter periodic pattern, i.e., daily, is observed for the system response to attacks compared to the weekly periodic pattern in passenger flow. From 1st Jan 2019 to 14th Jan 2019, the passenger flow has an extremely high mono-peak on weekdays, whereas a low mono-peak on weekends or public holidays (Fig. 11.12a). As the value of flow to stabilize its spatial distribution is relatively low and easy to reach (requires only 1~2 h from low flow and can last 18 h with high flow), the difference in the system performance under attacks between weekdays and weekends/public holidays diminishes, though, with dynamic passenger flow. Instead, a simple periodic pattern, with a periodicity of 1 day, is observed (Fig. 11.12b, c). The finding indicates that no separate policy about disaster prevention and repair is required to distinguish between weekdays and weekends/public holidays. Together with the finding of daily patterns, a well-designed unified solution should work for all days and any schedule, regardless of the peak flow value. A possible reason is that the quick onset of large passenger flow could stabilize the spatial distribution of flow in the network, i.e., LF.

11.6 Chapter Summary

A two-layer network approach is proposed to model the service disruption and recovery of the UTS under natural disaster shocks. The proposed method evaluates the change in functionality based on a new localized measure, i.e., LF. A customized fuzzy assignment method is used to obtain LF with uncertainty values. Comprehensive simulation of attacks under a portfolio of attack protocols and repairs, since varying severity of attacks enables transparency about the system response to service demand under adverse events, which is important and valuable for managers and decision-makers on the route of sustainable and green infrastructure under natural disaster shocks.

The proposed method is applied to the Hangzhou Metro system. The degradation curves under a portfolio of centrality-based attack protocols serve as a reference regime for the possible onsets of attacks in the real situation of natural disasters. The newly identified functionality measure, i.e., LF, provides a better locality in the identification of critical components. The optimized repair strategies relying on LF are 2~3 times faster and more robust (i.e., smaller normalized standard deviation) than those relying on the common RC metric, as tested on multiple attack protocols and varying severity of attacks. System response to attacks is not affected by passenger flow once the flow goes beyond a threshold value, which occupies the majority of UTS's operation hours. It indicates that high passenger flow enables a relatively stable flow distribution over the network. The results suggest that attention should be prioritized to optimizing one strategy for infrastructure resilience to defend against natural disaster shocks rather than customizing multiple ones targeted for flow variations for different clocks or days.

The contribution of this study can be summarized as follows. (1) A localized passenger flow estimation method based on smart card data is proposed to evaluate the disruptive service level. (2) A system resilience assessment method is developed in the absorptive and adaptive dimensions. (3) The identification of the vulnerable components of CIs with fine spatial resolution in emergence is performed. It can be potentially applied to UTS networks subjected to functionality degradation for enhanced resilience.

References

1. Xu Z, Chopra SS (2022) Network-based assessment of metro infrastructure with a spatial–temporal resilience cycle framework. Reliab Eng Syst Saf 223:108434
2. Martello MV, Whittle AJ, Keenan JM, Salvucci FP (2021) Evaluation of climate change resilience for Boston's rail rapid transit network. Transp Res Part D: Transp Environ 97:102908
3. Bi W, MacAskill K (2023) A network modelling-based approach for quantifying flood resilience of urban rail transit systems. In: EC3 Conference 2023: European Council on Computing in Construction
4. Yadav N, Chatterjee S, Ganguly AR (2020) Resilience of urban transport network-of-networks under intense flood hazards exacerbated by targeted attacks. Sci Rep 10:10350
5. Zheng M, Zuo H, Zhou Z, Bai Y (2023) Recovery strategies for urban rail transit network based on comprehensive resilience. Sustainability 15:15018
6. Zhu M, Chen J, Yin Y (2021) Repair strategies of an urban rail transit network based on optimal resilience. In: 2021 6th International conference on transportation information and safety (ICTIS), pp 871–876
7. Yang Y, Liu J, Shang P, Chen X, Cao J (2021) Temporal and spatial evolution of passenger flow in an urban rail transit network during station closure. IEEE Access 9:29623–29640
8. Mo X, Cao C, Li M, Wang DZW (2021) Predicting the impact of disruptions to urban rail transit systems. In: 2021 IEEE 37th international conference on data engineering (ICDE). IEEE, pp 2249–2254
9. Shao W, Zhang J, Zhao X, Liu W (2021) Robustness assessments of urban rail transit networks based on network utilization. IEEE Access 9:129161–129167
10. Wang Y, Zhao O, Zhang L (2024) Modeling urban rail transit system resilience under natural disasters: a two-layer network framework based on link flow. Reliab Eng Syst Saf 241:109619
11. Tan E, Liu B, Guo C, Ma X (2024) Restoration sequence optimization for vulnerable metro stations with limited budget: a case study of Beijing, China. Phys A Stat Mech Appl 653:130102
12. Zhang J, Ren G, Song J (2022) Resilience-based restoration sequence optimization for metro networks: a case study in China. J Adv Transp 2022:8595356
13. Xiao X-m, Jia L-m, Wang Y-h (2018) Correlation between heterogeneity and vulnerability of subway networks based on passenger flow. J Rail Transp Plan Manage 8:145–157

14. Lekha DS, Balakrishnan K (2020) Central attacks in complex networks: a revisit with new fallback strategy. Phys A Stat Mech Appl 549:124347
15. Meng Y, Tian X, Li Z, Zhou W, Zhou Z, Zhong M (2020) Exploring node importance evolution of weighted complex networks in urban rail transit. Phys A Stat Mech Appl 558:124925
16. Freeman LC (1977) A set of measures of centrality based on betweenness. Sociometry 40:35–41
17. Sun ED, Michaels TCT, Mahadevan L (2020) Optimal control of aging in complex networks. Proc Natl Acad Sci U S A 117:20404–20410
18. Wu J, Chen Z, Zhang Y, Xia Y, Chen X (2020) Sequential recovery of complex networks suffering from cascading failure blackouts. IEEE Trans Netw Sci Eng 7:2997–3007
19. Liu Y, McNeil S, Hackl J, Adey BT (2020) Prioritizing transportation network recovery using a resilience measure. Sustainable and Resilient Infrastructure 7:70–81
20. Xu Z, Chopra SS, Lee H (2022) Resilient urban public transportation infrastructure: a comparison of five flow-weighted metro networks in terms of the resilience cycle framework. IEEE Trans Intell Transp Syst 23:12688–12699
21. Pan X, Dang Y, Wang H, Hong D, Li Y, Deng H (2022) Resilience model and recovery strategy of transportation network based on travel OD-grid analysis. Reliab Eng Syst Saf 223:108483
22. Chen J, Liu J, Peng Q, Yin Y (2021) Strategies to enhance the resilience of an urban rail transit network. Transp Res Rec 2676:342–354

[illegible]

Chapter 12
Conclusions and Future Directions

12.1 Main Conclusions

Urban transit systems (UTS) play a critical role in ensuring the efficient movement of people in modern cities, although they face increasing threats from rapid urbanization, climate change, and unexpected disruptions. To tackle these challenges, this book presents a comprehensive framework that integrates system vulnerability modeling, multi-hazard vulnerability evaluation, and resilience-oriented planning and operations. By bridging theoretical modeling with practical applications, the book provides valuable insights and decision support for researchers, planners, and operators seeking to build robust and adaptive transit systems. To conclude, through its three main parts and the specific methods and analyses presented in each part, this book demonstrates its value and contributions in the following key aspects.

12.1.1 Systematic Vulnerability Modeling

A key contribution of this work is the development of systematic approaches to model the vulnerability of metro networks under diverse hazards. Unlike traditional methods that treat risks in isolation or rely on static assumptions, this book proposes a multi-scale, dynamic modeling framework that combines different complementary perspectives. Specifically, Chap. 2 develops a disaster chain model that maps how primary hazards can sequentially trigger related secondary events within metro systems, capturing the logical cause-and-effect pathways. Chapter 3 focuses on cascade failure modeling, simulating how an initial station or link disruption can propagate passenger load redistribution and cause widespread network breakdown under different attack strategies. Chapter 4 introduces an evolutionary dynamics model that couples system dynamics with Monte Carlo simulation to predict how

L. Zhang et al., *Resilience in Urban Transit Systems*, Advances in 21st Century Human Settlements, https://doi.org/10.1007/978-981-95-4322-9_12

vulnerability levels evolve over long-term operation, accounting for workforce, equipment aging, and management factors. By integrating these perspectives, we construct a layered and complementary modeling paradigm that captures spatial, temporal, and functional vulnerability features of urban metro systems. The resulting insights support targeted risk mitigation and provide a robust foundation for downstream resilience planning.

12.1.2 Multi-Hazard Risk Evaluation

Beyond modeling, the book contributes practical methods for evaluating the vulnerability of UTS against multiple hazards, specifically floods, fires, and evacuation scenarios. Chapter 5 presents a novel flood hazard analysis framework, employing evidential reasoning techniques to address uncertainties inherent in subjective assessment parameters. This approach quantifies risk expectations and identifies critical factors for flood mitigation, validated effectively through a detailed analysis of Shanghai's metro network. Chapter 6 shifts the focus to fire hazard analysis, introducing an advanced method that integrates 3D numerical simulations and multi-attribute decision analysis. This approach effectively captures the complexities of fire spread in confined metro spaces, enhancing the traditional ASET/RSET evaluation and identifying worst-case fire scenarios. Finally, Chap. 7 addresses passenger evacuation analysis through a hybrid framework combining agent-based simulations, random forest meta-modeling, and multi-objective optimization. By systematically evaluating evacuation strategies across diverse conditions, this approach improves evacuation performance indicators such as evacuation time, passenger density, and cost-effectiveness. Together, these chapters provide robust, multi-dimensional evaluation methodologies for metro system vulnerability, effectively guiding proactive risk management in UTS.

12.1.3 Resilience-Oriented Planning and Operations

Building upon the foundational understanding of system vulnerabilities and multi-hazard risks, this book further offers a series of resilience-oriented planning and operational strategies that enable metro systems to anticipate, absorb, and adapt to disruptions. Chapter 8 introduces a proactive strategy for pre-flood facility operation planning, focusing on station closure and protection decisions before a flood occurs. Chapter 9 shifts to emergency response optimization, particularly for metro fire incidents. It integrates Hierarchical Timed Colored Petri Nets (HTCPN) with multi-objective optimization to simulate and evaluate firefighting and passenger evacuation strategies in real time, balancing response time and resource deployment efficiency, and providing a responsive approach to fire emergency management. Chapter 10 extends resilience planning to the strategic deployment of flood control resources, proposing a multi-condition bilevel optimization model for warehouse location and inventory planning. Chapter 11 addresses the post-disruption recovery phase, presenting advanced methods for system repair planning. By constructing a two-layer network model and estimating dynamic link flow via a fuzzy-set-based method using smart card data, this chapter simulates service disruptions and optimizes repair strategies for urban rail transit systems under disaster shocks. Together, these four chapters provide comprehensive frameworks for resilience enhancement in UTS. By combining system simulation and optimization techniques, they offer practical solutions that strengthen preparedness, emergency responses, and recoverability, serving as vital tools for building a more resilient urban transit system under various hazards and disruptions.

12.2 Future Directions

UTS will continue to evolve rapidly in response to increasing urbanization, rising passenger demands, and heightened risks posed by extreme events and societal disruptions. As cities aim to build more resilient, efficient, and sustainable transit networks, advanced modeling, data integration, and smart technologies will play an important role in supporting this transformation. Despite the significant progress made through systematic vulnerability modeling, hazard evaluation, and resilience enhancement, several challenges remain in translating these frameworks into practical, real-time, and widely adaptable solutions. Future research and practice are expected to focus on the following directions, which are anticipated to shape the next generation of resilient and intelligent UTS.

12.2.1 Integration of Advanced Information Technologies

One promising direction for strengthening the resilience of UTS is the integration of advanced information technologies that enable real-time awareness, intelligent prediction, and coordinated emergency response. Among them, artificial intelligence (AI) offers a rich toolbox covering computer vision (CV), physics-informed machine learning (PIML), and large language models (LLMs), each bringing unique advantages. For instance, CV can automatically process CCTV footage to detect anomalies such as crowd congestion, unauthorized intrusions, or early signs of flooding inside tunnels and stations, enabling operators to respond faster than manual monitoring. Meanwhile, PIML combines data-driven learning with physical laws, providing accurate and explainable predictions for complex processes like underground water flow, smoke dispersion during fires, or structural deformation after earthquakes [1].

Beyond single-purpose models, the rise of general-purpose large language and multimodal models [2] opens new opportunities for integrated smart metro management. Such models can fuse textual guidelines, sensor data, and visual inputs to assist operators in complex decision-making—whether it is dynamically adjusting evacuation plans, generating real-time incident reports, or answering frontline staff queries. By serving as an intelligent assistant, large AI models can bridge information gaps across various subsystems and stakeholders, promoting more coordinated and adaptive responses during emergencies.

Equally important is the deployment of digital twin technology, which creates a real-time virtual replica of physical assets and processes [3]. Coupled with dense sensor networks and AI analytics, the digital twin continuously reflects the current state of metro facilities, simulates possible hazards, and tests response strategies. This digital system, when enhanced with AIoT (Artificial Intelligence of Things) [4], supports decentralized and automated actions such as activating flood barriers or rerouting trains when disruptions occur.

Finally, realizing the full benefits of these technologies calls for robust multi-scale human–computer interaction [5]. Intuitive dashboards, real-time alerts, and immersive tools like cloud AR/VR [4] enhance operators' situational awareness and support collaborative decision-making. Passengers can also benefit from AI-driven information services, receiving personalized guidance during disruptions or emergencies. Together, these advanced information technologies point to a future where resilient metro systems are not only technically robust but also smart, adaptive, and human-centered.

12.2.2 Model Portability, Robustness, Security and Safety

In the face of increasing system complexity and rapid urban dynamics, existing resilience models often fall short when applied beyond their original design conditions. To address this, it is necessary to enhance model portability, enabling seamless transfer and effective use of resilience models across different metro systems, operational contexts, and technological platforms. Equally important, model robustness must be strengthened to ensure reliable performance under noisy, incomplete, or rapidly changing data streams, providing stable guidance even during unexpected disruptions. Meanwhile, high standards of security are essential to protect sensitive passenger and operational data from breaches or misuse. Furthermore, ensuring operational safety requires that models can detect anomalies and respond to threats in real time. Together, these capabilities form the backbone of trustworthy resilience models that support continuous sensing, monitoring, and adaptive planning for safe and reliable urban transit operations.

To achieve high levels of portability, robustness, security, and safety, future resilience modeling should adopt several strategies. Enhancing portability requires building modular and adaptable model architectures that can be easily configured for different metro systems and seamlessly integrate varied data sources. Systematic cross-city validation is also vital to ensure that models perform consistently across diverse network structures and hazard scenarios, rather than being overly tailored to a single context. To reinforce robustness, hybrid modeling approaches that blend data-driven methods with physics-based principles and expert knowledge can be employed. Such approaches can help maintain reliable predictions even when data are incomplete or when facing unforeseen system disturbances. For security, federated learning stands out as an effective solution, allowing multiple devices or parties to collaboratively train a shared model without the need for centralized data storage.

In federated learning, local models are periodically aggregated and merged into a single global model, i.e., only updates to the model are transmitted back to the central server [6, 7]. This approach, combined with privacy-preserving techniques like differential privacy and secure computation, provides strong safeguards for sensitive passenger and operational information. Lastly, ensuring safety calls for embedding human oversight and clear traceability into every stage of the modeling process. Transparent decision logs, intuitive interfaces, and responsive feedback loops enable operators to monitor system behavior, identify anomalies, and intervene when needed. Together, these strategies will enable future resilience models to function as trustworthy, adaptable, and secure decision-support tools for complex UTS.

12.2.3 Policy and Stakeholder Collaboration

Beyond technical improvements, strengthening resilience in UTS depends heavily on effective policy frameworks and stakeholder collaboration. While advanced models and digital tools can offer powerful decision support, their real-world impact is limited if not backed by clear governance structures and coordinated multi-agency action. Currently, resilience planning often remains fragmented, with different departments, such as transport authorities, emergency services, urban planners, and utility providers. This lack of integration can delay critical response actions during disasters and hamper the smooth implementation of resilience strategies.

To overcome this, future efforts should prioritize setting up institutional mechanisms that encourage shared responsibility and continuous communication among different stakeholders. Developing standardized resilience guidelines and risk management protocols can help align goals across agencies. Public participation is also crucial: engaging communities in preparedness plans, evacuation drills, and real-time reporting can enhance system-wide situational awareness and foster a culture of shared resilience. Moreover, policymakers should consider embedding resilience requirements into metro planning codes, and then operators can adopt best practices in risk assessment and emergency management. Pilot projects and cross-city knowledge exchange can provide practical insights and help tailor global resilience concepts to local conditions. Ultimately, by integrating powerful policies, transparent governance, and active stakeholder collaboration, cities can transform technical resilience solutions into a sustained improvement in safety, reliability, and public trust for metro systems.

References

1. Chew AWZ, He R, Zhang L (2025) Physics informed machine learning (PIML) for design, management and resilience-development of urban infrastructures: a review. Arch Comput Methods Eng 32:399–439
2. Chang Y, Wang X, Wang J, Wu Y, Yang L, Zhu K et al (2024) A survey on evaluation of large language models. ACM Trans Intell Syst Technol 15:1–45
3. Pan Y, Zhang L (2023) Integrating BIM and AI for smart construction management: current status and future directions. Arch Comput Methods Eng 30:1081–1110
4. Pan Y, Zhang L (2021) Roles of artificial intelligence in construction engineering and management: a critical review and future trends. Autom Constr 122:103517
5. Ji A, He R, Chen W, Zhang L (2024) Computational methodologies for critical infrastructure resilience modeling: a review. Adv Eng Inform 62:102663
6. Banabilah S, Aloqaily M, Alsayed E, Malik N, Jararweh Y (2022) Federated learning review: fundamentals, enabling technologies, and future applications. Inf Process Manag 59:103061
7. Zhang C, Xie Y, Bai H, Yu B, Li W, Gao Y (2021) A survey on federated learning. Knowl-Based Syst 216:106775

FSC
www.fsc.org
MIX
Papier aus verantwortungsvollen Quellen
Paper from responsible sources
FSC® C105338

If you have any concerns about our products,
you can contact us on
ProductSafety@springernature.com

In case Publisher is established outside the EU,
the EU authorized representative is:
Springer Nature Customer Service Center GmbH
Europaplatz 3, 69115 Heidelberg, Germany

Printed by Libri Plureos GmbH
in Hamburg, Germany